Minor Millets

Dinesh Chandra Joshi
Amit Kumar Singh • Salej Sood
Manoj Prasad
Editors

Minor Millets

Cultivation, Breeding, Genomics and Uses

Springer

Editors

Dinesh Chandra Joshi
Division of Crop Improvement
ICAR-Vivekananda Parvatiya Krishi
Anusandhan Sansthan
Almora, Uttarakhand, India

Amit Kumar Singh
Division of Genomic Resources
ICAR-National Bureau of Plant Genentic
Resources
New Delhi, Delhi, India

Salej Sood
Crop Improvement Division
ICAR-Central Potato Research Institute
Shimla, Himachal Pradesh, India

Manoj Prasad
National Institute of Plant Genome
Research
New Delhi, Delhi, India

Department of Genetics
University of Delhi
New Delhi, India

ISBN 978-981-96-4264-9 ISBN 978-981-96-4265-6 (eBook)
https://doi.org/10.1007/978-981-96-4265-6

This Springer imprint is published by the registered company Springer Nature Singapore Pte Ltd.
The registered company address is: 152 Beach Road, #21-01/04 Gateway East, Singapore 189721, Singapore

Foreword

It is estimated that of the 80,000 plants discovered for potential economic utilization, roughly 30,000 are deemed edible, and around 7000 have been cultivated by humans at various times, of which just 158 are extensively utilized for food. Of these, 30 crops account for 90% of global food production, with only three crops rice, wheat, and maize supplying 60% of the world's total food need. Minor millets are a category of small-seeded grasses cultivated globally for human consumption, animal feed, and fodder. Minor millets have served as a vital source of sustenance for arid parts of Asia and Africa. They were the primitive crops domesticated by humans in Asia and Africa, subsequently disseminating worldwide as essential food sources for developing civilizations, owing to their versatile use and capacity to thrive in harsh growing conditions.

Minor millets constitute a crucial element of nutritional and livelihood security for resource-constrained farmers in India. They demonstrate extensive adaptability in tropical and subtropical climates, offering farmers optimal prospects for guaranteed yields, essential sustenance, necessary nutrition, and high-quality fodder. The promotion of millet production and value addition considerably contributes to the achievement of numerous Sustainable Development Goals (SDGs). The Food and Agriculture Organization of the United Nations (FAO) acknowledged the significant potential of minor millets by designating 2023 as the "International Year of Millets." The Government of India has just designated millets as "*Shri Anna*" because of their remarkable nutritional profile.

Millet production has significantly declined in recent decades, decreasing by 40% since the Green Revolution. Currently, an increasing number of individuals are acknowledging the health, lifestyle advantages, and environmental sustainability linked to millets. The availability of thoroughly organized and documented literature is crucial for achieving substantial advancements in science, technology, regulation, and policy to integrate small millets as sustainable and economically viable crops in modern agro-ecosystems. This book, *Minor Millets: Cultivation, Breeding, Genomics, and Uses*, will be valuable in creating ideas and solutions for the revival and long-term integration of small millets into modern cropping systems.

I am delighted to provide this foreword for the latest book edited by Drs. D C Joshi, Amit Kumar Singh, Salej Sood, and Manoj Prasad, who are familiar to me as partners to our All India Coordinated Research Project on Small Millets and National Agricultural Research System. This volume by the editors from Springer

Nature demonstrates their strong commitment to delivering extensive information on these sturdy minor millets. This book, comprising 16 chapters on various aspects of minor millets, will benefit scientists, students, FPOs, NGOs, startups, policy makers, and growers involved with millets.

ICAR-Indian Institute of Millets Research
Hyderabad, India

C. Tara Satyavathi

Preface

Almost 7000 plant species are edible to humans. Currently, only 12 major crop species account for over 75% of the food consumed globally. Indeed, rice, wheat, and maize account for nearly half of the world's dietary demands. As the number of crop species utilized in agriculture decreases, it is critical to address the consequences of this development on farmers and the environment. Furthermore, biodiversity loss can have a severe impact on the environment, reducing soil health, water quality, and ecosystem resilience. Prioritizing the utilization of a varied range of crops in agriculture is critical to ensuring long-term and resilient food systems. The resurgence of minor or small millets is one such adaptable solution for a wide range of biodiversity, climate change, nutritional, and health-related challenges. Small millets are a species of annual grasses that produce small seeds known as millets. Millets are primitive food crops that are presently grown in 131 countries. Minor millet species include barnyard millet (*Echinochloa* spp.), little millet (*Panicum sumatrense*), proso millet (*Panicum miliaceum* L.), kodo millet (*Paspalum scrobiculatum*), browntop millet (*Urochloa ramosa*), teff (*Eragrostis* tef), and fonio (*Digitaria exilis*).

These tiny grains constituted a significant element of several prevalent mixed farming practices across diverse agro-ecological zones worldwide. Nonetheless, agrobiodiversity diminished as modern agriculture prioritized a limited number of crops, leading to the replacement of traditional grains. In comparison to primary cereals, these crops exhibit resilience to diverse abiotic stress conditions. The grains are an especially important crop for regions that are vulnerable to famine since they can be kept for years without being damaged by storage pests. These small millets yield both food grains and high-quality fodder, particularly in rainfed regions. For the millions of underprivileged people dependent on starchy foods, these crops function as a reservoir of essential amino acids and micronutrients. Moreover, the gluten-free protein renders it an appealing option for individuals with celiac disease. The carbohydrate content of small millet grains possesses the distinctive characteristic of slow digestion, rendering them appropriate for diabetic nutrition.

In India, millets are grown in about 13.8 million hectares with an annual production of 17.3 million tons and contribute about 10% of the country's food grain basket. The FAO has declared 2023 to be the 'International Year of Millets' because of their nutrient-rich and climate-resilient qualities, which will increase their production and consumption and help achieve crop diversification in the face of climate change. This book is an attempt to compile the versatile aspects of small millets in a single volume while keeping in mind the global situation.

The edited book *Minor Millets: Cultivation, Breeding, Genomics, and Uses* comprised of 16 chapters will cover the latest advancements in the field of agronomy, genetic resources, and their utilization, quality seed production, plant protection, genome sequencing information, value added products, and improved technology dissemination. Chapter 1 of this book is basically a compilation of historical perspectives, botany, and taxonomic account of minor millets. Chapter 2 deals with the global production scenario of millets and their role in achieving global food and nutritional security. It also provides an account of major challenges and efforts made for genetic improvement of minor millets. Intake of diet poor in vital nutrients such as micronutrients, vitamins, and quality protein is a major cause of widespread global malnutrition. Minor millet grains are the reservoir of nutrients, protein, and fiber. All these aspects are narrated in Chap. 3. One of the major production constraint associated with minor millets is the unavailability of efficient crop management practices and agro-techniques. For instance, the cumbersome weed management practices in the absence of potent herbicides are the major factor for low productivity and high cost of cultivation of minor millets. Chapter 4 covering this aspect is the compilation of latest advancements in improved cultivation practices and efficient crop management practices in minor millets to realize the higher yield. Genetic resources of any crop are the building blocks for accelerating genetic gain and broadening the genetic base. Chapter 5 details the efforts made for collection, characterization, and utilization of genetic resources in minor millets. To some extent, these resources have been characterized systematically to identify trait-specific accessions and their utilization in hybridization program and trait mapping. One of the critical impediments hampering the progress of cultivar development in minor millets is the cumbersome floral biology and minute florets. Therefore, hand emasculation in minor millets is tedious and time and resource consuming. Chapter 6 summarized the latest advancements made in artificial hybridization techniques in small millets to create variability and facilitate recombination breeding. Although minor millets are tolerant to abiotic stresses, like other crop plants their productivity is affected by various biotic stresses including diseases and pests. Chapters 7 and 8 provide a comprehensive account of various diseases and pests of small millets and their efficient management practices. Chapter 9 provides the current status of breeding advancements and a detailed account of various cultivars of minor millets released for different agro-ecologies of the country. Good quality seed is the most critical input in enhancing productivity and large-scale dissemination of minor millets. Therefore, it is mandatory to follow the principles and procedures pertaining to seed production in minor millets. Chapter 10 of this book outlines the principles and procedures of quality seed production in minor millets. Compared to major crops, genomic resources are poorly documented in minor millets. Therefore, efforts are directed toward development of genomic resources in minor millets for dissecting the genetics of complex traits. Chapter 11 provides a detailed overview of genomic resources like molecular markers, linkage maps, transcriptomic, proteomic, and metabolic resources. Additionally, it also describes the role and possibility of deploying modern breeding tools to accelerate genetic gain in minor millets. Availability of well-assembled reference genome sequences will be a crucial step for utilizing genomic-assisted breeding for cultivar development in minor millets. Chapter 12 elaborates on the current national and international efforts employed for genome sequencing of minor millets. Owing to their exceptional nutraceutical potential, minor millets are being used in the development of various functional food products. Therefore, Chapters 13 and 14 of this book describe the technologies available for the development of value-added products and efforts for sustainable value chain development based on minor millets. Despite of modest progress in cultivar development, production gap exists between yields realized at farmer's fields because of poor technology dissemination. Frontline demonstrations (FLDs) at farmer's field have been very effective in popularizing high-yielding cultivars and improved crop management practices. Chapter 15 of this book describes the efforts made by the All India Coordinated Research Project on Small Millets for FLDs of improved cultivars and agronomic practices at farmers' fields in different agro-ecologies of the country. These efforts have played a vital role in technology dissemination and its impact on enhancing farm income. Recently, gene editing has emerged as one of the most potent technology for tackling major production constraints of agricultural crops. However, the most crucial step in utilizing gene editing in any crop plant is the availability of well-established regeneration protocols. Chapter 16, the final chapter of this book, describes the latest advancements made in regeneration protocols and transgenic development in minor millets. It also highlights the possibility and way forward for using gene editing in minor millets.

The chapters of this book have been authored by a team of scientists who are experts in various aspects of minor millets. We endeavored to prevent any overlap in the contents of the various chapters; however, some overlap in certain areas is inevitable.

We extend our sincere appreciation to all contributors for their chapters. We acknowledge their ongoing cooperation from the submission of draft chapters to the revision of chapters. We express our gratitude to our family members for their patience during the book editing process. This book covers basic and applied information pertaining to crop biology, breeding, genomics, agronomy, and uses of minor millets and will be useful to students, scientists, and policy makers.

Almora, Uttarakhand, India — Dinesh Chandra Joshi
New Delhi, Delhi, India — Amit Kumar Singh
Shimla, Himachal Pradesh, India — Salej Sood
New Delhi, Delhi, India — Manoj Prasad

Contents

Editors and Contributors

About the Editors

Dinesh Chandra Joshi is the Principal Investigator for the minor millets improvement program at ICAR-Vivekananda Parvatiya Krish Anusandhan Sansthan, Almora, India. He was involved in developing eight cultivars of minor millets and has registered several trait-specific genetic stocks of minor millets. He has decoded the genetics of economic traits, such as blast resistance in finger millet and root architecture in sorghum, utilizing genomics tools. He has published numerous research articles in reputable, peer-reviewed national and international journals. He is a recipient of the IACBG Fellowship from the Department of Biotechnology, as well as the Australian Centre of International Agricultural Research (ACIAR) Scholarship. Additionally, he received the Best Performing Centre Award from the All India Coordinated Research Project on Small Millets (AICRP-SM) for minor millets research.

Amit Kumar Singh is currently working as a Senior Scientist at ICAR-National Bureau of Plant Genetic Resources, New Delhi, India. He obtained his doctorate in Molecular Biology and Biotechnology from ICAR-Indian Agricultural Research Institute, New Delhi, India. He has over 18 years of experience in the field of plant genomics. His main research focus is on the development of novel genomic resources and trait mapping in cereal and underutilized crops by applying advanced genomics tools, such as transcriptomics and genome-wide association mapping. He has developed a novel promoter-based marker technique, known as CAAT box-derived polymorphism (CBDP), which is widely used by researchers worldwide for plant genome analysis. Dr. Singh has published many research publications in various international journals. He is a recipient of the prestigious Endeavour Award from the Australian Government and fellowships from many professional societies.

Salej Sood has worked with industry and the public sector for over 18 years in the field of plant breeding and genetics. He was involved in the development of multiple varieties and technologies for minor millets and other potential crops in India. He has worked in interdisciplinary teams and used both conventional and molecular tools for the genetic improvement of crops. His area of expertise includes marker-assisted selection, genome-wide association studies, and genomic selection. He has good experience in germplasm characterization and evaluation, varietal development, and development of mapping populations and association mapping panels. He has numerous peer-reviewed publications in reputable journals. He completed his postdoctoral research on genomic selection at La Trobe University, Melbourne, Australia, and his notable awards include the National Academy of Agricultural Sciences, India Associateship in 2019, the Endeavour Research Fellowship in 2018, the fellowship of the Indian Society of Genetics and Plant Breeding in 2021, and several other institutional and society research awards.

Manoj Prasad is currently working as a Professor and JC Bose National Fellow at the DBT-National Institute of Plant Genome Research (NIPGR), New Delhi, India. Prior to this, he served as an Alexander von Humboldt Fellow (2000–2003) at the Leibniz Institute of Plant Genetics and Crop Plant Research (IPK), Gatersleben, Germany. His research interests include plant molecular genetics, genomics and stress biology. Dr. Prasad has published more than 200 papers in high-impact journals, including Trends in Plant Sciences, Trends in Genetics, Trends in Microbiology, the Journal of Experimental Botany, and PNAS, USA. He is the recipient of prestigious awards and recognitions, including the Tata Innovation Fellow, National Bioscience Award, Birbal Sahni Medal, Golden Peacock Award, and the Hiralal Chakraborty Memorial Award.

Contributors

B. Amasiddha ICAR-Indian Institute of Millets Research, Hyderabad, India

N. Anuradha Agricultural Research Station, Acharya NG Ranga Agricultural University, Vizianagaram, India

Arthi Department of Plant Pathology, University of Agricultural Sciences, GKVK, Bengaluru, Karnataka, India

Lalit Arya ICAR, National Bureau of Plant Genetic Resources (NBPGR), New Delhi, India

Laxmi B College of Forestry, KSNUAHS, Ponnampet, India

D. Ratna Babu Agricultural College, Acharya NG Ranga Agricultural University, Bapatla, India

Yogesh Gangadhar Ben ICAR-AICRP on Small Millets, Kolhapur, Maharashtra, India

M. Bala Bharathi Agricultural College, Acharya NG Ranga Agricultural University, Bapatla, India

Sujata Bhat AICRP on Small Millets, University of Agricultural Sciences, GKVK, Bengaluru, India

Mahendar S. Bhinda ICAR-Vivekananda Parvatiya Krishi Anusandhan Sansthan, Almora, Uttarakhand, India

Stanislaus Antony Ceasar Division of Plant Molecular Biology and Biotechnology, Department of Biosciences, Rajagiri College of Social Sciences, Kochi, Kerala, India

C. Chaithra Project Coordinating Unit, ICAR-AICRP on Small Millets, GKVK, Bengaluru, Karnataka, India

M. Divya Agricultural College, Acharya NG Ranga Agricultural University, Bapatla, India

Rafi Dudekula ICAR-IIMR, Hyderabad, India

B. M. Dushyanthakumar KSNUAHS, Shivamogga, Karnataka, India

Suresha D. Ekbote Zonal Agricultural and Horticultural Research Station, Babbur Farm, Hiriyur, KSNUAHS, Shivamogga, Karnataka, India

Meghana G ICAR-IIMR, Hyderabad, India

K. N. Ganapathy ICAR-Indian Institute of Millets Research, Hyderabad, India

Prabhu C. Ganiger PC Unit, ICAR-AICRP on Small Millets, UAS, Bengaluru, India

Kamal Garg ICAR-Vivekananda Parvatiya Krishi Anusandhan Sansthan, Almora, India

ICAR-Indian Agricultural Research Institute, New Delh**i, India**

S. Gazala Parveen AICRP on Small Millets, University of Agricultural Sciences, GKVK, Bengaluru, India

Krati Goel Department of Food and Nutrition, College of Community Science, G.B. Pant University of Agriculture & Technology, Pantnagar, Uttarakhand, India

P. N. Guru ICAR-Central Institute of Post-Harvest Engineering and Technology, Ludhiana, Punjab, India

J. Jinu ICAR-Indian Institute of Millets Research, Hyderabad, India

Ayushi Joshi Department of Food and Nutrition, College of Community Science, G.B. Pant University of Agriculture and Technology, Pantnagar, Uttarakhand, India

Dinesh Chandra Joshi Division of Crop Improvement, ICAR-Vivekananda Parvatiya Krishi Anusandhan Sansthan, Almora, Uttarakhand, India

S. Kanika ICAR-Central Institute of Research in Cotton Technology, Matunga, Mumbai, India

Farooq Khan Department of Plant Pathology, University of Agricultural Sciences, GKVK, Bengaluru, Karnataka, India

Jyotsna Kiramnai ICAR-AICRP on Small Millets, Nandyala, Andhra Pradesh, India

Prasanna S. Koti Department of Plant Biotechnology, University of Agricultural Sciences, GKVK, Bengaluru, India

T. P. Ajeesh Krishna Division of Plant Molecular Biology and Biotechnology, Department of Biosciences, Rajagiri College of Social Sciences, Kochi, Kerala, India

Ajay Kumar ICAR-AICRP on Small Millets, Uttarakhand University of Horticulture and Forestry, Ranichauri, Uttarakhand, India

Anil Kumar Department of Plant Breeding and Genetics, Bihar Agriculture College, Bihar Agricultural University, Sabour, Bihar, India

Archana Kushwaha Department of Food and Nutrition, College of Community Science, G.B. Pant University of Agriculture and Technology, Pantnagar, Uttarakhand, India

Madhavi Latha Agricultural Research Station, Acharya NG Ranga Agricultural University, Ananthpuram, Andhra Pradesh, India

T. Maharajan Division of Plant Molecular Biology and Biotechnology, Department of Biosciences, Rajagiri College of Social Sciences, Kochi, Kerala, India

V. M. Malathi ICAR-Indian Institute of Millets Research, Hyderabad, India

Pavan Kumar Malav ICAR, National Bureau of Plant Genetic Resources (NBPGR), New Delhi, India

Rajendra Prasad Meena ICAR-Vivekananda Parvatiya Krishi Anusandhan Sansthan, Almora, Uttarakhand, India

Poojashree M. S. Department of Plant Pathology, University of Agricultural Sciences, GKVK, Bengaluru, Karnataka, India

Mehanathan Muthamilarasan Department of Plant Sciences, School of Life Sciences, University of Hyderabad, Hyderabad, Telangana, India

T. E. Nagaraja AICRP on Small Millets, University of Agricultural Sciences, GKVK, Bengaluru, India

Nagendra Kumar Nag ICAR-AICRP on Small Millets, Indira Gandhi Agricultural University, Raipur, Chhattisgarh, India

B. Nandini College of Horticulture, Kolar, University of Horticultural Sciences, Bagalkot, India

C. Nandini Zonal Agricultural and Horticultural Research Station, Babbur Farm, Hiriyur, KSNUAHS, Shivamogga, Karnataka, India

A. L. Narayanan Pandit Jawaharlal Nehru College of Agriculture and Research Institute, Govt. of Puducherry, Karaikal, Puducherry, India

Anusha M. Nayak Department of Plant Pathology, University of Agricultural Sciences, GKVK, Bengaluru, Karnataka, India

K. B. Palanna ICAR-AICRP on Small Millets, Project Coordinating Unit, University of Agricultural Sciences, Bengaluru, Karnataka, India

Anurag Panchal National Institute of Plant Genome Research, Aruna Asaf Ali Marg, New Delhi, India

Chithra Devi Pandey ICAR, National Bureau of Plant Genetic Resources (NBPGR), New Delhi, India

Sushil Pandey ICAR, National Bureau of Plant Genetic Resources (NBPGR), New Delhi, India

Amit Umesh Paschapur ICAR-Vivekananda Parvatiya Krishi Anusandhan Sansthan, Almora, Uttarakhand, India

H. E. Patil ICAR-AICRP on Small Millets, Waghai, Gujarat, India

T. S. S. K. Patro Agricultural Research Station, Acharya NG Ranga Agricultural University, Vizianagaram, Andhra Pradesh, India

T. Tharana Poonacha Department of Plant Pathology, University of Agricultural Sciences, GKVK, Bengaluru, Karnataka, India

Manoj Prasad National Institute of Plant Genome Research, New Delhi, Delhi, India

Department of Genetics, University of Delhi, South Campus, New Delh**i, India**

P. Kranthi Priya Agricultural Research Station, Acharya NG Ranga Agricultural University, Vizianagaram, Andhra Pradesh, India

Gutha Venkata Ramesh Department of Plant Pathology, Punjab Agricultural University, Ludhiana, Punjab, India

Y. Sandhya Rani Agricultural Research Station, Acharya NG Ranga Agricultural University, Vizianagaram, Andhra Pradesh, India

Sk. Abdul Salam Agricultural Research Station, Acharya NG Ranga Agricultural University, Vizianagaram, India

Ipsita Samal ICAR-National Research Centre on Litchi, Muzaffarpur, Bihar, India

Sangappa ICAR-IIMR, Hyderabad, India

K. Sathya ICAR-AICRP on Small Millets, Centre of Excellence in Millets, Athiyandal, Tamil Nadu, India

C. Tara Satyavathi ICAR-Indian Institute of Millets Research, Hyderabad, India

Neelam Shekhawat ICAR, National Bureau of Plant Genetic Resources (NBPGR), New Delhi, India

Manisha Shyam ICAR-AICRP on Small Millets, Dindori, Madhya Pradesh, India

Amit Kumar Singh Division of Genomic Resources, ICAR-National Bureau of Plant Genentic Resource, New Delhi, Delhi, India

Arjun Singh ICAR-Vivekananda Parvatiya Krishi Anusandhan Sansthan, Almora, India

ICAR-Indian Agricultural Research Institute, New Delhi, India

Ashish Kumar Singh ICAR-Vivekananda Parvatiya Krishi Anusandhan Sansthan, Almora, Uttarakhand, India

Badal Singh ICAR, National Bureau of Plant Genetic Resources (NBPGR), New Delhi, India

Mamta Singh ICAR, National Bureau of Plant Genetic Resources (NBPGR), New Delhi, India

Monika Singh ICAR-National Bureau of Plant Genetic Resources, New Delhi, India

Roshan Kumar Singh National Institute of Plant Genome Research, Aruna Asaf Ali Marg, New Delhi, India

T. S. Sukanya Project Coordinating Unit, ICAR-AICRP on Small Millets, GKVK, Bengaluru, Karnataka, India

B. S. Tomar ICAR-Indian Agricultural Research Institute, New Delhi, India

U. Triveni Agricultural Research Station, Acharya NG Ranga Agricultural University, Vizianagaram, Andhra Pradesh, India

K. Venkatesh ICAR, Indian Institute of Millet Research, Hyderabad, India

R. Venkateswarlu ICAR-Indian Institute of Millets Research, Hyderabad, India

Manjusha Verma ICAR-National Bureau of Plant Genetic Resources, New Delhi, India

Vishwanath Rohidas Yalamalle ICAR-Indian Agricultural Research Institute, New Delhi, India

History, Origin, Botany, and Taxonomy of Minor Millets

Badal Singh, Lalit Arya, Mamta Singh, Pavan Kumar Malav, K. Venkatesh, Chithra Devi Pandey, Sushil Pandey, Neelam Shekhawat, and Amit Kumar Singh

Abstract

The second green revolution must prioritize nutrition, focusing on diverse, environmentally friendly millets. These small-grained cereals, such as foxtail, kodo, little millet, etc., offer nutritional benefits and resilience to adverse conditions. Declining production of minor millets in India necessitates innovative approaches and policy support to meet growing demand and reverse the trend. Their gluten-free nature and sustainability align with global food security goals. Minor millets have been integral to India's agricultural heritage for millennia, with evidence of their cultivation dating back thousands of years. Historical records, including those from the Harappan civilization and the Iron Age, document the cultivation of various millet species such as foxtail, kodo, finger, little, native small millet, browntop, and bristly foxtail millet. These grains are mentioned in ancient Indian scriptures like the Samhita, Purana, and Veda, reflecting their significance in traditional culture and diets. While absent during the Paleolithic era, minor millets became prevalent during the transition to settled farming communities around the Neolithic period. Archaeological findings from sites such as Burzahom in Kashmir and Chirand in Bihar further confirm the historical presence and importance of minor millets in India's prehistoric era. Studying these historical sources provides valuable insights into the names, cultivation practices, and distribution of minor millets, essential for harnessing their potential as nutrient-rich and sustainable food sources. Taxonomically, minor millets belong to the Paniceae tribe

B. Singh (✉) · L. Arya · M. Singh · P. K. Malav · C. D. Pandey · S. Pandey · N. Shekhawat · A. K. Singh
Division of Genomic Resources, ICAR-National Bureau of Plant Genentic Resources, New Delhi, India
e-mail: badal.singh@icar.gov.in

K. Venkatesh
ICAR, Indian Institute of Millet Research, Hyderabad, India

D. C. Joshi et al. (eds.), *Minor Millets*,
https://doi.org/10.1007/978-981-96-4265-6_1

within the Panicoideae subfamily of Poaceae. Phylogenetic studies in millets are crucial for understanding their evolutionary relationships, genetic diversity, and breeding potential. Such research informs the development of improved varieties, conservation strategies, and sustainable agricultural practices, ensuring the resilience and productivity of these important crops in the face of environmental challenges.

Keywords

Minor millets · Agriculture heritage · Production · Conservation strategies · Origin · Taxonomy and history

1 Introduction

For our nation to expand and prosper in a way that is inclusive and equitable, the second green revolution must focus on nutrition, something that the first green revolution, which was production-oriented, overlooked. Though the green revolution contributed to India's self-sufficiency, the wheat-rice cropping system's productivity is currently on the verge of stabilization. Traditional grains which are both environmentally friendly and nutritionally sufficient need to be introduced for agricultural diversification. Agronomically diversified species of small grain cereal grasses, known as millets (particularly minor millets) or Nutri-cereals, are well suited to a variety of marginal growing circumstances where main cereals like maize, rice, and wheat are largely ineffectual.

Several small-grained cereal grasses are grouped together under the word millet. A number of small-grained millets, such as foxtail millet (kangni), kodo millet (kodo), proso millet (cheena), barnyard millet (sawan), little millet (kutki), and browntop millet, have been classified as minor millets based on grain size, whereas major millets include sorghum, pearl millet, and finger millet. Minor millets are short, slender, and grassy plants best suited to be grown in arid, semiarid, or mountain zones as rainfed crops in marginal and submarginal soils. These cereals may thrive in unfavorable soil and climate circumstances, and some of their cultivars have short growth seasons (60–80 days), yielding roughly 0.5 t ha^{-1}, which is an improvement over no crop (Verma et al. 2015). Nonetheless, low yield is more than made up for by their nutritional content when compared to many cereal crops (Saleh et al. 2013).

Nutritionally, millets might be a useful tool to address hidden hunger and malnutrition, which affect people all over the world. Additionally, millets are free of gluten, making them a suitable dietary choice for people with celiac disease or gluten sensitivity. This feature supports its applicability and accessibility across a range of dietary patterns and is consistent with the concepts of fair food security (Shukla and Bhise 2023). Emerging agri-production challenges, which include climate change, increased biotic and abiotic stresses and increased pressure on availability of irrigation resources demands sustainable approaches to ensure food security of global population. Millets are cultivated in accordance with more general sustainability objectives. Practices for growing millet promote the preservation of biodiversity, lessen the need for artificial fertilizers and pesticides, and save delicate ecosystems. Furthermore, millet production has enormous social and economic benefits, reviving rural communities and giving smallholder farmers a variety of revenue streams. Therefore, realizing millets' enormous potential and to increase domestic and international demand and to supply people with nutrient-rich meals, the United Nations (UN) declared the year 2023 as the "International Year of Millets (IYoM-2023)."

In India, historically, before agriculture reform (green revolution), minor millets were major part of diet in most of the part of country. However, their area and production declined with the onset of new agri-technologies. For instance, estimated production of minor millets was 2500 metric tons in 1953, which has declined to 500 metric tons in 2021 (approximately) and forecasted to 160 metric tons in 2026 (Prabhu et al. 2022). Innovative production technologies like improved varieties, seed production, standard agronomic practices, and better policy initiatives are needed to reverse the forecasted trend and to meet the national as well as global demand of these crops. New innovations require knowledge of diversity, distribution, botany, and next-generation biomolecular information of these promising minor millets which shall be covered in this chapter.

2 Historical Significance and Origin

India has a long history of minor millets; its cultivation and usage date back thousands of years. The records of cultivation of foxtail millet during the Harappan civilization; kodo millet and finger millet during early Iron age (1200–1000 BC); little millet, native small millet (*Setaria* spp.), and browntop millet during the later Harappan period (2500–2000 BC), browntop millet and bristly foxtail millet (*Setaria verticillata*) during the Neolithic-Chalcolithic period are some of the recorded proofs of Indian history. It is crucial to organize the material on minor millets because it is dispersed among several ancient scriptures, including the Samhita, Purana, and Veda. Compiling historical information about millets is crucial to realizing their promise as a nutrient-dense, sustainable food source. Historic evidences show the cultivation of minor millets that were absent during the Palaeolithic era (10,000 BCE); however, its cultivation coincides with the transition from nomadic lifestyles to settled farming communities around the Neolithic period (Lu et al. 2009). Archaeological evidence consistently confirms presence of minor millets in India. For instance, minor millet grains have been found at Neolithic sites in India, including Burzahom in Kashmir and Chirand in Bihar, demonstrating the crop's importance as a vital food source in the prehistoric era (Singh 2009). Archeological evidences as well as ancient literature of India provides insight of historic name, cultivation, and distribution of minor millets (Table 1).

Table 1 History and archeology of minor millets in India (Veerabadhran et al. 2023)

S. no.	Crop	Historic name	Literature source	Period: archeological records
1.	Finger millet	*Ragi* *Balbaja*	Yajurveda Atharveda	Neolithic (7000–1000 BC): Hallur in Karnataka, Senuwar in Bihar Chalcolithic (3000–700 BC): Ojiyana in Rajasthan, Ahar in Rajasthan, Daimabad, Inamgaon, Nevasa, Tuljapur Garhi, and Apegaon in Maharashtra Harappan (2500–1750 BC): Rojdi, Kuntasi, Surkotada, Shikarpur, Oriyo Timbo, Babar Kot in Gujarat; Harappa in Punjab, Hulas, Malhar, and Raja-Nal-ka-Tila in Uttar Pradesh Iron age (1500–500 BC): Raja-Nal-ka-Tila, Imlidih Khurd, Waina, and Hulas Khera in UP; Naikund, and Bhagimohari in Maharashtra
2.	Barnyard millet	Shyamaka	Yajurveda Atharveda	Neolithic (7000–10,000 BC): Shangshan site, China (Yang, 2015)
3.	Proso millet	Varaka, Chinaka	Kshudra Dhanya Dhanwantari Nighantu	Chalcolithic (3000–700 BC): Raja-Nal-ka-Tila in Uttar Pradesh Harappan (2500–1750 BC): Rojdi, Surkotada in Gujarat; Harappa in Punjab Iron age (1500–500 BC): Semthan in Jammu and Kashmir; Raja-Nal-ka-Tila in Uttar Pradesh
4.	Little millet	Chanaka	Kshudra Dhanya Dhanwantari Nighantu	Harappan (2500–1750 BC): Farmana in Haryana; Rojdi, Rangpur, Babar Kot, and Oriyo Timbo in Gujarat
5.	Kodo millet	Kodrava	Kshudra Dhanya Dhanwantari Nighantu Ayurveda	Chalcolithic (3000–700 BC): Senuwar in Bihar; Lahuradewa, Malhar, Waina, Imlidih Khurd and Narhan in Uttar Pradesh; Daimabad, Inamgaon, Nevasa, Tuljapur Garhi, and Apegaon in Maharashtra Harappan (2500–175 BC): Rojdi in Gujarat Iron age (1500–500 BC): Naikund and Bhagimohari, Maharashtra; Rupar in Punjab; Narhan in Uttar Pradesh; Hallur and Veerapuram in Karnataka
6.	Foxtail millet	Kangu	Kshudra Dhanya Dhanwantari Nighantu Ayurveda	Neolithic (7000–1000 BC): Hallur and Kurugodu in Karnataka; Tokwa in UP; Senuwar in Bihar Chalcolithic (3000–700 BC): Senuwar, Tokwa, Malhar, Raja-Nal-ka-tila in UP Harappan (2500–1750 BC: Farmana in Haryana; Kanmer in Shikarpur and Surkotada, Rojdi, Oriyo Timbo and Babor Kot in Gujarat
7.	Browntop millet	Aanava	Yajurveda	Neolithic India (7000–10,000 BC): Deccan (Boivin et al. 2014) Chalcolithic (300–700 BC): Orissa and Gangetic belt (Harvey 2006)

3 Finger Millet

The term "Eleusine" originates from the ancient Greek city of Eluesis, sacred to Demeter, the deity of agriculture. "Coracana" comes from "kurukkan," the Sinhalese term for the grain. Finger millet's origins have sparked debates across continents and centuries. De Candolle (1886) and Dixit and Dixit (1987) argued for its Indian roots. In Sanskrit texts, it's referred to as "ragi" or "rajika." Burkill (1935) suggested *Eleusine coracana* derived from the wild species *Eleusine indica*, with early selection in India. While Mehra (1963) viewed it as African, Vavilov (1951) proposed independent origins in India and Africa. Around 5000 years BC, domestication began in western Uganda and the Ethiopian highlands, reaching the Western Ghats of India by 3000 BC (Mehra 1963; Hilu and deWet 1976; Hilu et al. 1979). Two recognized races, the African highlands and Afro-asiatic lowland, reflect distinct patterns of variability, with African germplasm showing greater diversity. Despite this, India boasts over 5000 years of cultivation, leading to vast genetic diversity in local cultivars and landraces, positioning it as a secondary center of genetic diversity (Padulosi et al. 2009). de Wet (1995) suggests domestication occurred in an area stretching from western Uganda to the Ethiopian highlands, where *E. coracana* subsp. *africana* (syn. *Eleusine africana*) thrives.

4 Barnyard Millet

Barnyard millet (*Echinochloa* species) is an ancient millet crop widely cultivated in Asia, particularly India, China, Japan, and Korea. In India, barnyard millet was offered to deities in Vedic rituals and has been mentioned as "Shyamaka" in ancient Indian textbook like Charak Samhita, Puranas, and Vedas (Veerabadhran et al. 2023). *Echinochloa frumentacea*, also known as Indian barnyard millet, and *Echinochloa esculenta*, also known as Japanese barnyard millet, are two of the most widely used species of barnyard millet among the numerous cultivated and wild varieties (Sood et al. 2015). In a separate study, domestication of barnyard millet has been proposed to be initiated alongside rice and harvested and processed simultaneously (Yang 2015).

Role in Ancient Civilizations: In ancient China, it was considered one of the five sacred grains alongside rice, wheat, barley, and soybeans (Chen 2023). Similarly, in India's Vedic period, barnyard millet was revered as an important crop that provided sustenance and nourishment to the population (Bisht and Mukai 2000). In religious rituals and ceremonies, it symbolized fertility and abundance. The grain was often used in offerings to gods or deities during harvest festivals (Wu et al. 2022).

Contributions to Agriculture: With its adaptability to various climates and ability to thrive under low water availability and poor soil conditions, barnyard millet plays a vital role in ensuring food security in regions with challenging agricultural environments. Moreover, its high nutritional content makes it an excellent source of dietary fiber, protein, vitamins (such as B-complex), and minerals like iron and calcium (Saleh et al. 2013 and Ugare et al. 2014). These attributes contribute not only to sustaining rural livelihoods but also promoting agro-biodiversity while addressing malnutrition challenges. The versatility of barnyard millet extends beyond human consumption; it can be used for animal feed production due to its rich nutrient profile.

Culinary and Nutritional Importance: Barnyard millet stands out for its superior nutritional profile compared to both major and minor millets, as noted by Singh et al. (2010), Saleh et al. (2013), Chandel et al. (2014), and other researchers. These studies highlight barnyard millet grains as rich sources of dietary fiber, iron, zinc, calcium, protein, magnesium, fats, vitamins, and various essential amino acids. Notably, barnyard millet boasts a lower average carbohydrate content compared to other millets, typically ranging from 51.5 to 62.0 g/100 g (Saleh et al. 2013).

Furthermore, barnyard millet contains a higher proportion of crude fiber than any other cereal, ranging from 8.1% to 16.3%, according to findings by Ugare et al. (2014). This high ratio of carbohydrates to crude fiber is beneficial for managing blood sugar levels, as it facilitates a slower release of glucose into the bloodstream. Research indicates that the resistant starch present in barnyard millet may contribute to reducing serum cholesterol, triglycerides, and blood sugar levels, as evidenced in studies involving mice (Krishna Kumari and Thayumanavan 1998). In a clinical trial involving human volunteers, Ugare et al. (2014) observed a reduced glycemic index (GI) among type 2 diabetes groups following regular consumption of barnyard millet meals. At the time, research indicated that barnyard millet exhibited a relatively higher protein content (11.2–12.7%) compared to other primary cereals and millets. Barnyard millet's iron content was notably higher than that of other cereals and millets, despite similarities in overall mineral, ash, fat, and amino acid contents. For instance, studies by Saleh et al. (2013), Renganathan et al. (2017), and Vanniarajan et al. (2018) consistently reported higher iron concentrations in barnyard millet grains, ranging from 15.6 to 18.6 mg/100 g. This logically positions barnyard millet as a superior source of iron compared to major cereals and millets. Moreover, the dehulling process has significantly reduced phytic acid levels, resulting in a lower phytate content in barnyard millet grains (3.30–3.70 mg/100 g), as highlighted by Panwar et al. (2016). This reduction in phytic acids enhances the bioavailability of minerals present in the grains. Consequently, barnyard millet emerges as an ideal dietary option for individuals suffering from anemia, particularly women in underdeveloped regions and those with lifestyle disorders. Additionally, barnyard millet exhibits twice the amount of polyphenols and carotenoids compared to finger millet. These compounds, as emphasized by Panwar et al. (2016), offer numerous potential health benefits to humans, further underscoring the nutritional value of barnyard millet.

5 Proso Millet

This crop is proposed to be domesticated first in China and likely to be cultivated preceding wheat. Basically, it is used for human consumption, livestock, and bird feed. This is one of the best adapted rotational crops for semiarid regions, mainly used for its dietary importance toward human health. Traditional use of this millet spans in Russia, Germany, and China and dates to 8000–10,000 BP mainly in cuisines. Russians used it as an ingredient for preparing sweet dishes and supplementing meat- and vegetable-based diets. Chinese people used to consume it without milk and sugar, frequently mixed with different vegetables like beans, potato, or squash. In Germany, people also used to take it as a sweet dish added with apples and honey (Santra and Rose 2013).

Role in Ancient Civilizations: Proso millet (PM) is one of the ancient millets in many countries in the world mainly in semiarid and drought-prone areas. PM is mainly consumed as food in those cultures from ancient times, and it had a pivotal role in the early development of multi-crop agricultural systems and settling of the farming societies (NPR millet).

Contributions to Agriculture: PM and other millets were replaced by higher-yielding crops like wheat, corn, and soybean in agricultural system of those countries during the twentieth century. As a result, the significance of PM and other millets in rural socio-economy was greatly reduced in the last century agrarian system. Proso millet is one of the oldest cultivated and among the first domesticated crops, mainly distributed in the semiarid region of Asia (East Asia, West Asia, and South Asia) and Europe. The probable reason of proso millet in the first league of domesticated crops is its short duration, minimum crop management, resilience to abiotic stress, and being nutritionally rich. These could be the important attributes that also made proso millet spread over Asia and European agricultural systems (Singode et al. 2023). It can grow in range of soils that are light to deep black and sandy to clay. Proso millet is grown in 0.77 million ha area, with a production of 1.19 million tons, 1535 kg per hac yield. A large proportion (89.7%) of the area under proso millet is located in five countries, viz., Russian Federation, the USA, Ukraine, North Korea, and Kazakhstan, which produce 86.4% of global proso millet. In India it is grown 0.41 lakh ha area, with 0.22 lakh tons production and 531 kg/hac yield. The major growing states are Tamil Nadu, Bihar, Karnataka, Andhra Pradesh, Madhya Pradesh, and northeastern states. This crop can be used as the best alternative for winter wheat-based dryland grain production systems for crop diversification and intensification. This also serves as a good water conservation crop in deep soils for wheat and also provides protection from weeds and insect pests.

Culinary and Nutritional Importance: The current and past decades are known for the use of proso millet as bird feed mainly. In in Russian, German, and Chinese dishes, proso millet porridge is quite popular. The nutritional value of proso millet can be compared with wheat (Table 1). Due to its mild flavor, light color, and gluten-free nature, it is mostly preferred in Europe and North America. Since it has a health benefit in celiac disease, several food industries use this millet in their gluten-free recipes such as pasta, bread, multigrain mixes, etc. The other culinary usage of these millets known is tongba, a millet-based alcoholic brew in Nepal and India, white millet soup, millet with mushrooms, and candied millet puffs, a specialty of Osaka, Japan (candied millet puffs, a specialty of Osaka, Japan (Santra and Rose 2013).

Geographical Distribution: Evidences from archaeological surveys report that proso millet was first domesticated about 10,000 BP in Northern China. It is distributed from Northern China to Himachal Pradesh of India, Nepal, Ukraine, Russia, Belarus, the Middle East, Turkey, Romania, and the Great Plains states of the US. After its domestication about 10,000 years ago, it moved from China to the Black Sea region of Europe by 5000 B.C., and in 1875 German-Russian immigrants introduced it to North America (Santra and Rose 2013). Weedy forms of proso millet covering a widespread area from the Caspian Sea. *Panicum miliaceum* is primarily a crop of temperate regions, but it is very adaptable and can be grown in climates which are too hot and dry and on soils which are too shallow and poor for successful cultivation of other cereals. In India cultivation as a grain crop occurs up to an altitude of 3000 m in the Himalayas (Tran 2017).

6 Foxtail Millet

Foxtail millet, based on archeological evidence and research, has been proposed to be originated in China (Hunt et al. 2008); however, its geographical origin remains controversial (Fukunaga and Kawase 2024). Kawase and Sakamoto (1987) proposed that foxtail millet originated in the region spanning from Afghanistan to India, in contrast to the theories of a Chinese origin and multiple origins, because accessions with less specialized genetic makeup and primitive morphological traits were discovered in that region. It was common for people in ancient China to combine foxtail with broomcorn grains along with rice (Tsang et al. 2017). Foxtail millet can survive high temperatures, droughts, disease and insect infestations, and low soil nutrition.

Culinary and Nutritional Importance: According to Bandyopadhyay et al. (2017) and Sharma and Niranjan (2018), foxtail millet is high in antioxidant compounds, minerals (3.3%), vitamins (VB1, 0.59 mg/100 g; VB2, 0.11 mg/100 g; VB3, 3.20 mg/100 g; VB5, 0.82 mg/100 g), fiber (6.7%), proteins (12.3%), amino acids, and antioxidants. According to Moharil et al. (2019), foxtail millet shares comparable amino acid compositions as corn; it has comparatively high levels of glutamic acid, alanine, aspartic acid, leucine, methionine, and lysine. Hou et al. (2022) gathered samples of foxtail millet from Jincheng City and Datong, China, and discovered that genotypic and environmental variables affected the protein levels. According to Li et al. (2021), the range of amylose contents in various foxtail millet varieties was 5.25% to 32.25%. Asharani et al. (2010) reported that the primary vitamin E isomers in finger millet were γ and α tocopherols, while δ and γ tocopherols were important in small and proso millets, and γ tocopherol was the isomer in foxtail millet. Virtamo et al. (2014) found that α tocopherol was the most physiologically active of these isomers, since it was shown to reduce prostate cancer mortality at a moderate dose. Moreover, foxtail millet possesses hypoglycemic and wound-healing qualities, increases the proliferation of intestinal flora, and raises blood cholesterol levels (Sharma and Niranjan 2018).

7 Kodo Millet

Early Utilization by Indigenous Communities: Traditionally integral to the diets of indigenous communities across India, kodo millet (*Paspalum scrobiculatum*) holds particular significance for two of Madhya Pradesh's seven tribal groups, the Gond and Baiga (Yadava and Jain 2006). Thriving in damp habitats within tropical and subtropical regions worldwide, kodo millet serves diverse purposes. While it functions as a valuable pasture crop in arid areas spanning Asia, New Zealand, and the USA, it retains indigenous status in India (Yadava and Jain 2006). Predominantly cultivated in Madhya Pradesh, Chhattisgarh, Uttar Pradesh, Tamil Nadu, Maharashtra, Karnataka, and certain parts of Andhra Pradesh, kodo millet is known by various names across the subcontinent (Ayyangar and Rao 1934; Deshpande et al. 2015). In Hindi, it's referred to as kodo, while in Tamil it's Varagu, in Telugu Arika, in Kannada Harka, and in Gujarati, Marathi, and Punjabi Kodra, among others. Moreover, it carries several English aliases, including bastard millet, creeping paspalum, ditch millet, Indian paspalum, koda grass, scrobic, water couch, and Kodohirse in German (Knees and Gupta 2013).

Role in Ancient Civilizations: In various regions worldwide, including India, Mexico, China, and Africa, millets feature prominently among the earliest archaeological plant discoveries. Despite this, these small millets have received notably less attention from archaeologists and botanists compared to the more prominent cereals like rice, wheat, barley, and maize. Archaeological findings from sites such as Harappa of the Indus Civilization have documented the recovery of small millet seeds, indicating their widespread cultivation and use during the period from 3300 to 1900 B.C. However, specific mention of kodo millet is lacking in these records (Weber and Fuller 2008). Kodo millet, on the other hand, was domesticated approximately 3000 years ago in southern Rajasthan and Maharashtra (Kajale 1977).

Contributions to Agriculture: In India, kodo millet cultivation spans approximately 2.44 lakh hectares (2005–2006), predominantly in southern, western, central, and northern states. This cultivation yields roughly 0.73 lakh tons of grains, with productivity averaging around 312 kg/ha. Madhya Pradesh and Chhattisgarh collectively encompass nearly 80% of the cultivated area, while Tamil Nadu, Maharashtra, Uttar Pradesh, and Gujarat also make substantial contributions.

Culinary and Nutritional Importance: Kodo millet stands out for its nutritional profile, rivaling popular cereals such as rice, wheat, barley, or bajra. It offers a remarkable 37–38% dietary fiber content, the highest among cereals, alongside elevated levels of polyunsaturated fatty acids (PUFA) and minerals compared to rice or wheat. Additionally, it is easily digestible and rich in essential B vitamins like niacin, B6, and folic acid, as well as essential minerals such as calcium, iron, potassium, magnesium, and zinc. Being gluten-free, it serves as an excellent alternative for individuals with gluten intolerance.

Regular consumption of kodo millet brings notable health advantages, particularly for postmenopausal women grappling with cardiovascular diseases, high blood pressure, and elevated cholesterol levels. The crop boasts valuable traits including high herbage, branched ears, numerous seeds per raceme, high fertility, and exceptional storage capability, rendering it one of the hardest among small millets. Its production potential is particularly noteworthy in marginal, low-fertility soils and regions characterized by chronic moisture deficits, thereby enhancing its resilience in modern agricultural systems (Yadava and Jain 2006).

8 Little Millet

Little millet (*Panicum sumatrense*) was originated in Indian peninsula (Weber and Fuller 2008). The species name is based on a specimen collected from Sumatra (Indonesia) (de Wet et al. 1983a, b). It is mainly cultivated in the Caucasus, China, East Asia, India, and Malaysia. Little millet was domesticated in the Eastern Ghats of India around 2000 years ago. It is primarily grown in peninsular Indian states such as Andhra Pradesh, Karnataka, Tamil Nadu, and Kerala today. The crop is adaptable to both dry and humid climates, allowing for cultivation in both drought-prone and water-logged places. It grows quickly and can resist severe conditions.

Contributions to Agriculture: Tribal and poor farmers cultivate the crop in low-fertile soils with few or no financial resources for food and feed. It outperforms other cereal crops in terms of rejuvenation. In India, the crop is cultivated on 291 thousand hectares with an annual production of 102 thousand tons and a productivity of 349 kg per hectare, which is significantly lower than other cereal crops.

Culinary and Nutritional Importance: These millets have very tiny seeds, yet they are highly nutritious and simple to harvest. They can be consumed raw or powdered and used to bake cakes and other baked goods. Moreover, they can be sprouted and added to salads. Although the seed can be cooked whole and turn extremely gelatinous, portion of the seed will pass straight through the digestive system without being digested since it is difficult to smash all of the tiny seeds in the mouth. The blossoms are added to ceremonial maize bread as a culinary coloring.

9 Browntop Millet

Browntop millet (*Brachiaria ramosa* (L.) Stapf; *Panicum ramosum* L.) was present in the staple-crop fields as a weed alongside other crops in southern India from beginning of the third millennium BCE (Fuller 2006). During the second millennium BCE, it reached to Gujarat and Tamil Nadu (Cooke et al. 2005). The grains of browntop millet have been observed in scattered sites in Odisha and the Gangetic plains. However, literature does not provide any authenticate information about commercial farming of this millet (Harvey et al. 2006). Up to the seventh century, existence of browntop millet was recognized at certain areas in Maharashtra (Paithan). However, with the passage of time, cultivation of this crop got substituted by more productive millets, including pearl millet, sorghum, finger millet, as well as foxtail millet.

Contributions to Agriculture: Among different small millets, the importance of browntop millet has been recognized recently as it has huge potential to make faring in resource-poor and fragile ecological conditions and thus can ensure economic and nutritional security as well as production sustainability of small-holders. The small millets can play an important role in this regard as these are ecologically sound, belong to C4 group of plants, tolerant to different adverse climatic conditions including drought, and can produce a moderate yield for food and nutritional sustainability. The crop can be utilized by people from economically weaker sectors for food purposes.

Cultural and Culinary Importance: In Karnataka, farmers popularly/locally called browntop millet as branched or chaduru korale and round panicle/dundu korale. Normally branched one has low pests and diseases, whereas round panicle browntop millet gives higher yields. They use the traditional drillers for sowing the seeds one inch below the topsoil. This method is recommended for a better yield. Five kg seeds are required per acre. Seeds germinate by the fifth day of sowing. Farmers get seven to eight quintal grains per acre and four tractor loads of good-quality fodder. They consume grains by making roti or as rice. The crop matures within 60–70 days. The shelf life of the seeds is about 5 to 6 years, while the shelf life of browntop rice is only about 25–30 days (Mallikarjuna 2016). Browntop millets are suitable for use as cover crop for erosion control, with the ability to accumulate significant amounts of lead and zinc in shoot and root tissues making it an important plant for remediation of contaminated soils and pasture and food for wildlife. Browntop millet tends to be ground into flour and used to make flat breads (roti, dosa) or polished and boiled to make gruel (anna, kheer). Some of these foods are used in religious rituals, which may partly account for its persistence in cultivation (Klmata et al. 2000).

Geographical Distribution: Browntop millet also known as pedda-sama known as signal grass, and it is one of the rarest millets. Being native to India, it grows well in the dryland tracts of Karnataka-Andhra Pradesh border areas, covering regions of Tumkur, Chitradurga, and Chikkaballapur districts in Karnataka and Anantapur districts in Andhra Pradesh. It is named differently in Indian languages as *korale* in Kannada and *Andakorra* in Telugu, and the crop is additionally grown and consumed in limited quantities in Bundelkhand region (Niyogi 2018).

Although its distribution is highly relict today, restricted to remote parts of Andhra Pradesh, Karnataka, and Tamil Nadu states in southern India (Klmata et al. 2000), it appears to have been a major staple crop in the late prehistory of the wider region of the Deccan (Fuller et al. 2004).

Native Regions: Browntop millet, which goes by the scientific name *Brachiaria ramosa* (L.) Stapf. or Urochloa ramosa (L.) R.D. Webster, is known locally as pedda-sama and korne and has a limited cultivation largely confined to southern India. Domestic and wild/weedy forms of browntop millet are found in agricultural systems, often within the same field. It is used as both a human food crop and fodder. Being domestic in India, it grows well in the dryland tracts of Karnataka-Andhra Pradesh border areas, Tamil Nadu, and Maharashtra (Bhat et al. 2018), covering regions of Tumkur, Chitradurga, Mandya, and

Chikkaballapur districts in Karnataka and Anantapur districts in Andhra Pradesh, and the crop is additionally grown and consumed in limited quantities in Bundelkhand region (north central India) (Reddy and Prasad 2017). In South India, *Brachiaria ramosa* appears to have been the first staple millet (Fuller et al. 2004) and has also turned up as an early crop/resource in the Neolithic Ganges (Harvey et al. 2006). In several parts of India, browntop millet is known by local names which translate to "illegal wife of little millet [*Panicum sumatrense*]," reflecting its tendency to grow within fields of little millet as a mimic weed (Sakamoto 1987).

Global Spread: The crop is grown in Africa, Arabia, China Australia, Bangladesh, Bhutan, Cambodia, India, Malawi, Myanmar; Nepal, Senegal, South Africa, Yemen, and Zimbabwe (Clayton et al. 2006). In 1915, it was introduced to the USA from India (Oelke et al. 1990), especially to feed game birds.

10 Teff

Teff (*Eragrostis tef*), a gluten-free cereal, originated in Ethiopia between 4000 and 1000 BC (Spaenij-Dekking et al. 2005). Ethiopia's Amhara region stands out as one of the leading producers of teff. Notably, the Gojjam zones (East Gojjam and West Gojjam) and Awi zones contribute significantly to teff production, representing the majority of the region's output. Despite its longstanding status as a staple food in Ethiopia and Eritrea, teff has only recently gained recognition as a food crop in other parts of the world.

11 Fonio Millet

Fonio holds significant cultural importance in West African traditions, notably in the creation myth of the Dogon people of Mali (National Research Council 1996). Its cultivation has been documented in historical records dating back to the fourteenth century in Mali and later in the Fouta Djallon highlands of Guinea (Cruz et al. 2016; Portères 1976), where it continues to be used as a symbolic dowry in weddings. Despite being domesticated in West Africa, the exact origins of fonio remain unclear. Some scholars argue it to be an ancient crop (Gestrich et al. 2021), while others suggest it may have been domesticated around 2000 years ago (Champion et al. 2021; Champion and Fuller 2018). Archaeobotanical analyses in West Africa are limited, with none conducted to date in Guinea.

12 Botanical Characteristics and Taxonomy

Within the grass family (Poaceae), there are approximately 8000 species spanning 600 genera. Historically, around 300 species were utilized, yet only 35 species across 20 genera have undergone domestication (de Wet 1986). Notably, finger millet (*Eleusine coracana*), proso millet (*Panicum miliaceum*), foxtail millet (*Setaria italica*), little millet (*Panicum sumatrense*, formerly *Panicum miliare*), barnyard (sawa) millet (*Echinochloa colona*, formerly *Echinochloa frumentacea*), kodo millet (*Paspalum scrobiculatum*), teff (*Eragrostis tef*), and fonio millet (*Digitaria exilis* and *Digitaria iburua*) are among the most significant and are highlighted in this chapter (Table 2). Taxonomically minor millets largely fall to taxon Paniceae (tribe) of subfamily Panicoideae under family Poaceae of plant kingdom (Vetriventhan et al. 2020) and have been summarized in Table 2.

13 Finger Millet

Growth Habit and Morphology: Finger millet, a sturdy annual grass, typically reaches heights ranging from 30 to 150 cm and completes its seed cycle within 75 to 160 days, as noted by DeWet (2006). Its leaves and culms exhibit a characteristic green color, while boasting a resilient fibrous root system that anchors firmly in the soil, making it resistant to uprooting. Finger millet comprises two subspecies: *E. coracana* subsp. *africana* (wild form) and *E. coracana* subsp. *coracana* (cultivated form). While *E. coracana* subsp. *africana* bears resemblance to *E. indica*, the latter is distinguishable by smaller spikelets and oblong grains.

The inflorescence of finger millet consists of digitally arranged spikes known as fingers, with each spikelet typically containing four to ten florets serially arranged. These spikelets measure around 5–8 mm in length and 3–4 mm in width. With the exception of terminal ones, all florets within the finger millet inflorescence are fertile. Notably, the seed pericarp is distinct from the kernel and can be readily separated from the seed coat. Finger millet grains exhibit varying shapes, ranging from oblong to round and oval, with color variations spanning from reddish-brown to white. In wild forms, seed shattering at maturity is common, while this trait has been eliminated in cultivated forms through domestication (De Wet et al. 1984). Finger millet seeds display seasonal dormancy, remaining viable in the soil and germinating only during the crop season (Chuah et al. 2004).

Reproductive Biology and Flowering Mechanism: The inflorescence of finger millet is a panicle composed of spikes, each containing numerous spikelets arranged sequentially, with four to ten florets per spikelet. Most florets within the finger millet inflorescence are perfect flowers, except for a few terminal ones. In each spike, spikelets open from the top downward, while within each spikelet, florets open from the bottom upward, with one floret opening per day. These florets are enveloped by a pair of scales known as palea and lower flowering glumes called lemma. Near the base of the ovary, two small scales known as lodicules are present. Florets possess a unilocular bicarpellary gynoecium with a superior ovary and a single ovule, while the androecium comprises three hypogynous stamens. The androecium completely surrounds the stigma, facilitating self-pollination (Gupta et al. 2012). Cross-pollination in finger millet is minimal due to the feathery stigma being covered by dehiscing anthers upon floret opening (Dodake and Dhonukshe 1998). Reported natural crossing estimates generally do not exceed 1% in finger millet (Rachie and

Table 2 Minor millets with there taxonomic classification, Growth Habit, Origin and Botany

S. no.	Crop name	Genus	Species	Subspecies	Race	Subrace	Growth habit	Native place	Inflorescence type	Pollination type (self/ cross)	Chromosome no.	Seed size	Seed shape
1.	Finger millet	Eleusine	*Coracana*	*Africana coracana*	Africana Spontanea Elongata Reclus a Sparsa Plana Compacta Vulgans	Laxa Senata Confusa grandigluma Lihacea, stellata, incurvata, digitata	Annual Herb	India and Africa	Panicle (fingers)	Self	$2n = 4x = 36$		Oblong to round and oval
2.	Barnyard millet	*Echinochloa*	*Echinochloa crus-galli*	*Echinochloa crus-galli* subsp. *utilis*, *Echinochloa colonum* subsp. *edulis* and *E. oryzicola*			Annual herb	S. and E. Europe to Asia, W., E,. and S. Tropical Africa to S. Africa, Madagascar	Compound ascending racemes (terminal panicle)	Self	$2n = 6x = 36$	0.39 × 3.05 mm	Ovate to lanceolate or elliptic
3.	Proso millet	*Panicum*	*Panicum miliaceum* L.	*Panicum miliaceum* subsp. *ruderale* (Kitag.) Tzvelev	Miliaceum	Miliaceum Patentissimum Contractum Compactum Ovatum	Annual herb	Indian subcontinent to Myanmar	Panicle, spikelets	Self	$2n = 4x = 36$	3 x 2 mm	Broadly ovoid,
4.	Little millet	*Panicum*	*Panicum sumatrense*		Nana robusta	Laxa, erecta Laxa, compacta	Annual herb	Tropical and subtropical Asia	Panicle, spikelets	Self	2n = 6x = 36	2.95 x 3.10 mm	Elliptical, oval
5.	Kodo millet	*Paspalum*	*Paspalum scrobiculatum*	NA	Regularis Regulans Vanabilis		Annual herb	Tropical and subtropical Old World to N. and E. Australia	Raceme	Self	$2n = 4x = 40$	0.5 x 2 mm	Ellipsoidal

6.	Foxtail millet	*Setaria*	*Setaria italica* (L.) P. Beauv.	*Viridis (wild) italica* (L.) Briq. (cultivated)	Moharia Maxima Assamense Indica	Anstata, fusiformis, glabra Compacta, spongiosa Erecta, glabra, nana, Profusa	Annual herb	Southern Asia	Terminal spike	Self	$2n = 2x = 18$	2 × 1.5 mm	Oval, convex
7.	Browntop millet	*Urochloa*	*Urochloa ramosa* (L.) T.Q.Nguyen	*Brachiaria ramosa* (L.) Stapf.)			Annual herb	Africa to tropical and subtropical Asia	Panicle, spikelets	Self	$2n = 4x = 38$	1 × 2 mm	Grains elliptic

Peters 1977). Anthesis and pollination in *Eleusine coracana* have been extensively studied and described by various authors, including Ayyangar (1932), Ayyangar and Wariar (1934), Chavan et al. (1955), Chavan and Shendge (1957), and Coleman et al. (1921). it efficient in making use of occurs abruptly. During the study, it was observed that the stigma emerged from the lemma covered by a thick cloud of pollen dust, with the elongation of the style and filaments closely associated with the bursting of anthers onto the stigma as it emerged. This limited the opportunity for cross-pollination, as anthers dehisced while still inside the flower, minimizing pollen loss. Following anther dehiscence, the flower closed, with no visible stigma, only empty and dehisced anthers protruding. Interestingly, under conditions of high humidity and low temperature, both the stigma and anthers were exposed when the flower closed. These behavioral sequences predispose finger millet to chasmogamous and cleistogamous reproduction (Dodake and Dhonukshe 1998).

Adaptations to Environmental Conditions: Finger millet exhibits adaptability to a variety of soil types, often thriving on reddish-brown lateritic soils with good drainage, particularly in tropical regions. It prefers free-draining sandy loam soils that receive consistent rainfall throughout the growing season. While finger millet demonstrates better tolerance to drought compared to major cereals, extended dry periods can markedly diminish both forage and grain productivity. Therefore, while finger millet may endure dry spells better than other crops, prolonged periods of drought still pose significant challenges to its growth and yield.

14 Barnyard Millet

Growth Habit and Morphology: Barnyard millet demonstrates a broad adaptability, thriving at altitudes of up to 2000 m above mean sea level during the summer season (Gupta et al. 2009). This tall, robust annual can reach heights of up to 220 cm, boasting a notably short generation time, with the fastest growth among all small millets, completing its life cycle from seed to seed in 45–60 days (depending on accession and growth environment) (Denton 1987; Padulosi et al. 2009). However, growth may take longer under the northern hill ecosystem, as reported by Hulse et al. (1981), who noted that *Echinochloa* millets may require 3–4 months to mature at high elevations. The leaf blades of barnyard millet are flat and wide, lacking ligules. Its inflorescence is a terminal panicle measuring 10- to 25-cm-long, featuring dense racemes of 3- to 4-mm-long spikelets (Napper 1965). These inflorescences are typically erect, occasionally drooping, with shapes varying from cylindrical, pyramidal, and globose to elliptic. Racemes, ranging from few to numerous, are densely crowded with spikelets arranged in four irregular rows on the triquetrous rachis and are loosely or tightly appended to the rachis. The spikelets exhibit colors ranging from green, brown to purple, and are crowded on one side of the rachis. They are typically two-flowered, either awnless or awned, with red or green awns, and positioned on short, rough pedicels subtended by two glumes (Mitich 1990). The lower floret is neuter (sterile), possessing a lemma and small palea, while the upper floret is bisexual (Gupta et al. 2010). The sterile lemma is five-veined, while the fertile lemma is plano-convex, elliptic, smooth, and shiny, abruptly sharp-pointed or cuspidate, with margins inrolled below over the palea and the apex of the palea not enclosed. The palea is flat, with a surface texture similar to that of the fertile lemma (Napper 1965). The ovary is superior and contains two distinct styles with plumose stigma, while the grain measures 2- to 3-mm-long and 1- to 2-mm-wide, enclosed in a white, shining, hardened lemma and palea.

Reproductive biology and flowering mechanism: Barnyard millet displays variability in various traits such as flowering time, inflorescence shape, morphological features, pigmentation of spikelets, and plant type, as noted by Obara (1938). Flowering typically begins at the top of the inflorescence and progresses downward over a period of 10–15 days. Flowers generally open between 5 to 10 am, with the highest number of flowers opening between 6 and 7 am (Sundararaj and Thulasidas 1976; Jayaraman et al. 1997). Within individual racemes, flowering initiates at the marginal ends before proceeding to the middle.

The flowers of barnyard millet are hermaphrodite, possessing both male and female reproductive organs. Prior to anther dehiscence, the stigmatic branches spread, facilitating flower opening (Seetharam et al. 2003). Late-season florets often remain cleistogamous, not opening (Maun and Barrett 1986). Barnyard millet primarily undergoes self-pollination and is self-compatible, although some degree of outcrossing has been recorded, facilitated by wind pollination (Maun and Barrett 1986; Potvin 1986, 1991).

Distinguishing between the two cultivated millet species, Indian and Japanese barnyard millets, is possible based on panicle morphology, texture of the glumes, and lower lemma (Yabuno 1971, 1987; de Wet et al. 1983a, b). *Echinochloa frumentacea* exhibits smaller, awnless spikelets with membranous glumes, contrasting with the larger, usually awned spikelets and chartaceous upper glumes and lower lemma in *E. esculenta*. Based on inflorescence morphology, *E. frumentacea* has been classified into four races: Stolonifera, Intermedia, Robusta, and Laxa. Similarly, *E. utilis* (Syn. *E. esculenta*) has been classified into two races: utilis and intermedia.

15 Proso Millet

Growth Habit and Morphology: Proso millet is an annual plant with usually erect plant type having 30- to 90- to 100- to 120-cm-tall plant height with few tillers and an adventitious root system. Slight hairs are present on leaves and stems. Stem are slender and glamorous with hollow internodes. However, Indian genotypes with heavy branching and tillering are reported. No ligules but short auricles are present in the leaves which are nearly 30-cm-long. The stem is terminated by a drooping panicle 10- to 45-cm-long that may be open or compact. Proso ($2n = 36$) is considered a self-pollinated crop, but natural cross-pollination may exceed by 10% (Popov 1946). Physiological maturity starts from top to bottom of the panicle. Since, the seed maturity in panicles are not uniform, delayed harvesting may lead to yield losses. The inflorescence of drooping panicle may have different degrees of compactness.

Reproductive Biology and Flowering Mechanism: The main rachis of panicle is glamorous. Spikelets with two flowers are about half centimeter long. Plumes partially close the flowers. The five-nerved outer glume is short and broad. However, the inner glume is almost as long as the spikelet and is a broad, pointed, many-nerved. The lower flower of the spikelet is sterile, consisting of a lemma and a very much reduced palea only, while the upper flower is perfect with a stiff, broad, shining lemma and an equally stiff, broad palea, which also becomes shining as the grain matures. The perfect flower contains two lodicules, three stamens, and an ovary with two long styles and feathery stigmas. The flowers are almost completely self-fertilized and produce a nearly globular grain, enclosed tightly in the persistent lemma and palea. In proso millet the flowering starts from the top of the panicle and proceeds toward the bottom of the panicle. Floral anthesis starts from 10 AM in the morning hours and remains effective till 12 in the noon. It takes about 10–15 days to complete the flowering of all the florets. It has been observed that the pollen shedding from anthers coincides with stigma receptivity. Anthers remain sticky when the florets are open, but the moment anthesis commences, the anthers shed pollen and become dry. The duration of flower opening spans from 10 to 15 min. Elevated temperatures, dry weather, and intense sunshine may enhance the flowering.

Seed Formation and Maturity: Domestic proso millet has yellow or light brown seeds, while the weedier wild proso has brown to black seeds. Seed maturity is the crucial and most important factor determining the seed quality under successful seed production program. Physiological maturity of the seeds is widely defined as the stage at which the seed attains its maximum dry weight and related physical characters (Shaw and Loomis 1950). This can be assessed by visualizing changes in seed or floret color. Seed maturity is not uniform in panicles. Fixation of the optimum stage of physiological as well as harvestable maturity will ensure better quality seed and reduce shattering loss. Harvesting of proso millet seed crop within a week after physiological maturity will enable to get seeds of best quality and minimize the seed losses.

Adaptations to Environmental Conditions: Proso millet is a C4 plant, which makes it efficient in making use of limited water resources. Also, its short life cycle (60–90 days) makes it a suitable crop for escaping drought conditions, thus, suitable for dryland crop rotations in arid and semiarid regions. Growing this crop with winter wheat or peas may help increasing the economic benefit as well as provide healthy crop rotation. Being a C4 cereal crop, it has more capacity to convert carbon dioxide into oxygen that helps reduce CO2 emission. Producing proso millet is estimated to result in 25% fewer carbon emissions as compared to corn making it a grain that helps with mitigating climate change (drylandgenetics.com). Its fibrous root system gains its nutrients and water from the topsoils and leaves the deeper water available for deep-rooted crops. Also, it reduces soil erosion. Oftentimes it does not need nitrogen application and thus reduces the agricultural input cost. To an extent it provides nutritional security too in resource-poor areas and indulges in a protein-rich grain. Proso millet has the highest percentage of protein across all millet species with 12.5% protein. However, there is a need to assess the yield projections and climate sensitivity under changing climate conditions. However, a few studies such state that proso millet is an excellent candidate for low-input farming systems under changing climate (Wimalasiri et al. 2023).

16 Foxtail Millet

Growth Habit and Morphology: The C4 annual monocot foxtail millet has thin, upright, leafy stems that can reach heights of 90–180 cm. Its adventitious roots are often slender and lanky, and its root system is thick. The culms are vertical and thin and have hollow internodes; the leaves are arc-broad and hairless. Its architecture is reminiscent of a unique domesticated plant with a single stalk or few tillers, as well as big inflorescences that mature in a very consistent manner (Doust et al. 2009). The inflorescence is a narrow panicle with tiny branches that resemble spikes and frequently nod at the top. The panicle takes on the appearance of a foxtail due to the dense packing and combination of stiff bristles (usually long sterile branches, frequently reddish or purplish in color) with the spikelets (one yellow pistilled flower per spikelet) (http://www.plantguide.org/foxtail-millet.html).

Reproductive Biology and Flowering Mechanism: It can generate hundreds of seeds per flower and has a comparatively short generation time (up to 15 weeks) (Reddy et al. 2006; Doust et al. 2009). The tiny seeds (about 2 mm in diameter) are covered in a fragile, thin hull that easily separates during the threshing process. Additionally, there are notable differences in seed color among foxtail millet cultivars, including yellow, red, white, and black.

17 Kodo Millet

Growth Habit and Morphology: Kodo millet is a robust, annual tufted grass featuring a shallow root system and typically reaching heights of 60–90 cm. Its stems are characterized by being glabrous, with nodes that swell and turn purple as they mature. The leaves are simple, alternate, and either erect or suberect, measuring up to 40 cm in length, with surfaces that are usually glabrous but may occasionally bear soft hairs. Maturation of kodo millet occurs within 3–4 months, with average yields ranging from 250 to 1000 kg/ha and a potential yield reaching 2000 kg/ha (Hulse et al. 1981; Harinarayana 1989). The 1000 grain weight of kodo millet is approximately 6.7 g. While the crop thrives in full sunlight, it can also tolerate partial shading. Kodo millet primarily undergoes self-pollination, with flowers that are predominantly cleistogamous.

Reproductive Biology and Flowering Mechanism: The inflorescences of kodo millet comprise three to five racemes arranged alternately, each measuring up to 13 cm in length. These racemes bear spikelets in two to three rows along one side of a flattened rachis. The grains themselves are elliptic-orbicular, ranging from 1.5 to 2.5 mm in length, and are enclosed in hard, corneous husks that can vary in color from light red to dark grey (de Wet et al. 1983a, b). Flower opening typically occurs between 2:30 am to 11:30 am, although this timing can vary by region, with certain lines exhibiting protogyny (Youngman and Roy 1923; Ayyangar and Rao 1934; Verma 1989; Yadava 1997).

Adaptations to Environmental Conditions: The extensive cultivation of minor millets over centuries and their adaptation to diverse regions characterized by harsh farming conditions have resulted in significant genetic variability within these crops. Among them, kodo millet stands out for its exceptional hardiness and highest drought resistance, enabling it to yield well even in growing periods lasting between 80 and 135 days.

18 Little Millet

The plant is 5–2.5 m tall, with sturdy stems that are somewhat branched and covered in long hair. Its color is primarily reddish or purplish. Stipules are absent; leaves are arranged spirally, simple, and entire; the petiole can grow up to 8-cm-long; the blade is broadly ovate to rhomboid-ovate; and the inflorescence can grow up to 1.5 m in length and is complex, with many agglomerated cymes, axillary, terminal, and pendant spikes and bracts that are 2- to 4-mm-long, pale, membranous, and adorned with a long awn. Little millet has sessile, unisexual blooms with five mucronate tepals and filaments approximately 1 mm in length; three stigmas crown the female flowers. Fruit is ovoid, long, circumscissile, nearly smooth or slightly wrinkled, and it has one seed. It narrows suddenly to a short, thick beak.

19 Browntop Millet

Growth Habit and Morphology: Browntop millet is an annual warm-season species that grows 1–3 ft. tall. The smooth stems have pubescent nodes and may stand erect or ascend from a decumbent base. The leaves are 2.2- to 18-cm-long and 6- to 18-mm-wide; both surfaces are smooth. The inflorescence is indeterminate, open, spreading with simple axis and stalked flowers. It has 3–15 inflorescences and white flowers. Seeds are ellipsoid and tan in color; they mature in approximately 60 days (Sheahan 2014).

Like other millets, it is ecologically hardy and prominently considered for cultivation in drylands. This millet features a wider adaptability and is grown in a variety of soils and climates, having the quality to tolerate a moderate level of environmental stress as C4 plants. Being a facultative upland plant, browntop millet prefers sandy loam soils with a pH 5–6.5 under full sun. Unlike other millets, browntop millet has a unique quality as it can be grown in the partial shade which ensures wider choice of adoption even in fruit orchards or in agro-forestry system. Browntop millet grows in rocky, shallow soils from sea level up to 2450 m (FAO 2007), and it is adaptable to almost all upland soil (Mitchell and Tomlinson Jr. 1989), but it does not survive in temperature less than 11 °C (Sheahan 2014).

The identification of browntop millet grain and spikelets can be difficult due to its similarity to *Setaria italica.* Although the panicle is distinct from Setaria by being looser and non-bristly, the grains themselves are very similar. Grains are ovate to round and have a long embryo, roughly two thirds to three fourths of the length of the grain. They tend to be smaller than *Setaria italica* and squatter in cross section. The surface of well-preserved grains can be used for identification as these have a distinctive undulating pattern, although this again has similarities to *S. italica* (Fuller et al. 2004). The husk has a fine-beaded and rugose pattern, which again has some resemblance to that of *Setaria* spp., but it is somewhat coarser than *S. italica* and finer than *S. verticillata*. Morphotypes or races are not known in guinea millet and browntop millet.

Pollination and Seed Formation: Browntop millet is a warm season annual grass that is a heavy seed producer. This millet seed is grown in a variety of soils and climates. Browntop millet is used as a wildlife food plot crop and livestock summer grazing crop, for erosion control and hay production and as a food grain crop.

20 Teff

Teff stands out as an underutilized crop and the sole cultivated cereal species within the genus *Eragrostis* of the tribe Eragrostidae, which encompasses approximately 350 species worldwide (Tefera et al. 2003). In the flora of Ethiopia and Eritrea alone, there are 44 species representing this genus (Phillips 1995). Eragrostis is the largest genus in the subfamily Chloridoideae, a lineage that has received comparatively less research attention than other grass subfamilies like Pooideae, Panicoideae, and Bambusoideae (Zhu et al. 2012). The grains of teff are remarkably small, measuring between 1.0 and 1.7 mm in length and 0.6 and 1.0 mm in width (Bedane et al. 2015). While its seed color can range from white (ivory) to dark brown (black), the most common colors are white, red, and brown. Teff, like many C4 crop plants, demonstrates versatility in its ability to thrive across various agroecological zones in Ethiopia, including arid and semiarid regions prone to drought and heat where maize, wheat, and rice struggle (Teferra et al. 2000; Gressel 2008; Araya et al. 2010). Its capacity to adapt to colder climates enables cultivation within altitudinal ranges spanning from 800 to 3200 meters above sea level (a.s.l) (Tefera 2011).

21 Fonio Millet

The genus *Digitaria* comprises approximately 230 species, among which *Digitaria exilis*, commonly known as "hungry rice" or "true fonio," and *Digitaria iburua*, black fonio, are two domesticated species. True fonio is a tetraploid species with $2n = 4x = 36$. Notably, it is renowned for its short-growing season, ranging from 40 days to 3.5 months. This agronomic characteristic is likely attributed to its rapid germination, occurring within 3–4 days after sowing, and early vigorous growth (Purseglove 1972). The seeds of true fonio are exceedingly small and surrounded by glumes and lemmas, resembling the appearance of finger millet, barley, and rice (Ibrahim Bio Yerima and Achigan-Dako 2021).

21.1 Phylogeny

21.1.1 Finger Millet

The cultivated species *E. coracana* is an allotetraploid with a chromosome count of $2n = 4x = 36$, exhibiting morphological similarities to both *E. indica* ($2n = 18$) and *E. africana* ($2n = 36$). Cytological, biochemical, and molecular evidence suggests that *E. indica* contributed one of the genomes (designated as AA) to cultivated *E. coracana* and also serves as the maternal genome donor to *E. africana* (Hilu 1988, 1995; Werth et al. 1994; Mallikharjun et al. 2005). *E. africana*, also tetraploid, shares significant morphological features with *E. coracana*, indicating a genetic relationship and natural gene flow between them. This suggests that *E. coracana* potentially originated from *E. africana* through selection and further mutation toward larger grains (Chennaveeraiah and Hiremath 1974; Hilu and deWet 1976).

Studies on ribosomal DNA (rDNA) sites on chromosomes of diploid species like *E. indica* and *Eleusine floccifolia*, as well as tetraploid species like *E. africana*, have shown similarities. This suggests that the two diploid species might have contributed two genomes to *E. africana*. However, later evidence disapproved *E. floccifolia* as the B genome donor based on nuclear internal transcribed spacer (ITS) and plasmid trnT-trnF sequences (Bisht and Mukai 2000; Bisht and Mukai 2001; Neves et al. 2005).

Biochemical and genetic evidence indicates a sister relationship between *E. indica* and *Eleusine tristachya*, as well as between *E. floccifolia* and *Eleusine jaegeri* (Liu et al. 2011; Hiremath and Chennaveeraiah 1982; Hiremath and Salimath 1991; Hilu and Johnson 1992). In plastid phylogeny, three polyploid species, namely, *E. africana*, *E. coracana*, and *Eleusine kigeziensis*, share a common ancestor with the *E. indica*-*E. tristachya* clade, which represents a source for the maternal parents. The paternal parents of these three tetraploids are believed to no longer exist (Liu et al. 2014).

21.1.2 Barnyard Millet

Members of the *Echinochloa* genus encompass both minor cereal crops and major weeds in fields. Understanding their interspecific relationships has been challenging due to high morphological variation, leading to ambiguous phylogenetic understandings among taxa (Hilu 1994; Michael 1994, 2001). The most comprehensive insights into this genus have stemmed from a series of studies by Yabuno (1962, 1984, 1996, 2001).

Two cultivated species, Japanese barnyard millet (*E. esculenta*) and Indian barnyard millet (*E. frumentacea*), share close relationships with their wild counterparts, *Echinochloa crus-galli* and *E. colona*, respectively. Both cultivated species and their progenitors are hexaploid with a chromosome count of $2n = 6x = 54$, where $x = 9$ (Yabuno 1962, 1966). Interspecific hybrids resulting from crosses between *E. crus-galli* × *E. esculenta* and *E. colona* × *E. frumentacea* exhibit normal meiotic division with 27 bivalents. However, crosses between these two groups, involving cultivated species and their ancestors, result in meiotic irregularities such as univalents, laggards, and micronuclei. These cytogenetic findings suggest that the hexaploid wild species *E. colona* and *E. crus-galli* are potential progenitors of *E. frumentacea* and *E. esculenta*, respectively, and that the two cultivated species have distinct genomic compositions (Yabuno 1966). Crosses between *E. crus-galli* and *Echinochloa oryzoides* reveal 18 bivalents and 9 univalents, suggesting that 2 of the 3 genomes of *E. crus-galli* are homologous to *E. oryzoides*. This indicates that *E. crus-galli* is an allohexaploid formed through natural hybridization between the tetraploid *E. oryzoides* and an, as yet, undiscovered diploid species of *Echinochloa* (Yabuno 1966, 1984).

Crossing experiments and observations of meiotic chromosomal behavior in interspecific hybrids led to the classification of *Echinochloa* species into cross-compatible groups, indicating polyploidization and amphidiploidization in the evolutionary diversification process of this genus (Yabuno 1966, 1984, 2001). However, reliance on conventional taxonomic traits such as morphology, reproduction, and ecology may yield ambiguous results due to domestication and adaptive syndromes (Yabuno 1966, 1984, 2001; Yamaguchi et al. 1996).

Subsequent efforts to understand the genetic relationships among Eurasian annual *Echinochloa* species involved the use of molecular markers by various researchers (Yasuda et al. 2002; Yamaguchi et al. 2005; Ruiz-Santaella et al. 2006). Molecular analysis confirmed that Asian annual weedy *Echinochloa* species, including their domesticated counterparts, can be grouped into three cross-compatible groups: the *Echinochloa oryzicola* group, *Echinochloa crus-galli* complex (group), and *Echinochloa colona-frumentacea* group (Yamaguchi et al. 2005). Species divergence is evident from nucleotide sequencing of specific regions of cpDNA, with the three groups showing distinct cytoplasmic lineages (Yamaguchi et al. 2005).

The tetraploid *E. oryzicola* includes wild representatives such as *E. oryzicola* (Vasing.) and *Echinochloa phyllopogon* (Stapf) Koss., along with Mosuo barnyard millet. The hexaploid *E. crus-galli* group comprises four wild varieties of *E. crus-galli* (L.) P. Beauv. and the cultivated species *Echinochloa esculenta* (A. Braun) Scholz (Yabuno 1984, 1996). The hexaploid *E. colona-frumentacea* group includes the wild species *E. colona* (L.) Link and the cultivated *Echinochloa frumentacea* (Roxb.) Link.

The hypothesis that *E. crus-galli* originated from a hybrid between *E. oryzicola* and an unknown diploid species (Yabuno 1966) was confirmed using internal transcribed spacer and chloroplast DNA sequences by Aoki and Yamaguchi (2008). They found the same nuclear lineage between *E. oryzicola* and *E. crus-galli*, indicating that *E. oryzicola* is the paternal donor of *E. crus-galli*. Additionally, phylogenetic analysis of the sh4 gene in *Oryza* detected homoeologue copies in *E. oryzicola*, *E. crus-galli*, *Echinochloa stagnina*, *E. colona*, and *E. crus-pavonis*. This indicated a genomic relationship between Asian *Echinochloa* species, supporting that the allohexaploid *E. crus-galli* shares two genomes with its parental donor, *E. oryzicola*. The Asian perennial tetraploid species *E. stagnina* shares one genome with *E. oryzicola* and possesses an unknown genome, *E. crus-pavonis* from the New World, which shows a close affinity of two genomes with *E. crus-galli* and *E. oryzicola*. Meanwhile, *E. colona* exhibited distant affinities in all homoeologous copies (Aoki and Yamaguchi 2009).

21.1.3 Proso Millet

Proso millet is a tetraploid species with basic chromosome number x = 18. The species is reported to be an allotetraploid that appeared as a consequence of wide hybridization between two different parental ancestors. A close relative *Panicum capillare* is supposed to be a contributor of one of the two genomes in Panicum miliaceum. The second subgenome was contributed by allotetraploid species *Panicum repens* that received one of its genomes from an unknown ancestor (Hunt et al. 2014). Diversion of the two subgenomes in proso millet occurred 5.6 million years ago. However, little fractionation has occurred in its genome, and an appreciable quantum of genes has been retained in both the subgenomes. A genome size of 920 mega base pairs was provided by Zou et al. 2019.

21.1.4 Foxtail Millet

There is ongoing debate on the foxtail millet's origins. The majority of evolutionary research suggested that foxtail millet was domesticated in China from *Setaria italica* ssp. Viridis; however, a small number of studies suggested that it was domesticated independently in other regions, such as Central and West Asia to Eurasia. Phylogenetic study reveals the domestication/natural selection of foxtail millet involves evolution of several genes viz., waxy controlling amylose content in the endosperm, the polyphenol oxidase (PPO) gene for the phenol color reaction (Phr) in grains, two genes involved in heading date (heading date 1 [HD1], pseudo-response regulator 37 [PRR37]), shattering genes (qSh1, Sh1, and SvLes1), and the C gene involved in leaf sheath pigmentation (Fukunaga and Kawase 2024). However, detailed study is still required to confirm its origin and evolution.

21.1.5 Kodo Millet

Taxonomically, kodo millet belongs to the family Poaceae, subfamily Panicoideae, tribe Paniceae, and genus *Paspalum*. The genus *Paspalum* encompasses approximately 250 species, among which are *Paspalum scrobiculatum* L, *Paspalum notatum*, *Paspalum conjugatum*, *Paspalum compactum*, and *Paspalum dilatatum*, cultivated for their grain or as fodder. The wild progenitor of paspalum species is *Paspalum sangunale*. Historical records from the Hassan district of Karnataka during 1772–1782 indicate the presence of *P. compactum*, *P. conjugatum*, *Paspalum orbiculare*, and *P. scrobiculatum* (Saldanha and Nicolson 1978).

21.1.6 Little Millet

Little millet was domesticated in India 5000 years ago. *Panicum silopodium* is a weed that has been domesticated into little millet. Tetraploid ($2n = 4x = 36$) plants belong to the Poaceae family, including little millet. The hybrid chromosomes of *P. silopodium* and *Panicum sumatrense* show just one quadrivalent and a very complete pairing, indicating the divergence of the two species. Based on panicle morphology, little millet is classified into two races: nana and robusta, each of which has two subraces: nana's laxa and erecta and robusta's laxa and compacta (Patil et al. 2024).

21.1.7 Browntop Millet

There is no conclusive phylogenetic study regarding browntop millet. Due to its propensity to develop as a mimic weed in tiny millet fields, browntop millet is known by regional names that roughly translate to "illegal wife of little millet [*Panicum sumatrense*]" in various parts of India (Sakamoto 1987). Browntop millet has indehiscent or shattering spikelets and can develop in an open or compact panicle. On the other hand, the domestic form is compact and somewhat indehiscent, and it behaves similarly to other domestic grains (Klmata et al. 2000). Browntop millet is typically managed as an insurance crop in areas where it grows as a weed in other millet fields.

21.1.8 Teff

Teff, an allotetraploid with a chromosome count of $2n = 4x = 40$ (Ayele et al., 1996), traces its origins to two diploid progenitors (Jones et al. 1978). Although the precise diploid progenitors of teff remain unidentified, numerous studies focusing on morphological, anatomical, cytological, and biochemical traits indicate a close relationship between teff and *Eragrostis pilosa*, a wild allotetraploid known for its early maturity and seed shattering. Interestingly, *E. pilosa* is the only species known to be cross-compatible with modern teff varieties (Costanza et al. 1979). This evolutionary link is further supported by DNA sequence data from the nuclear waxy gene and plastid locus rps16, as reported by Ingram and Doyle (2003). Finger millet (*Eleusine coracana*) stands out as the closest cultivated species to teff, with teff's closest subfamily being Panicoideae, which includes maize, sorghum (*Sorghum bicolor*), and pearl millet (*Pennisetum glaucum*) (Cannarozzi et al. 2014).

21.1.9 Fonio Millet

The precise ancestral species of *Digitaria exilis* remains uncertain. However, the closest wild relative of true fonio is believed to be *Digitaria longiflora* (Hatz) Pers., which thrives in hot regions across Africa and Asia (Harlan Jr 1992). Additionally, *Digitaria fuscescens* (Presl) Henr., a species morphologically akin to *D. longiflora*, has been suggested as another potential ancestor for true fonio (Porteres 1955).

22 Future Prospects and Conservation

Minor millets are important for maintaining food security even though they are often overlooked in favor of major cereal crops like rice, wheat, and maize. This is especially true in areas that are susceptible to environmental degradation and climate change. Minor millets are very nutrient-dense and full of important elements like dietary fiber, protein, and vitamins and minerals (Saleh et al. 2013). Minor millets add variety to diets and encourage better eating practices, which improves nutrition and well-being for groups that are more vulnerable. Minor millets can thrive in a variety of agroclimatic environments, such as arid and semiarid areas with poor soil fertility and unpredictable rainfall patterns. The production of minor millets improves agricultural systems' resilience to climate change and water constraint by diversifying cropping systems and lowering reliance on water-intensive crops like rice and wheat. In contrast to commodity crops, which are frequently prone to price swings and market instability, minor millets provide farming communities with a reliable source of income and a secure means of subsistence. Furthermore, their value-added goods, such as millet flour, snacks, and drinks, have become well-liked health foods in metropolitan markets, giving rural women and other marginalized groups new chances for entrepreneurship and revenue production (Saleh et al. 2013). Despite of its potential of significant contributions to food security, nutrition, and sustainable agriculture, minor millets face threats such as genetic erosion, habitat loss, and changing agricultural practices. Adequate genetic diversity is essential for the adaptation and resilience of crop species to changing environmental conditions, pests, and diseases. Conserving diverse germplasm collections, through traditional farming practices, seed saving, local knowledge systems, community seed bank, and long-term seed gene banking is crucial for preserving valuable genetic traits and ensuring the long-term sustainability of minor millet cultivation. By including minor millet conservation into more general sustainable development objectives, we may improve food systems' resilience, encourage biodiversity conservation, and deal with issues like food insecurity and climate change. Exploiting inherent property of low agri-input requirement of minor millets along with developing high-yielding varieties without compromising the nutrient composition will be the breakthrough for future sustainable food and nutritional security goals.

23 Conclusion

Minor millets are among the first crops to be domesticated in India, and they have been cultivated by ancient agricultural communities in many different parts of the world. Millets have provided populations with essential nutrition and ecological benefits throughout Indian history. The domestication and cultivation of millets have not only contributed to agricultural biodiversity but also to cultural diversity, as evidenced by their integral roles in various cuisines and traditions across different regions. Historical evidence underscores their cultural significance and provides valuable insights for revitalizing traditional agricultural practices. The renaissance of minor millets in India represents a pivotal shift toward sustainable, nutritious agriculture. Rooted in ancient traditions yet embracing modern scientific advancements, millets offer a promising solution to contemporary challenges of food security and environmental sustainability. Through phylogenetic studies, we can uncover their genetic diversity, enabling targeted breeding efforts for enhanced resilience. Understanding the origin and evolution of these crops through generating information related to phylogeny, biosystematics and genomic make up are crucial for enhanced utilization and improvement of these crops. Taxonomic studies of minor millets illuminate their genetic heritage, aiding conservation and breeding efforts. It has been disputed how millets spread throughout the world after being domesticated, especially in India, which grows the greatest variety of millet crop species as a staple diet. Despite uncertainties surrounding the exact progenitors of certain millet species like teff and fonio, ongoing research continues to unravel their genetic relationships and evolutionary histories. Minor millets like proso millet, kodo millet, and foxtail millet were brought to India either in the medieval or colonial eras, according to common literature. However, new analyses based on data from archaeobotany show that they existed considerably earlier, in antiquity. More evolutionary research is required to trace the genetic diversity and cultural legacies of accessions from around the world using conventional and molecular techniques, given the divergent opinions regarding the spread of millets and the growing interest in them as nutritious cereals worldwide. As research into millets progresses, it promises to uncover more about their genetic diversity, evolutionary pathways, and potential contributions to sustainable agriculture and food systems in the future. Mining genotypes best suited for enhancing productivity, postharvest processing, and value-added products are the important future thrust area to work on.

References

Aoki D, Yamaguchi H (2008) Genetic relationship between Echinochloa crus-galli and Echinochloa oryzicola accessions inferred from internal transcribed spacer and chloroplast DNA sequences. Weed Biol Manag 8(4):233–242

Aoki D, Yamaguchi H (2009) Oryza sh4 gene homologue represents homoeologous genomic copies in polyploid Echinochloa. Weed Biol Manag 9(3):225–233

Araya A, Keesstra SD, Stroosnijder L (2010) A new agro-climatic classification for crop suitability zoning in northern semi-arid Ethiopia. Agric For Meteorol 150(7–8):1057–1064

Asharani, V. T., Jayadeep, A., & Malleshi, N. G. (2010). Natural antioxidants in edible flours of selected small millets. International Journal of Food Properties, 13(1), 41–50.

Ayyangar GNR (1932) The inheritance of characters in ragi, Eleusine coracana Gaertn. Madras Agric J 20:1–9

Ayyangar GN, Rao VP (1934) Studies in Paspalum Scrobiculatum, L. The Kodo Millet. Madras Agric J 22:419–425

Ayyangar RG, Wariar UA (1934) Anthesis and pollination in Ragi, Eleusine Coracana (Gaertn.): the finger millet part-II. Indian J Agric Sci 4:386

Ayele, M., Doleže, J., Van Duren, M., Brunner, H., & Zapata-Arias, F. J. (1996). Flow cytometric analysis of nuclear genome of the Ethiopian cereal tef [Eragrostis tef (Zucc.) Trotter]. Genetica, 98, 211–215.

Bandyopadhyay, T., Jaiswal, V., & Prasad, M. (2017). Nutrition potential of foxtail millet in comparison to other millets and major cereals. The foxtail millet genome, 123–135

Bedane GM, Saukuru AM, George DL, Gupta ML (2015) Evaluation of teff ('Eragrostis tef'[Zucc.] trotter) lines for agronomic traits in Australia. Aust J Crop Sci 9(3):242–247

Bhat S, Ganiger PC, Nandini C, Thippeswamy V (2018) Browntop millet—a review. Agri Res and Tech. Open Access J 14:555937. https://doi.org/10.19080/ARTOAJ.2018.14.555937

Bisht MS, Mukai Y (2000) Mapping of rDNA on the chromosomes of Eleusine species by fluorescence in situ hybridization. Genes Genet Syst 75(6):343–348

Bisht, M. S., & Mukai, Y. (2001). Identification of genome donors to the wild species of finger millet, Eleusine africana by genomic in situ hybridization. Breeding Science, 51(4), 263–269.

Boivin N, Fuller DQ, Korisettar R, Petraglia M (2014) The south deccan prehistory project. Karnatak Univ., School of Oxford, University College London. http://www.homepages.ucl.ac.uk/~tcrndfu/web_project/home.html

Burkill IH (1935) A dictionary of economic products of Malaya peninsula. Crown Agents of the Colonies, London

Cannarozzi G, Plaza-Wüthrich S, Esfeld K, Larti S, Wilson YS, Girma D, de Castro E, Chanyalew S, Blösch R, Farinelli L, Lyons E (2014) Genome and transcriptome sequencing identifies breeding targets in the orphan crop tef (Eragrostis tef). BMC Genomics 15:1–21

Champion L, Fuller DQ (2018) New evidence on the development of millet and rice economies in The Niger River Basin: Archaeobotanical results from Benin. Plants and people in the African past: Progress in African archaeobotany 529–47

Champion L, Fuller DQ, Ozainne S, Huysecom É, Mayor A (2021) Agricultural diversification in West Africa: an archaeobotanical study of the site of Sadia (Dogon Country, Mali). Archaeol Anthropol Sci 13:1–21

Chandel G, Meena RK, Dubey M, Kumar M (2014) Nutritional properties of minor millets: neglected cereals with potentials to combat malnutrition. Curr Sci 107:7. (00113891)

Chavan, V. M., Gopalkrishna, N and Khadilkar, B. T. 1955. Blooming and anthesis in ragi (Eleusine coracana (Linn) Gaertn.) Poona Agricultural College Magazine, pp. 75–79

Chavan, V. M., & Shendge, P. Y. (1957). Plant breeding in south Asia with reference to millets.

Chen, H. (2023). Animals and plants in Chinese religions and science. Anthem Press.

Chennaveeraiah MS, Hiremath SC (1974) Genome analysis of Eleusine coracana (L.) Gaertn. Euphytica 23:489–495

Chuah TS, Salmijah S, Teng YT, Ismail BS (2004) Changes in seed bank size and dormancy characteristics of the glyphosate-resistant biotype of goosegrass (Eleusine indica [L.] Gaertn.). Weed Biol Manag 4(2):114–121

Clayton WD, Vorontsova MS, Harman KT, Williamson H (2006) GrassBase-the online world grass flora [internet]

Coleman, L. C. (1921). The collection of ragi in Mysore. Bulletin, Department of Agriculture., Mysore, Gen. Ser, 60–62.

Cooke M, Fuller DQ, & Rajan K (2005) Early historic agriculture in southern Tamil Nadu: Archaeobotanical research at Mangudi, Kodumanal and Perur . In U. Franke-Vogt, & J. Weisshaar (Eds.), South Asian archaeology: proceedings of the European Association for South Asian Archaeology Conference (pp. 329-334). Bonn, Germany., 7th-11th July 2003

Costanza SH, Dewet JM, Harlan J (1979) Literature review and numerical taxonomy of Eragrostis tef (T'ef). Econ Bot 33:413–424

Cruz JF, Béavogui F, Dramé D, Diallo TA (2016) Fonio, an African cereal. CIRAD:154. ISBN 978–2–87614-720-1

De Candolle A (1886) Origin of cultivated plants. Hafner, New York

de Wet JMJ (1986) Origin, evolution and systematics of minor cereals. Pages 19·30. In: Seethararna A, Riley KW, Harinarayana G (eds) Small millets in global agriculture. Oxford and IBH Publ. Co., New Delhi, India, pp 19–30). Proc. Int. Small Millets Workshop, 1st, Bangalore, India

de Wet JMJ (1995) Minor cereals. In: Smartt J, Simmonds NW (eds) Evolution of crop plants, 2nd edn. Longman, Harlow, pp 202–207

de Wet JMJ, Rao KEP, Brink DE (1983a) Systematics and domestication of Panicum sumatrense (Graminae). Journal d'agriculture TraditBot appliquée 30:159–168

de Wet JM, Prasada Rao KE, Mengesha MH, Brink DE (1983b) Domestication of mawa millet (Echinochloa colona). Econ Bot 37(3):283–291

De Wet JM, Rao KP, Brink DE, Mengesha MH (1984) Systematics and evolution of Eleusine coracana (Gramineae). Am J Bot 71(4):550–557

Denton DC (1987) Food crops for waterfowl. Fireside waterfowler: fundamentals of duck and goose. Ecology 352

Deshpande SS, Mohapatra D, Tripathi MK, Sadvatha RH (2015) Kodo millet-nutritional value and utilization in Indian foods. J Grain Process Storage 2(2):16–23

DeWet JMJ (2006) Eleusine coracana (L.) Gaertn., record from Protabase. In: Brink M, Belay G (eds) PROTA (plant resources of tropical Africa/Ressources végétales de l'Afrique tropicale), Wageningen, Netherlands. http://www.prota4u.org Accessed 15 July 2018

Dixit AS, Dixit SS (1987) The occurrence of Eleusine africana Kennedy O'Byrne in India and its significance in the origin of Eleusine coracana. Proc Plant Sci 97:1

Dodake SS, Dhonukshe BL (1998) Variability in floral structure and floral biology of finger millet (Eleusine coracana (L.) Gaertn.). Indian J Genet Plant Breed 58(01):107–112
Doust, A. N., Kellogg, E. A., Devos, K. M., & Bennetzen, J. L. (2009). Foxtail millet: a sequence-driven grass model system. Plant physiology, 149(1), 137–141.
FAO (2007) Eco-crop data sheet: Brachiaria ramosa. Food and Agriculture Organization of the UN. http://ecocrop.fao.org/ecocrop/srv/en/dataSheet?id=3814
Fukunaga K, Kawase M (2024) Crop evolution of foxtail millet. Plants 13(2):218
Fuller DQ (2006) Agricultural origins and frontiers in South Asia: a working synthesis. J World Prehist 20:1–86
Fuller D, Korisettar R, Venkatasubbaiah PC, Jones MK (2004) Early plant domestications in southern India: some preliminary archaeobotanical results. Veg Hist Archaeobotany 13:115–129
Gestrich N, Champion L, Keïta D, Coulibaly N, Fuller DQ (2021) Evidence of an eleventh-century AD Cola nitida trade into the middle Niger region. Afr Archaeol Rev 38:403–418
Gressel J (2008) Tef—the crop for dry extremes. In: Genetics glass ceilings: Transgenics for crop biodiversity. John Hopkins University Press, USA, pp 241–256
Gupta A, Mahajan V, Kumar M, Gupta HS (2009) Biodiversity in the barnyard millet (Echinochloa frumentacea Link, Poaceae) germplasm in India. Genet Resour Crop Evol 56:883–889
Gupta A, Mahajan V, Gupta HS (2010) Genetic resources and varietal improvement of small millets for Indian Himalaya. Biodiversity potentials of the Himalaya 305–16
Gupta N, Gupta AK, Gaur VS, Kumar A (2012) Relationship of nitrogen use efficiency with the activities of enzymes involved in nitrogen uptake and assimilation of finger millet genotypes grown under different nitrogen inputs. Sci World J 1:625731
Harinarayana G (1989) Breeding and varietal improvement of small millets in India. Small Millets in Global Agriculture, 59–70
Harlan Jr (1992) Crops & man, 2nd edn. American Society of Agronomy and Crop Science Society of America, Madison. Wisconsin
Harvey EL (2006) Early agricultural communities in northern and eastern India : an archaeobotanical investigation. Unpublished Ph.D. dissertation,. University College, London
Hilu KW (1988) Identification of the 'A' genome of finger millet using chloroplast DNA. Genetics 118:163–167
Hilu KW (1994) Evidence from RAPD markers in the evolution of Echinochloa millets (Poaceae). Plant Syst Evol 189:247–257
Hilu KW (1995) Evolution of finger millet: evidence from random amplified polymorphic DNA. Genome 38:232–238
Hilu KW, DeWet JMJ (1976) Racial evolution of finger millet, Eleusine coracana. Am J Bot 63:1311–1318
Hilu KW, Johnson JL (1992) Ribosomal DNA variation in finger millet and wild species of Eleusine (Poaceae). Theor Appl Genet 83:895–902. https://doi.org/10.1007/BF00226713
Hilu KW, DeWet JMJ, Harlan JR (1979) Archaeobotany and the origin of finger millet. Am J Bot 66:330–333
Hiremath SC, Chennaveeraiah MS (1982) Cytogenetical studies in wild and cultivated species of Eleusine (Gramineae). Caryologia 35:57–69
Hiremath SC, Salimath SS (1991) Quantitative nuclear DNA changes in Eleusine (Gramineae). Plant Syst Evol 178:225–233
Hou, S., Men, Y., Wei, M., Zhang, Y., Li, H., Sun, Z., & Han, Y. (2022). Total protein content, amino acid composition and eating-quality evaluation of foxtail millet (Setaria italica (L.) P. Beauv). Foods, 12(1), 31.
Hulse, J. H., Laing, E. M., & Pearson, O. E. (1981). Sorghum and millets: their composition and nutritive value (p. 997pp).
Hunt HV, Vander Linden M, Liu X, Motuzaite-Matuzeviciute G, Colledge S, Jones MK (2008) Millets across Eurasia: chronology and context of early records of the genera Panicum and Setaria from archaeological sites in the Old World. Veg Hist Archaeobotany 17:5–18

Hunt HV, Badakshi F, Romanova O, Howe CJ, Jones MK, Heslop-Harrison JP (2014) Reticulate evolution in Panicum (Poaceae): the origin of tetraploid broomcorn millet, P miliaceum. J Exp Bot 65(12):3165–3175
Ibrahim Bio Yerima AR, Achigan-Dako EG (2021) A review of the orphan small grain cereals improvement with a comprehensive plan for genomics-assisted breeding of fonio millet in West Africa. Plant Breed 140(4):561–574
Ingram AL, Doyle JJ (2003) The origin and evolution of Eragrostis tef (Poaceae) and related polyploids: evidence from nuclear waxy and plastid rps16. Am J Bot 90(1):116–122
Jayaraman N, Suresh S, Nirmala A, Ganeshan NM (1997) Genetic enhancement and breeding strategies in small millets. In National seminar on small millets pp. 23–24
Jones BM, Ponti J, Tavassoli A, Dixon PA (1978) Relationships of the Ethiopian cereal t′ ef (Eragrostis tef (Zucc.) Trotter): evidence from morphology and chromosome number. Ann Bot 42(6):1369–1373
Kajale, M. D. (1977). Ancient grains from excavations at Nevasa, Maharashtra. Geophytology, 7(1), 98–106.
Kawase M, Sakamoto S (1987) Geographical distribution of landrace groups classified by hybrid pollen sterility in foxtail millet, Setaria italica (L.) P. Beauv Japanese. J Breed 37(1):1–9
Klmata M, Ashok EG, Seetharam A (2000) Domestication, cultivation and utilization of two small millets, Brachiaria ramosa and Setaria glauca (Poaceae), in South India. Econ Bot 54:217–227
Knees SG, & Gupta AK (2013) Paspalum scrobiculatum. The IUCN Red List of Threatened Species
Krishna Kumari S, Thayumanavan B (1998) Characterization of starches of proso, foxtail, barnyard, kodo, and little millets. Plant Foods Hum Nutr 53:47–56
Li, S., Zhao, W., Min, G., Li, P., Zhang, A., Zhang, J., ... & Liu, J. (2021). Effects of different amylose contents of foxtail millet flour varieties on textural properties of chinese steamed bread. Processes, 9(7), 1131.
Liu Q, Triplett JK, Wen J, Peterson PM (2011) Allotetraploid origin and divergence in Eleusine (Chloridoideae, Poaceae): evidence from low-copy nuclear gene phylogenies and a plastid gene chronogram. Ann Bot 108(7):1287–1298
Liu Q, Jiang B, Wen J, Peterson PM (2014) Low-copy nuclear gene and McGISH resolves polyploid history of Eleusine coracana and morphological character evolution in Eleusine. Turk J Bot 38(1):1–2
Lu H, Zhang J, Liu KB, Wu N, Li Y, Zhou K, Ye M, Zhang T, Zhang H, Yang X, Shen L (2009) Earliest domestication of common millet (Panicum miliaceum) in East Asia extended to 10,000 years ago. Proc Natl Acad Sci 106(18):7367–7372
Mallikarjuna H (2016) Korale's growing popularity. Deccan Herald News Paper
Mallikharjun DR, Hiremath SC, Rao SR, Kumar A, Sheelavanthmath SS (2005) Genome interrelationship in the genus Eleusine (Poaceae) as revealed through heteroploid crosses. Caryologia 58(4):300–307
Maun MA, Barrett SC (1986) The biology of Canadian weeds.: 77. Echinochloa crus-galli (L.) Beauv. Can J Plant Sci 66(3):739–759
Mehra KL (1963) Consideration of the African origin of Eleusine coracana (L.) Gaertn. Curr Sci J 32:300–301
Michael P (1994) Distribution and taxonomy of Echinochloa—a world view with a key to the species occurring in China. Proceedings of the 5th Weed Science Conference of China (Kunming, China, 25–30 November 1994), 161—166
Michael P (2001) The taxonomy and distribution of Echinochloa species (barnyard grasses) in the Asian-Pacific region, with a review of pertinent biological studies. In Proceedings of the 18th APWSS conference (Beijing, China, 28 May-2 June 2001) 2001 (pp. 57–66). Standard Press of China
Mitchell WA, & Tomlinson WH Jr. (1989) Browntop millet (Panicum ramosum): section 7.1.5. USACE wildlife resources management manual. Technical report EL—89-12. US Army Engineer Waterways Experiment Station, Vicksburg, Miss., U. S. a.
Mitich LW (1990) Intriguing world of weeds: Barnyardgrass. Weed Technol J 4:918–920

Moharil, M. P., Ingle, K. P., Jadhav, P. V., Gawai, D. C., Khelurkar, V. C., & Suprasanna, P. (2019). Foxtail millet (Setaria italica L.): potential of smaller millet for future breeding. Advances in Plant Breeding Strategies: Cereals: Volume 5, 133–163.
Napper DM (1965) Grasses of Tanganyaika: with keys for identification, bulletin 18, 146 pp. Ministry of Agriculture Forest and Widlife, Tanzania
National Research Council, Global Affairs, Office of International Affairs, Board on Science, & Technology for International Development. (1996). Lost crops of Africa: volume I: grains (Vol. 1). national academies press.
Neves SS, Swire-Clark G, Hilu KW, Baird WV (2005) Phylogeny of Eleusine (Poaceae: Chloridoideae) based on nuclear ITS and plastid trnT–trnF sequences. Mol Phylogenet Evol 35(2):395–419
Niyogi D (2018) Millets back in our fields and plates, The Millenium Post, 4 September. Available at: http://www.millenniumpost.in/opinion/milletsback-in-our-fields-and-plates-317237
NPR Millet (2018) How a trendy ancient grain turned nomads into farmers. Available online: https://www.npr.org/sections/thesalt/2015/12/23/460559052/millet-how-a-trendy-ancient-grain-turned-nomads-into-farmers. Accessed 19 March 2025
Obara T (1938) Studies of barnyard millet: I. Classification of varieties by agricultural traits. Japanese J Crop Sci 9:471–518
Oelke EA, Oplinger ES, Putnam DH, Durgan BR, Doll JD, Undersander DJ (1990) Millets. University of Wisconsin, Madison, WI. Alternative field crops manual. https://hort.purdue.edu/new crop/afcm/millet.html
Padulosi S, Mal B, Ravi SB, Gowda J, Gowda KT, Shanthakumar G, Yenagi N, Dutta M (2009) Food security and climate change: role of plant genetic resources of minor millets. Indian J Plant Genet Resour 22(1):1–6
Panwar P, Dubey A, Verma AK (2016) Evaluation of nutraceutical and antinutritional properties in barnyard and finger millet varieties grown in Himalayan region. J Food Sci Technol 53:2779–2787
Patil HE, Pali V, Sao A, Patil GB, Patel UN (2024) Floral biology, pollination, genetics, origin and diversity in little millet (Panicum sumatrense L. Roth ex. Roem. and Schultz). In Genetic improvement of small millets (pp. 555–570). Singapore: Springer Nature Singapore
Phillips S (1995) Poaceeae (Gramineae). In: Hedberg I, Edwards S (eds) Flora of Ethiopia and Eritrea, vol 7. University of Addis Ababa, Addis Ababa, Ethiopia, pp 1–429
Popov GI (1946) The importance of diversity in millet. Agrobiologiya 2:28–43
Porteres R (1955) Les ccreales mineurcs du genre Digitaria en Afrique et en Europe. Jour. d'Agric. Trup. et de Bot. Appl. 2
Portères R (1976) African cereals: Eleusine, Fonio, Black Fonio, Teff, Brachiaria, paspalum, Pennisetum, and African Rice. In: Harlan J, Wet JMJ, Stamler ABL (eds) Origins of African plant domestication. Mouton, pp 409–463. https://doi.org/10.1515/9783110806373.409
Potvin, C. (1986). Biomass allocation and phenological differences among southern and northern populations of the C4 grass Echinochloa crus-galli. The Journal of Ecology, 915–923.
Potvin, C. (1991). Temperature-induced variation in reproductive success: field and control experiments with the C4 grass Echinochloa crus-galli. Canadian journal of botany, 69(7), 1577–1582.
Prabhu R, Gowri MU, Gayathri R, Govindaraj M, Manikandan G (2022) Growth dynamics and forecasting of minor millets in India: a time series analysis. J Appl Nat Sci 14(SI):145–150
Purseglove JW (1972) Tropical crops: monocotyledons l. Longman Group Limiled, London
Rachie KO, Peters LV (1977) The Eleusines: a review of the world literature. ICRISAT, Hyderabad, p 179
Reddy, V. G., Upadhyaya, H. D., & Gowda, C. L. L. (2006). Characterization of world's foxtail millet germplasm collections for morphological traits. International Sorghum and Millets Newsletter, 47, 107–109.
Reddy A, Prasad KG (2017) Return of the forgotton crop: browntop millet. LEISA India 19:27–28
Renganathan VG, Vanniarajan C, Nirmalakumari A, Raveendran M, Thiyageshwari S (2017) Cluster analyses for qualitative and quantitative traits in barnyard millet Echinochloa frumentacea (Roxb. Link) germplasm. Bioscan 12:1927–1931

Ruiz-Santaella JP, Bastida F, Franco AR, De Prado R (2006) Morphological and molecular characterization of different Echinochloa spp. and Oryza sativa populations. J Agric Food Chem 54(4):1166–1172

Sakamoto S (1987). A preliminary repost of the studies on millet cultivation and its agro-pastoral culture complex in the Indian subcontinent (1985). Kyoto: Kyoto University Research Team for the studies on millet cultivation and its agro-pastoral culture complex in the Indian subcontinent

Saldanha CJ, Nicolson DH (1978) Flora of Hassan District, Karnataka, India. Amerind Pub, Co. Pvt. Ltd., New Delhi, pp 746–747

Saleh AS, Zhang Q, Chen J, Shen Q (2013) Millet grains: nutritional quality, processing, and potential health benefits. Compr Rev Food Sci Food Saf 12(3):281–295

Santra DK, Rose D (2013) Alternative uses of proso millet. University of Nebraska-Lincoln Neb Guide G 2218:3–6

Seetharam A, Gowda J, Halaswamy JH (2003) Small millets. In: Chowdhury SK, Lal SK (eds) Nucleus and breeder seed production manual. Indian Agricultural Research Institute, New Delhi, India, pp 54–67

Sharma, N., & Niranjan, K. (2018). Foxtail millet: Properties, processing, health benefits, and uses. Food reviews international, 34(4), 329–363.

Shaw RH, Loomis WE (1950) Bases for the prediction of corn yields. Plant Physiol 25(2):225

Sheahan CM (2014) Plant guide for browntop millet (Urochloa ramosa). USDA-Natural Resources Conservation Service, Cape May Plant Materials Center, Cape May, NJ

Shukla P, Bhise S (2023) Mighty millets: bespoke for multi nutrients. Curr J Appl Sci Technol 42(32):54–62. https://doi.org/10.9734/cjast/2023/v42i324220

Singh U (2009) A history of ancient and early medieval India: from the stone age to the 12th century *(PB)*. Pearson Education India

Singh KP, Mishra HN, Saha S (2010) Moisture-dependent properties of barnyard millet grain and kernel. J Food Eng 96:598–606. https://doi.org/10.1016/j.jfoodeng.2009.09.007

Singode A, Chapke RR, Domathoti B, Padmaja PG, Bhat VB, Malathi VM, Venkateswarlu R, Nayak S, Ratnavathi CV (2023) Proso millet: importance and cultivation. Indian Farming 73(1):110–1

Sood S, Khulbe RK, Kumar A, Agrawal PK, Upadhyaya HD (2015) Barnyard millet global core collection evaluation in the submontane Himalayan region of India using multivariate analysis. Crop J 3(6):517–525

Spaenij-Dekking L, Kooy-Winkelaar Y, Koning F (2005) The Ethiopian cereal tef in celiac disease. N Engl J Med 353(16):1748–1749

Sundararaj DP, Thulasidas G (1976) Botany of field crops. Macmillan Publisher

Tefera MM (2011) Land use/land-cover dynamics in Nonno District, Central Ethiopia. J Sustain Dev Africa 13:123–141

Tefera H, Assefa K, Belay G (2003) Evaluation of interspecific recombinant inbred lines of Eragrostis tef x E. Pilosa. J Genet Breed 57:21–30

Teferra T, Tefera H, Simane B, Tuinstra M (2000) The influence of moisture stress on growth, leaf water loss rate and phonological development of tef (Eragrostis tef). Trop Sci 40:100–107

Tran G (2017) Proso millet (*Panicum miliaceum*), forage. Feedipedia, a programme by INRAE, CIRAD, AFZ and FAO https://www.feedipedia.org/node/409

Tsang CH, Li KT, Hsu TF, Tsai YC, Fang PH, Hsing YI (2017) Broomcorn and foxtail millet were cultivated in Taiwan about 5000 years ago. Bot Stud 58:1

Ugare R, Chimmad B, Naik R, Bharati P, Itagi S (2014) Glycemic index and significance of barnyard millet (Echinochloa frumentacae) in type II diabetics. J Food Sci Technol 51:392–395

Vanniarajan C, Anand G, Kanchana S, Veeranan V, Giridhari A, Renganathan VG (2018) A short duration high yielding culture-barnyard millet ACM 10145. Agric Sci Digest Res J 38(2):123–126

Vavilov NI (1951) The origin, variation, immunity and breeding of cultivated plants. The Roland Press Co., New York

Veerabadhran V, Palakuru HR, Pratap GP, Rajeshwari B, Prasad GP, Srikanth N, Acharya R (2023) A critical review of the history and practices of millets in India. J Drug Res Ayurvedic Sci 8(Suppl 1):S18–S25
Verma SNP (1989) Research on small millets at JNKVV. College of Agriculture, JNKVV, Rewa, Madhya Pradesh, India, p 126
Verma S, Srivastava S, Tiwari N (2015) Comparative study on nutritional and sensory quality of barnyard and foxtail millet food products with traditional rice products. J Food Sci Technol 52(8):5147–5155
Vetriventhan M, Azevedo VC, Upadhyaya HD, Nirmalakumari A, Kane-Potaka J, Anitha S, Ceasar SA, Muthamilarasan M, Bhat BV, Hariprasanna K, Bellundagi A (2020) Genetic and genomic resources, and breeding for accelerating improvement of small millets: current status and future interventions. Nucleus 63:217–239
Virtamo, J., Taylor, P. R., Kontto, J., Männistö, S., Utriainen, M., Weinstein, S. J., ... & Albanes, D. (2014). Effects of α-tocopherol and β-carotene supplementation on cancer incidence and mortality: 18-Year postintervention follow-up of the Alpha-Tocopherol, Beta-Carotene Cancer Prevention Study. International Journal of Cancer, 135(1), 178–185.
Weber SA, Fuller DQ (2008) Millets and their role in early agriculture. Pragdhara 18(69):e90
Werth CR, Hilu KW, Langner CA (1994) Isozymes of Eleusine (Gramineae) and the origin of finger millet. Am J Bot 81(9):1186–1197
Wimalasiri EM, Ashfold MJ, Jahanshiri E, Walker S, Azam-Ali SN, Karunaratne AS (2023) Agro-climatic sensitivity analysis for sustainable crop diversification; the case of Proso millet (Panicum miliaceum L.). PLoS One 18(3):e0283298
Wu D, Shen E, Jiang B et al (2022) Genomic insights into the evolution of Echinochloa species as weed and orphan crop. Nat Commun 13:689. https://doi.org/10.1038/s41467-022-28359-9
Yabuno T (1962) Cytotaxonomic studies on the two cultivated species and the wild relatives in the genus Echinochloa. Cytologia 27:296–305
Yabuno T (1966) Biosystematic study of the genus Echinochloa. J Japanese Bot 19:277–323
Yabuno T (1971) A note on barnyard millet. SABRAO Newslett 3:43–45
Yabuno T (1984) A biosystematic study on Echinochloa oryzoides (Ard.) Fritsch. Cytologia 49:673–678
Yabuno T (1987) Japanese barnyard millet (Echinochloa utilis, Poaceae) in Japan. Econ Bot 41:484–493
Yabuno T (1996) Taxonomy and phylogeny of the genus Echinochloa. In: Yabuno T, Yamaguchi H (eds) Natural history of genus Echinochloa. Daw Elanco/Dow Chemical Company, Tokyo (in Japanese), pp 16–28
Yabuno T (2001) Taxonomy and phylogeny of the genus Echinochloa. In: Yabuno T, Yamaguchi H (eds) Natural history of genus Echinochloa. Zennokyo Shuppan, Tokyo (in Japanese), pp 15–30
Yadava HS (1997) Gamma rays induced variation in Kodo millet. Adv Plant Sci 9(II suppl):165–169
Yadava HS, Jain AK (2006) Advances in kodo millet research. Directorate of Information and Publications of Agriculture, Indian Council of Agricultural Research
Yamaguchi H, Umemoto S, Masanaga Y (1996) Studies on barnyard grasses, especially on the non-shattering form of Echinchloa oryzicola Vasing [in Yungui plateau, China]. Weed Res Japana J 41:111–115
Yamaguchi H, Utano AY, Yasuda K, Yano A, Soejima A (2005) A molecular phylogeny of wild and cultivated Echinochloa in East Asia inferred from non-coding region sequences of trnT-L-F. Weed Biol Manag 5(4):210–218
Yang, X., Fuller, D. Q., Huan, X., Perry, L., Li, Q., Li, Z., ... & Lu, H. (2015). Barnyard grasses were processed with rice around 10000 years ago. Scientific Reports, 5(1), 16251. https://doi.org/10.1038/srep16251
Yasuda K, Yano A, Nakayama Y, Yamaguchi H (2002) Molecular identification of Echinochloa oryzicola Vasing. and E. Crus-galli (L.) Beauv. Using a polymerase chain reaction–restriction fragment length polymorphism technique. Weed Biol Manag 2(1):11–17

Youngman W, Roy SC (1923) Pollination methods amongst the lesser millets. Agric J India 18:580–583

Zhu Q, Smith SM, Ayele M, Yang L, Jogi A, Chaluvadi SR, Bennetzen JL (2012) High-throughput discovery of mutations in tef semi-dwarfing genes by next-generation sequencing analysis. Genetics 192(3):819–829

Zou C, Li L, Miki D, Li D, Tang Q, Xiao L, Rajput S, Deng P, Peng LI, Jia W, Huang R (2019) The genome of broomcorn millet. Nat Commun 10(1):436

Nutritional Composition and Nutraceutical Properties of Minor Millets

V. M. Malathi, S. Kanika, R. Venkateswarlu, and J. Jinu

Abstract

Minor millets are a group of small-grained cereals, extensively cultivated across the arid and semiarid parts of the world. These grains exhibit excellent agronomic traits and are highly nutritious. Minor millet grains are a good source of protein, dietary fibre, as well as micronutrients. These grains also house abundant phytochemicals including phenolic acids and flavonoids and phytosterols. Various processing methods including fermentation, soaking, germination, and malting enhance grain nutrient content and hence influence its bioavailability. Owing to their excellent nutritional and phytochemical composition, minor millets exhibit potential nutraceutical properties. Consumption of minor millet grains can mitigate micronutrient deficiencies and ameliorate diabetes. They also exhibit hypolipidaemic as well as prebiotic potential. Furthermore, minor millets exhibit several health-beneficial effects owing to the bioactivities of phenolic phytochemicals. More research on the health benefits and nutrient bioavailability of minor millets would pave way for optimal utilization of these grains ensuring sustainable nutrition.

Keywords

Minor millets · Proximate composition · Micronutrient composition · Phenolics · Bioavailability · Bioactivities

---

V. M. Malathi (✉) · R. Venkateswarlu · J. Jinu
ICAR-Indian Institute of Millets Research, Hyderabad, India
e-mail: malathi@millets.res.in

S. Kanika
ICAR-Central Institute of Research in Cotton Technology, Mumbai, India

D. C. Joshi et al. (eds.), *Minor Millets*,
https://doi.org/10.1007/978-981-96-4265-6_2

Abbreviations

ABTS	2,2′-azino-bis (3-ethylbenzothiazoline-6-sulfonic acid) diammonium salt
COX-1	Cyclooxygenase 1
COX-2	Cyclooxygenase 2
DPPH	2,2′-diphenyl-1-picrylhydrazyl radical
FRAP	Ferric reducing antioxidant power
HDL	cholesterol- High-density lipoprotein cholesterol
LOX	Lipoxygenase
ORAC	Oxygen radical absorbance capacity
TFC	Total flavonoid content
TPC	Total phenolic content

1 Introduction

Millets include a diverse group of small-grained cereals, belonging to *Poaceae*. They are widely cultivated across the arid, semiarid, and tropical parts of the world, where they assume significance as staple crops. Apart from their food use, millets are also utilized as bird feed (in Western countries), fodder, and fuel. Millets are regarded as climate-smart crops as they exhibit adaptability to high temperature, drought, etc. and can grow well even with low agricultural inputs (Rooney 1996; Taylor 2017). These crops have a shorter crop duration and are less prone to biotic stress as well. Most importantly, millet grains are nutritionally superior and exhibit nutraceutical properties and hence are popularly known as nutri-cereals.

Millets encompass sorghum (*Sorghum bicolor*), pearl millet (*Pennisetum glauccum*), finger millet (*Eleusine coracana* (L.) Gaertn.), foxtail millet (*Setaria italica* (L.) P. Beauv.), little millet (*Panicum sumatrense* Roth. ex. Roem. and Schult.), barnyard millet (*Echinochloa* spp.), kodo millet (*Paspalum scrobiculatum* (L.)), proso millet (*Panicum miliaceum* (L.)), and browntop millet (*Brachiaria ramosa* (*L.*)). Fonio (*Digitaria exilis*) and teff (*Eragrostis tef*) are also millets, and their cultivation is predominantly confined to Nigeria and Ethiopia, respectively (Malleshi et al. 2021). Sorghum, pearl millet, as well as finger millet constitute the major millets, often due to the relatively larger area under cultivation among other millets. The other millets, viz., foxtail, little, barnyard, kodo, proso, and browntop, constitute the minor millets (Goswami et al. 2023). However, in some literature, the minor millets and finger millet are grouped together as small millets, owing to the comparatively smaller grain size among millets.

Unlike the major cereal crops, minor millets are regionally important crops; they assume significance in terms of imparting food security in the region they are widely cultivated. Minor millets serve as staple in some parts of China (foxtail), India (foxtail, little, barnyard, kodo, browntop), and other Asian countries (Vetriventhan et al. 2020). Recently, minor millets are receiving considerable attention from researchers and consumers across the globe owing to their climate-smartness, excellent nutritional value, and gluten-free nature. They hold immense potential to tackle the global challenges of lifestyle-associated diseases and malnutrition. There is a recent upsurge in the utilization of minor millets in food industry, especially in the preparation of gluten-free functional foods. This chapter describes the composition of nutrients, antinutrients, and phytochemicals and bioavailability of nutrients of various minor millet grains. It further describes the nutraceutical properties of minor millets in terms of the bioactivities of the phenolic compounds as well as peptides based on the available studies.

2 Nutritional Composition of Minor Millets

The minor millet grains possess excellent nutritional value. The grain nutritional composition depends on the type of minor millet as well as the environment. Genotypic variation in nutritional composition may also exist. The minor millets including foxtail, little, barnyard, proso, kodo, and browntop millet possess an outer hull layer which has to be removed for making them edible. Upon decortication, the nutritional value of the grain often varies as many constituents including the fibre, micronutrients, and phenolic compounds are localized in the outer regions of the grain. The proximate composition of minor millets is presented in Table 1.

As with most other cereals, carbohydrates form the major macronutrient constituent of minor millet grains; most of the grain carbohydrates are present as starch, while, the soluble sugars constitute only less than 2% carbohydrates. Also, the minor millets consist of good amounts of dietary fibre. The dietary fibre includes both soluble and insoluble fractions; in minor millets, the insoluble fractions are often high. Minor millets are a good source of protein. They are gluten-free and are hence a viable protein source for celiacs too. Minor millet proteins are composed of good amounts of essential amino acids; nevertheless, lysine is limiting. Fat content is generally low with little and foxtail millet having higher fat content among the minor millets. Minor millet fat contains good amounts of unsaturated fatty acids including the essential fatty acids. Furthermore, minor millets are abundant sources of micronutrients too (Tables 2 and 3).

The nutritional composition of the minor millets is described in detail hereunder:

Table 1 Proximate composition of minor millets (g/100 g)

Minor millet	Carbohydrate[a] (%)	Dietary fibre (%)	Protein (%)	Fat (%)	Ash (%)
Foxtail millet	51–73	9.4–10.5	6–17	1.2–5.2	1.5–4.3
Little millet	65.5–78.5	7.7	10.1–13.4	3.5–5.4	2.9–3.4
Barnyard millet (*E. frumentacea*)	51.5–62.0	12.6	6.7–12.7	2.5–6.3	2.0–5.0
Kodo millet	72–79	6.4–10.4	6.213.1	2.5–4.9	1.7–4.1
Proso millet	56.1–76	11.2	6.4–16	1.7–4.1	0.8–8.8
Browntop millet	71.3	11.3	8.9	1.9	3.9

References: Veena et al. (2005); Ugare et al. (2014); Longvah et al. (2017); Sharma et al. (2017); Taylor (2017); Serna-Saldivar and Espinosa-Ramírez (2019); Malleshi et al. (2021). Sharma et al. (2021)

[a]Carbohydrates essentially reflect the starch content

Table 2 Important minerals in minor millets (mg/100 g)

Minor millet	Ca	Mg	Fe	Zn	P	K	Reference
Foxtail millet	10–40	70–140	2.8–24	2.1–5.1	70–360	250–580	Taylor (2017); Serna-Saldivar and Espinosa-Ramírez 2019; Malleshi et al. (2021); Sharma et al. (2021);
Little millet	16	91.4–133	1.3	1.8–3.7	130–220	105–129	Panwar et al. (2016); Sharma et al. (2017); Longvah et al. (2017); Malleshi et al. (2021)
Barnyard millet	22–27.1	38.2	15.6–22.98	4.2–5.9	–	–	Serna-Saldivar and Espinosa-Ramírez (2019)
Kodo millet	10–40	122–147.5	1.3–13	0.7–1.6	101–332	94–170	Longvah et al. (2017); Taylor (2017) Serna-Saldivar and Espinosa-Ramírez (2019); Malleshi et al. (2021)
Proso millet	8–40	60–180	0.8–31	1.4–6.8	210–340	113–590	Taylor (2017); Serna-Saldivar and Espinosa-Ramírez (2019); Malleshi et al. (2021);
Browntop millet	28	94.5	7.7	2.7	276	408	Malleshi et al. (2021)

Table 3 Important B-vitamins in minor millets (mg/100 g)

Minor millet	Thiamine	Riboflavin	Niacin	Reference
Foxtail millet	0.48–0.59	0.11–0.12	3.2–3.7	Taylor (2017); Serna-Saldivar and Espinosa-Ramírez (2019)
Little millet	0.26–0.41	0.05–0.28	1.3	Longvah et al. (2017); Taylor (2017)
Barnyard millet	0.36	0.11	4.2	Serna-Saldivar and Espinosa-Ramírez (2019)
Kodo millet	0.17–0.5	0.05–0.20	1.5	Longvah et al. (2017); Serna-Saldivar and Espinosa-Ramírez (2019)
Proso millet	0.41–0.68	0.17–0.28	1.8–5.1	Taylor (2017); Serna-Saldivar and Espinosa-Ramírez (2019)
Browntop millet	0.42	0.29	–	Malleshi et al. (2021)

2.1 Foxtail Millet

Foxtail millet grain is ovoid with 1–2 mm diameter (Fig. 1) (Taylor 2017). Like in most other cereals, in foxtail millet, starch forms the major component of the carbohydrate with content ranging from 51 to 73% (w/w) depending on the genotypes (Serna-Saldivar and Espinosa-Ramírez 2019; Kaimal et al. 2021; Kalsi and Bhasin 2023). The starch granules were reported to be mostly small spherical and also large polygonal (Serna-Saldivar and Espinosa-Ramírez 2019). The amylose/amylopectin content of foxtail millet varies according to the genotypes; amylose content up to 31% has been reported in foxtail millet grains (Nakayama et al. 1998). Further, a range of amylose content from 0.17% to 16.4% has been reported in foxtail land races from Taiwan (Yin et al. 2019). While the amylose of foxtail millet grains with a regular type endosperm varies from 17.4% to 24%, the waxy genotypes (high amylopectin; 1.3% to 2.7% amylose) have also been reported in foxtail millet (Fujita et al. 1989; Nakayama et al. 1998).

As per various studies, the total dietary fibre (TDF) content of foxtail millet genotypes may vary. Taylor (2017) reported a TDF content of 9.4% in the foxtail millet. More recently, Malleshi et al. (2021) summarized a TDF content of 10.5% in foxtail millet grains with a higher content of insoluble dietary fibre (IDF) (8.7%) than soluble dietary fibre (SDF) (1.8%) as in other minor millet grains. Earlier, Geervani and Eggum (1989) reported that the TDF of whole foxtail millet grain amounted up to 19.11% and that of the dehusked grains amounted up to 2.40%. Protein content as high as 17% has been reported in foxtail millet (Taylor 2017). Prolamin forms the major storage fraction followed by glutelins (Serna-Saldivar and Espinosa-Ramírez 2019). In foxtail millet, the content of the limiting amino acid, lysine, is about 1.9 g/100 g. While studying the chemical and physiochemical properties of two varieties (yellow and white) of foxtail millet, Kamara et al. (2009) reported that most essential amino acids (except lysine) are present in the defatted flour of foxtail millet, with leucine, phenylalanine, and tyrosine present in higher amounts. The fat content of foxtail millet genotypes varies from 1.2 to 5.2%, and the total ash content in the grains varies from 1.5 to 4.3% (Taylor 2017).

a b

Fig. 1 Foxtail millet (**a**) whole grain (**b**) dehulled grain

Regarding the micronutrients, foxtail grains are a good source of mineral nutrients including Zn, Mg, K, etc. and B-vitamins including thiamine, riboflavin, and niacin (Tables 2 and 3).

2.2 Little Millet

Little millet grains are elliptical to oval with about 1.8–1.9 mm length (Fig. 2) and vary in colour from grey, pale to straw. Little millet has a carbohydrate content as high as 78.5%, while Longvah et al. (2017) reported a carbohydrate content (by difference method) of about 65.5%. The TDF content of little millet was found to be 7.7% with IDF of 5.4% and SDF of 2.3%. Protein content ranging from 10.1 to 13.4% has been reported in little millet grains. The lysine content of little millet is about 1.1 g/100 g (Taylor 2017). Further, the fat content of little millet is quite high, ranging from 3.5% to 5.4% with good amounts of unsaturated fatty acids. The ash content varies from 2.9% to 3.4%. Regarding the micronutrient composition, the little millet grains are good source of Zn, Mg, P, as well as other minerals and B-vitamins.

2.3 Barnyard Millet

Barnyard millet grains are pale to straw, tiny elliptical to oval grains of length about 2 mm (Fig. 3). *Echinochloa frumentacea* (Indian barnyard millet) and *Echinochloa esculenta* (Japanese barnyard millet) are among the widely cultivated barnyard species. The composition of *E. frumentacea* is included in this chapter. Barnyard millet genotypes contain up to 51.5–65.5% starch (Veena et al. 2005; Serna-Saldivar and Espinosa-Ramírez 2019; Malleshi et al. 2021). Starch granules reported in the Indian barnyard millet grains are large polygonal as well as large and small spherical types. Krishna Kumari and Thayumanavan (1998) found that the amylose

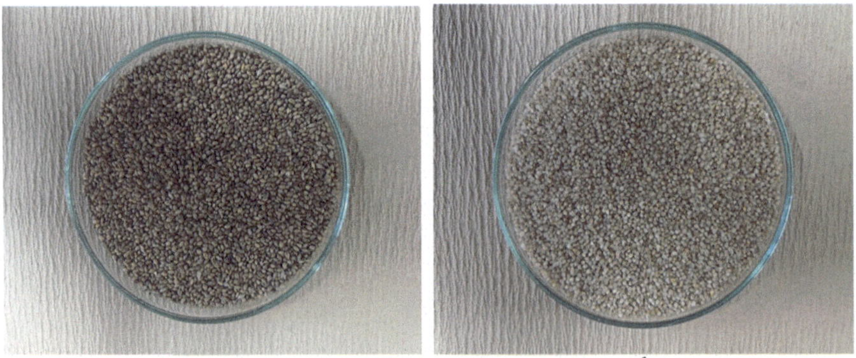

a b

Fig. 2 Little millet (**a**) whole grain (**b**) dehulled grain

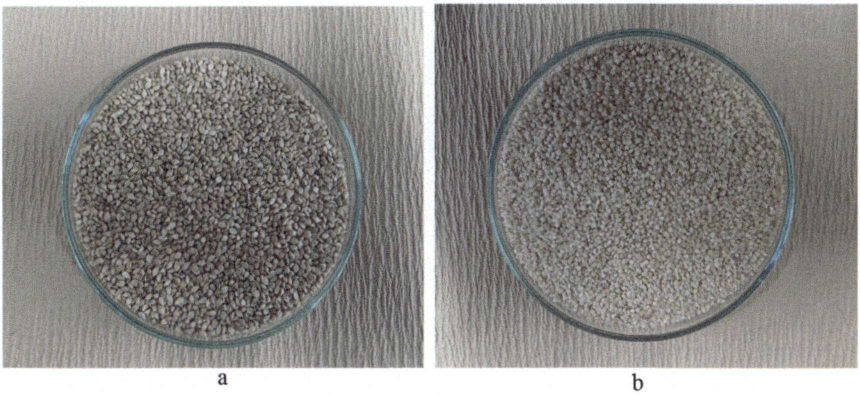

Fig. 3 Barnyard millet (**a**) whole grain (**b**) dehulled grain

content of barnyard millet is 20%. Barnyard millet has a dietary fibre content of about 12.6% (Ugare et al. 2014) which is the highest reported among the minor millets. They reported that the insoluble dietary fibre content amounted up to 8.4% and soluble fraction was 4.2%, while, in a previous study, Indian barnyard millet was found to have an insoluble dietary fibre content of 6.1–10.5% and soluble dietary fibre content of 3.5–4.6% (Veena et al. 2005). Protein content ranging from 11.2 to 12.7% was reported in barnyard millet (Serna-Saldivar and Espinosa-Ramírez 2019), while protein content as low as 6.7% has also been reported (Malleshi et al. 2021). Like most cereals, lysine (1.4 g/100 g) is the limiting amino acid in barnyard grains too (Taylor 2017). Fat content as high as 6.3% has been previously reported (Serna-Saldivar and Espinosa-Ramírez 2019). Linoleic acid and oleic acid are among the predominant fatty acids reported in barnyard millet. The ash content of barnyard millet grains was found to be ranging from 2.0% to 5.0%. Barnyard millet is a rich source of mineral nutrients including Fe and Zn and B-vitamins especially niacin.

2.4 Kodo Millet

Kodo millet grains are darkish brown to black and are elliptical to oval in appearance with a length of about 2.5 mm (Fig. 4). Kodo millet has a macronutrient composition of 72–79% carbohydrates, 6.2–13% protein, 3.2–4.9% fat, 1.72–5.0% ash, and 6.4–10.4% dietary fibre. The starch granules are often large polygonal types. Kodo millet starch contains about 19.6% amylose (Krishna Kumari and Thayumanavan 1998). The TDF content of kodo millet grains ranges from 6.4 to 10.4%. As per Longvah et al. (2017), kodo millet grains contain TDF of 6.4% with about 4.3% IDF and 2.1% of SDF. Up to 13.1% protein has been reported in kodo millet grains. Glutelin followed by prolamin forms the major protein fractions of kodo millet (Sudharshana et al. 1988). Also, kodo millet contains better lysine

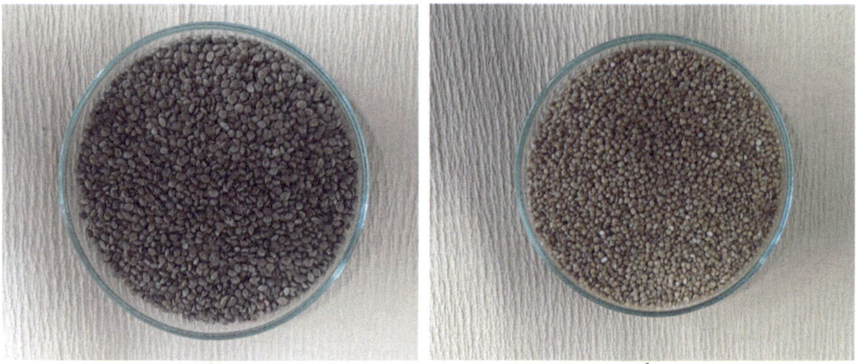

a b

Fig. 4 Kodo millet (**a**) whole grain (**b**) dehulled grain

content (2.4 g/100 g) among the minor millets (Taylor 2017; Serna-Saldivar and Espinosa-Ramírez 2019). The grain fat content of kodo millet ranges up to 4.9%; linoleic and oleic acids were found to be the predominant fatty acids. Kodo is a good source of micronutrients (Tables 2 and 3) especially Mg, Ca, etc.

2.5 Proso Millet

Proso millet grains are oval in shape with a length of close to 3 mm and about 2 mm diameter (Fig. 5). They vary in colour from white and yellow to orange in some genotypes (Taylor 2017). Most of the carbohydrates in proso millet, as mentioned for other minor millets, are contributed by starch (56–58%) (Serna-Saldivar and Espinosa-Ramírez 2019); some studies also report up to 76% starch in proso millet. Krishna Kumari and Thayumanavan (1998) reported a bimodal distribution of starch granules in proso millet grains: small spherical and large polygonal. Proso millet grains have an amylose content of about 14.9–17.4% (Wen et al. 2014), while high amylose content of up to 33.9% has been reported in proso millet grains from Canada (Annor et al. 2014). High amylopectin, waxy lines are also reported in proso millet (Serna-Saldivar and Espinosa-Ramírez 2019). The dietary fibre content is about 11.2% with a higher content of IDF (9.3%) as compared to SDF (1.9%) (Malleshi et al. 2021). Devisetti et al. (2014) evaluated the dietary fibre content in whole, dehusked (brown grain) and debranned proso millet grain. They reported a TDF content of 19.4%, 11.1%, and 5.5%, respectively, in the whole, dehusked, and debranned grains.

Proso millet grains contain abundant amounts of protein, up to 16% (Taylor 2017). A slight variation in the protein content between whole grain (13.6%), dehusked (14.8), and debranned grains (14.1) was reported (Devisetti et al. 2014). Prolamins constitute the major protein fraction in proso millet as in foxtail millet (Serna-Saldivar and Espinosa-Ramírez 2019). Lysine (1.5 g/100 g) is the limiting amino acid. Up to 4.1% fat was reported in proso millet grains (Taylor 2017).

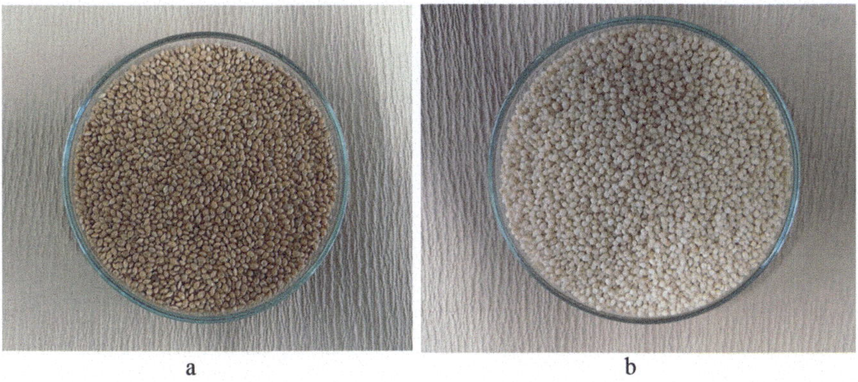

Fig. 5 Proso millet (**a**) whole grain (**b**) dehulled grain

A wide variation in the ash content was reported in proso millet grains, 0.8–8.8% (average content of 2.7%) (Taylor 2017). Furthermore, like all other minor millets, proso millet is also rich in micronutrients.

2.6 Browntop Millet

Browntop millet grains are oval- to round-shaped and pale to straw-coloured (Fig. 6). The available studies show that the browntop grains contain about 71.3% carbohydrates, 8.98% protein, 1.89% fat, and 3.9% ash content (Malleshi et al. 2021). The TDF content in browntop millet is 11.3% with an IDF of 9.4% and SDF of 1.7%. These grains are a good source of micronutrients including Fe, Mg, etc. (Table 2).

3 Phytochemical Composition of Minor Millets

Along with the macro- and micronutrients, most cereal grains contain phytochemicals, which are often plant secondary metabolites, with protective functions. These compounds exhibit potential bioactivities with human health benefits. They are mainly distributed in the outer layers of the grains, especially in the bran, along with other constituents (Sharma et al. 2021). Millet grains contain abundant amounts of phytochemicals, the phenolic compounds and phytosterols being the important ones.

3.1 Phenolic Compounds

Phenolic compounds are among the major classes of phytochemicals in millet grains; also, millets are known to possess higher amounts of phenolic compounds among most other cereals (Taylor and Duodu 2015). Phenolic acids and flavonoids

Fig. 6 Browntop millet (**a**) whole grain (**b**) dehulled grain

are the major phenolics present in both major and minor millets. Tannins are polyphenolic compounds reported mainly from some genotypes of sorghum as well as finger millet (Dykes and Rooney 2006); tannins are detected in comparatively lower amounts in the minor millets. As the condensed tannins bind to proteins interfering with their digestibility, they are regarded as antinutritional factors. The occurrence of condensed tannins in minor millets is discussed under antinutrients.

The composition, abundance, and distribution in grains of phenolic compounds may vary with different minor millets and are also influenced by other factors including the environmental conditions (Shahidi and Chandrasekara 2013; Sharma et al. 2021).

Phenolic acids and flavonoids in minor millets exist in the soluble (which includes the free phenolics as well as those which are found in ester or ether linkage with the soluble carbohydrates) and bound (which includes those phenolics which are esterified to the cell wall polysaccharides) forms (Chandrasekara and Shahidi 2011a). Phenolic acids reported from minor millets include hydroxy benzoic acids, viz., gallic, protocatechuic, *p*-hydroxy benzoic acid, and hydroxy cinnamic acids, viz., caffeic, *trans*-cinnamic, *p*-coumaric, chlorogenic acid, and their derivatives (Mattila et al. 2005; Chandrasekara and Shahidi 2011a; Zhang et al. 2014; Zhang and Liu 2015; Kumari et al. 2017). Chandrasekara and Shahidi (2010) evaluated the total phenolic content (TPC) of minor millets, viz., kodo millet, little millet, foxtail millet, and proso millet grains as well as the major millets including pearl and finger millet. The authors reported that kodo millet extracts displayed the highest, while proso millet extract showed the lowest TPC of soluble fractions. Higher phenolic contents in the bound extracts as compared to the soluble extracts were reported in kodo millet. Almost similar phenolic contents in both soluble and bound extracts were reported for foxtail millet. In little and proso millet, the TPC of soluble extracts was higher as compared to the bound fraction (Chandrasekara and Shahidi 2010). Also, recently, a TPC of 19.6–20.9 and 3.0–4.5 μmol FAE/g in the soluble and bound phenolic fractions, respectively, was reported in the little millet cultivars, showing that soluble fractions possessed higher phenolic content (Pradeep and Sreerama 2018). Evaluation of phenolic contents in the soluble and bound phenolic extracts from whole grain and various milled fractions (dehulled, hulled, branned, and pearled) of foxtail and little millet showed that the phenolic contents reduced in the order hull, bran, and pearled grains imply that the phenolics are concentrated on the outermost regions of the grains (Pradeep and Sreerama 2017).

Flavonoids identified in minor millets include luteolin, myricetin, quercetin kaempferol, naringenin, and daidzein among others (Chandrasekara and Shahidi 2011a; Pradeep and Sreerama 2018; Ofosu et al. 2020). Chandrasekara and Shahidi (2010) studied the total flavonoid content (TFC) of the soluble and bound phenolic fractions of minor millet whole grains including foxtail, kodo, little, and proso along with other millets. They reported that there was a significant difference between the flavonoid content of soluble and bound fractions, with the soluble fractions showing higher contents. The TPC, TFC, and the phenolic compounds identified in minor millets are presented in Table 4.

3.1.1 Phytosterols

Phytosterols are among the phytochemicals that form integral part of plant cell membrane providing structural stability. They also serve as precursors of other biologically important molecules including brassinosteroids, steroidal saponins, phytoecdysteroids, etc. Phytosterols are known to exhibit bioactivities including antioxidant, anticarcinogenic, and hypocholesterolaemic effects. Among the minor millets, the foxtail millet bran oil reportedly contains phytosterols, viz., β-sitosterol, campesterol, and stigmastanol at a concentration of 56.26%, 15.91%, and 15.12%, respectively. Also, stigmasterol, fucosterol, and ergostanol were found to be present at concentrations, 5.58%, 4.16%, and 2.97%, respectively (Pang et al. 2014). In another study, Ji et al. (2019a) also detected phytosterols including campesterol, stigmasterol, β-sitosterol, and sitostanol in the oils extracted from coarse bran, skin bran, polished bran, and mixed bran of foxtail millet.

4 Antinutrients

Most cereal grains as well as legumes are known to contain antinutrients, the compounds that interfere with the accessibility and availability of nutrients. Phytic acid, condensed tannins, trypsin inhibitors, oxalates, etc. are among the antinutritional factors present in minor millets. Phytic acid (*myo*-inositol hexakisphosphate) is the storage form of phosphorus (P) in most cereals and other seeds; in grains, phytic acid is commonly found as its mixed mineral form, phytate (Raes et al. 2014). This antinutrient is majorly located on the aleurone layer of the grain and also in the germ layer to a slighter extent (Serna-Saldivar and Rooney 1995). Phytic acid chelates mineral ions, mainly the divalent metal ions including Fe^{2+}, Zn^{2+}, Ca^{2+}, Mg^{2+}, etc. reducing their bioavailability. Moreover, phytic acid tends to bind and precipitate proteins, thus reducing the activities of enzymes (Sharma et al. 2021). Estimation of

Table 4 Phenolic phytochemicals of minor millets

Minor millet	Total phenolic content	Total flavonoid content	Phenolic compounds and their derivatives identified	References
Foxtail millet	10.79 ± 0.82 µmol FAE/g defatted meal (soluble) 11.59 ± 0.23 µmol FAE/g defatted meal (bound)	1.26 ± 0.03 CE/g of defatted meal (soluble) 0.47 ± 0.09 µmolCE/g of defatted meal (bound)	Gallic, protocatechuic, *p*-hydroxy benzoic, gentisic, vanillic, syringic, caffeic, *trans*-cinnamic, *p*-coumaric, chlorogenic acid, sinapic, *trans*-ferulic acid, *N′*,*N″*-dicaffeoylspermidine, *N′*, *N″*-di-*p*-coumaroylspermidine, *N′*-*p*-coumaroyl-*N″*-feruloylspermidine hexoside, *N′*, *N″*-diferuloylspermidine, *N′*-caffeoylspermidine, *N′*-*p*-coumaroyl-*N″*-feruloylspermidine,, *N′*-*p*-coumaroyl -*N″*-caffeoylspermidine, *N′*-caffeoyl-*N″*- feruloylspermidine, *N′* *N″*-diferuloyl spermidine-dihexoside Catechin, myricetin, luteolin, daidzein, quercetin, apigenin, naringenin and kaempferol, kaempferol-C-hexoside, kaempferol-C, O-dihexoside, apigenin-C-dihexoside	Chandrasekara and Shahidi (2010, 2011a); Zhang and Liu (2015); Kumari et al. (2017); Pradeep and Sreerama (2017); Xiang et al. (2019)
Little millet	19.6–20.9 µmol FAE/g (soluble) 3.0–4.5 µmol FAE/g (bound)	9.92–13.64 µmol catechin equivalents/g (soluble) 0.79–1.45 µmol catechin equivalents/g (soluble)	Gallic, dihydroxy benzoic acid, vanillic, caffeic, chlorogenic, ferulic, sinapic and *p*-coumaric acid Apigenin, kaempferol and luteolin, catechin, myricetin, daidzein, quercetin, naringenin	Chandrasekara and Shahidi (2011a); Pradeep and Sreerama (2018)

Barnyard millet	129.5 ± 4.95 mg FAE/100 g DW	101.3 ± 10.4 mg CE/100 g DW	Luteolin, tricin, catechin, kaempferol, apigenin, isorhamnetin and 3,7-dimethylquercetin	Watanabe 1999; Ofosu et al. (2020)
Kodo millet	32.39 μmol FAE/g defatted meal (soluble) 81.64 μmol FAE/g defatted meal (bound)	33.71 μmol CE/g defatted meal (soluble) 4.53 μmol CE/g defatted meal (bound)	Protocatechuic, *p*-hydroxy benzoic, gallic, syringic, vanillic caffeic, *p*-coumaric acid, chlorogenic, sinapic, *trans*-ferulic, cinnamic acid Vitexin, isovitexin, apigenin, luteolin, and quercetin	Chandrasekara and Shahidi (2010, 2011a)
Proso millet	7.19 μmol FAE/g defatted meal (soluble) 2.21 μmol FAE/g defatted meal (bound)	1.18 μmol CE/g of defatted meal (soluble) 0.44 μmol CE/g of defatted meal (bound)	Vanillic acid, *p*-hydroxy benzoic acid, syringic, caffeic acid, ferulic acid, *p*-coumaric acid, chlorogenic acid and its derivatives; protocatechuic acid, ferulic acid dehydrodimers Rutin and kaempferol *N*-caffeoyl *N″*feruloyl spermidine; *N*,*N′* dicaffeoyl spermidine; *N*-*p*-coumaroyl serotonin	Mattila et al. (2005); Chandrasekara and Shahidi (2010, 2011a); Seo et al. (2011); Zhang et al. (2014); Yuan et al. (2022)

phytic acid in the whole flour of different millets, viz., foxtail, barnyard, little, proso, kodo, and finger millets, in comparison with whole wheat, showed that proso and barnyard millets recorded the highest contents (Sharma and Gujral 2019). Panwar et al. (2016) reported a content of about 3.37–3.70 mg/g phytic acid in the barnyard millet varieties from the Himalayan region, while a lower phytic acid content of 1.2–1.4 mg/g was reported in kodo millet (Shyam and Singh 2018). The phytic acid levels of whole proso millet grains were found to be 7.2 mg/g, while the content in the dehusked grain was 8.3 mg/g. Also, the polished grain (debranned grain) recorded a phytic acid content as low as 2.2 mg/g (Devisetti et al. 2014). In the same study, the authors also showed that foxtail millet showed a similar trend, up to 9.9 mg/g (whole), 11.0 mg/g (dehusked/brown grain), and 4.9 mg/g (polished grain), suggesting that the phytic acid is concentrated in the outer layers of the grain.

Condensed tannins, also referred as proanthocyanidins, are polyphenolic compounds formed of flavan-3-ol units, viz., catechin, epicatechin, and epigallocatechin. They bind and form complexes with dietary protein as well as digestive enzymes thus interfering with protein digestibility (Taylor 2017). They are also known to bind and lower bioavailability of minerals (Sharma et al. 2021). Condensed tannins are reported from some genotypes of sorghum and finger millet, and its distribution in the minor millets is in general low. A low tannin content of 1.06–1.2 mg/g was reported in the germplasm lines of kodo millet (Shyam and Singh 2018). Contrastingly, Sharma and Gujral (2019) determined the tannin contents in the whole flours of minor millets and found that kodo millet recorded the highest tannin content. They found that kodo millet had the highest levels of tannins among the studied minor millets, while the lowest was in proso millet. Low tannin content of 0.067 g/100 g (DM) was reported in the dehusked little and barnyard millet grains.

Protease inhibitors are known to cause pancreatic hyperplasia along with interference in amino acid utilization thereby hampering protein digestion. Protease inhibitors such as trypsin inhibitors have been previously reported in barnyard millet grain (Pattabiraman 1986; Panwar et al. 2016).

5 Bioavailability

Bioavailability pertains to the amount of nutrients consumed that can be absorbed and utilized in normal metabolic processes. It refers to the extent at which a substance is absorbed and becomes available for use or storage in the body. It involves multiple stages from their release during digestion to their absorption into the bloodstream and eventual utilization within the body's tissues. Beyond nutritional worth, the physical and functional traits, specifically digestibility and bioavailability, influence how effectively the body assimilates and utilizes a food product in the gastrointestinal tract. These properties are contingent on various elements such as food origin, architecture, granularity, composition, matrix, processing techniques, and nutrient interplays (Galanakis 2021).

The bioavailability of nutrients relies on their accessibility, which, in turn, depends on their discharge from the intricate food matrix. This liberation process is influenced by the complex components embedded within the matrix. Ultimately, the rate at which these nutrients traverse the intestinal mucus layer and undergo molecular changes significantly impacts their availability (Macierzanka et al. 2019). Researchers have evaluated diverse in vitro methods to unravel the intricate molecular interactions governing the relationship between food components and micronutrients. These methods aim to decode the complex molecular dynamics guiding the interaction between food elements and micronutrients. Thus, when assessing the adequacy of dietary intake, the focus shifts to the bioavailability of nutrients rather than solely relying on the nutrient content of the food (Galanakis 2021; Rodríguez et al. 2020; Kumar et al. 2022). Various other factors, including the individual's genetic makeup, age, gender, overall health, and even the composition of their gut bacteria, can also significantly impact how well the body absorbs and uses the macro/micronutrients.

5.1 Factors Affecting Bioavailability of Minor Millet Nutrients and Phenolics

In addition to its balanced nutritional composition, millets contain tannins, phytates, and trypsin inhibitory factors, impacting nutrient bioavailability (Devisetti et al. 2014). These anti-nutritional compounds hinder nutrient absorption and accessibility in humans, compromising the nutritional value of millet-based diets. Both dietary and physiological aspects contribute to altering the availability of nutrients from millets. Hence, the goal should prioritize diminishing antinutrient quantities and amplifying nutrient levels (Samtiya et al. 2020). Sharma and Gujral (2019) found a negative correlation of in vitro protein digestibility, total dietary fibre, tannin, as well as phytic acid and phenolic content in minor millets, as these components bind proteins causing its precipitation and further breakdown. Proso millet proteins showed lower in vitro digestion than that of wheat proteins (Gulati et al. 2017).

Similarly, polyphenols of millet grains serving as rich reservoirs of antioxidants are predominantly concentrated within the bran fraction through covalent bonding with indigestible polysaccharides. However, these bound phenolic compounds exhibit limited bioavailability due to reduced accessibility to digestive enzymes in the gastrointestinal tract. The health advantages and biological functions shown by phenolic compounds in living organisms rely not just on their concentration in food but also on their bioaccessibility and bioavailability. Bioaccessibility, pertaining to phenolics, represents the fraction of ingested phenolics available for absorption (Hedren et al. 2002). This measure is crucial in understanding how much of these compounds are actually absorbed and utilized by the body after digestion. Chandrasekara and Shahidi (2012) reported that the in vitro enzymatic digestion and fermentation of various types of millet grains (finger, kodo, proso, pearl, foxtail) liberated 2–15% more total phenolic and flavonoids from millets compared to those obtained from identical millet grain samples extracted using 70% acetone. The obtained compounds were considered accessible and showed considerable antioxidant activity.

Hence, appropriate processing becomes vital to maximize the favourable nutritional attributes. Various processing methods including soaking, fermentation, germination, autoclaving, and enzyme addition can be employed to enhance nutrient bioavailability (Tharifkhan et al. 2021). Preprocessing of these grains demonstrates a substantial reduction in the impact of antinutrients, resulting in improved nutrient bioavailability and enhanced bioactive compounds, corroborated by various scientific studies (Hithamani and Srinivasan (2014)).

5.2 Strategies to Enhance Bioavailability of Nutrients

Various processing methods including fermentation, soaking, malting, and germination, significantly affect the bioavailability of protein in millets. Fermentation aids in breaking down antinutrients like phytate and enhances the conversion of insoluble proteins into soluble forms through the action of microorganism-synthesized proteolytic enzymes, thereby improving protein accessibility for digestion. Soaking, a simple technique, triggers the release of stored nitrogen, leading to increased protein levels in millets. Malting, involving both soaking and germination, instigates biochemical changes that notably boost protein content. Additionally, germination during malting activates enzymes that break down complex proteins into simpler forms, further enhancing digestibility. These processing methods collectively contribute to augmenting the protein content and improving the digestibility and bioavailability of proteins in millets including foxtail millet and proso millet among others, offering practical approaches to combat protein deficiencies, especially in regions where such nutritional concerns prevail (Amadou et al. 2014; Morah and Etukudo 2017).

Despite its lower presence compared to other grains like wheat, brown rice, barley, and corn, reducing phytic acid levels in millets is essential for enhancing the availability of vital nutrients. As mentioned previously, phytic acid tends to bind to the cations like zinc, iron, and calcium compromising their digestibility and absorption. Sheethal et al. (2022) reported a decrease in phytic levels by 34.1%, 62.9%, and 29.35% in finger millet, sorghum, and pearl millet, respectively, by using traditional household processing method involving fermentation. The processing decreased the phytic acid levels substantially in the studied millets. Moreover, protein and ash content increased, while water-soluble vitamin retention was high. Also, positive impacts of various processing methods such as germination and fermentation on mineral bioavailability are reported in finger millet (Tatala et al. 2007). Decreased phytate and enhanced minerals and dietary fibre have been reported in barnyard millet grains upon germination (Sharma et al. 2016).

Breeding or genetic modification is another possible approach for improving the nutrient bioavailability. Kruger et al. (2013) examined whether a genetic reduction of phytate by about 80–86% could enhance the bioaccessibility, uptake, and absorption of iron and zinc in the major millet, sorghum. This evaluation was carried out using a dialysis assay to measure bioaccessibility, a Caco-2 cell assay to evaluate uptake, and a suckling rat pup model to analyse absorption, aiming to determine the potential impact of reduction in the phytic acid levels on the availability and absorption of iron. They found that the Caco-2-based method could be more useful than the dialysability assay for Zn absorption studies. The substantial phytate reduction (80–86%) notably enhanced the uptake and absorption of zinc in the rat pup model compared to the control. Genetic reduction of phytate led to a noteworthy increase in iron bioaccessibility. However, this alteration did not consistently translate into increased iron uptake and absorption across all the cases.

5.3 Strategies to Enhance Phenolics Bioavailability

Kumari and colleagues investigated the impact of diverse cooking techniques (steaming, open boiling, cooking, dry roasting) on the phenolics in finger millet flour-based foods. Their study utilized a simulated in vitro digestion model to analyse the bioaccessibility and the absorption of phenolics. By assessing phenolic extracts and their antioxidant activity throughout digestion, they found that open-pan boiling preserved higher phenolic content and displayed better antioxidant activity compared to other cooking methods. Moreover, the study highlighted that the release of phenolic compounds escalated progressively from the gastric to intestinal phases across all food types, emphasizing the influence of specific food preparation techniques on the bioaccessibility and potential absorption of phenolic compounds (Kumari et al. 2019). Germination was found to enhance TPC, TFC, as well as antioxidant activity in barnyard millet and proso millet (Sharma et al. 2016; Xiang et al. 2023).

Another approach for enhancing the bioavailability involves the bioavailability enhancers which don't act independently but serve to amplify the bioavailability and consequent effectiveness of active compounds. Piperine, an alkaloid found in black pepper, has been noted for its ability to heighten the absorption of various medicinal compounds and phytochemicals. When orally administered with piperine, the bioavailability of polyphenols from finger millet increased notably in rats (Hithamani and Srinivasan 2017). In minor millets such studies are limited.

6 Nutraceutical Properties of Minor Millets

Millet grains are known for their nutraceutical value. Several studies have demonstrated the health-beneficial properties of minor millets including amelioration of micronutrient deficiencies as well as some lifestyle-associated diseases through in vitro and in vivo methods. Some of these studies demonstrate the beneficial role of the minor millet grain supplementation and the macro/micronutrients associated with it, while there are other studies that demonstrate the nutraceutical properties of minor millet phenolics as well as peptides. These studies are presented herewith.

Regular consumption of minor millets including kodo, little, and foxtail millets along with other millets could mitigate iron deficiency anaemia and improve the haemoglobin as well as serum ferritin levels as reported in a systemic review and meta-analysis (Anitha et al. 2021). The role of proso millet protein in curing liver injury was reported by Nishizawa et al. (2002). They devised rat models with liver injury induced by D-galactosamine or carbon tetrachloride to assess the effect of dietary supplementation of proso millet protein. Upon feeding diet with 20% proso millet protein for 2 weeks, significant suppression in the serum levels of aspartate aminotransferase, alanine aminotransferase, and lactate dehydrogenase (which were elevated during injury) was found in the treated rats as compared to their counterparts rats fed with casein-based diet. Further, the therapeutic role of proso millet protein concentrate against type 2 diabetes was reported (Park et al. 2008). Feeding the protein concentrate from proso millet grains was found to enhance plasma levels of high-density lipoprotein (HDL) cholesterol as well as adiponectin and reduces the levels of glucose and insulin in the obese, type 2 diabetic rats fed with a high-fat diet as compared to rats fed with control diet. Roles of minor millets including foxtail and proso millets in lowering serum triglycerides have been reported as demonstrated by studies on rat models (Lee et al. 2010; Sireesha et al. 2011). More recently, Ito et al. (2022) found the potential diabetic ameliorative role of bran from Japanese barnyard millet. They reported that when rats were supplemented with barnyard millet bran, the blood glucose levels could be maintained. Furthermore, the bran-based diet could reduce diabetic polyuria, water intake, and HbA1C levels. They also observed a higher expression of the antioxidant enzyme, heme oxygense-1, in diabetic rats supplemented with bran-based diet, demonstrating its antidiabetic role.

Furthermore, studies have shown that the hypoglycaemic effect associated with foxtail millet grains can be ascribed to the insoluble fibres (Bangoura et al. 2013). Dietary fibre from foxtail millet exhibits prebiotic effect as demonstrated by Farooq et al. (2013). The foxtail millet dietary fibre serves as substrates for fermentation by the probiotic bacteria including *Lactobacillus rhamnosus*, *Lactobacillus acidophilus*, *Bifidobacterium longum,* and *Bifidobacterium bifidum* producing short-chain fatty acids such as acetate, propionate, and butyrate which are implicated in the promotion of gut health. More recently, Ren et al. (2021) studied the glucose-lowering role of foxtail millet in high-fat diet as well as streptozotocin-induced diabetic rats. They found that foxtail millet supplementation enhanced the relative abundance of gut bacteria (*Lactobacillus* and *Ruminococcus*), which in turn correlated negatively with the blood glucose level. Further, foxtail millet supplementation reversed gene expression in diabetic rats thereby promoting reduced blood glucose levels as well as reduced inflammation.

6.1 Nutraceutical Properties of Minor Millet Phenolics

6.1.1 Radical Scavenging Activity

A number of studies have demonstrated the radical scavenging role of phenolic phytochemicals from small millets. Hegde and Chandra (2005) evaluated the

2,2′-diphenyl-1-picrylhydrazyl radical (DPPH) radical quenching of methanolic extracts of minor millets including kodo, little, foxtail, barnyard, and other millets and their white varieties by electron spin resonance (ESR) spectroscopy. They found that the kodo millet extracts exhibited much higher (70%) radical quenching as compared to the other millets studied. Also, the white varieties of millets including foxtail millet showed lower quenching compared to their coloured counterparts, suggesting the role of phenolic compounds in seed coat in radical scavenging. Nevertheless, no correlation was observed between phenolic content and the antioxidant activity in the study. Chandrasekara and Shahidi (2010, 2011a, b) demonstrated the radical scavenging/antioxidant activity of phenolic extracts of minor millets including foxtail, kodo, proso, little, and other millets in terms of reducing power, trolox equivalent antioxidant capacity, β-carotene linoleate system and ferrous ion chelation, and oxygen radical absorbance capacity. Kumari et al. (2017) reported the antioxidant activity of soluble and bound phenolic compounds of the minor millets, foxtail, and proso millet along with other millets. The study observed that the antioxidant oxidant activity was affected by the variety as well as location of cultivation as demonstrated by the higher antioxidant activity of varieties from dry zone.

Recent studies also showed the antioxidant role of phenolic compounds of foxtail millet and other minor millets. Pradeep and Sreerama (2018) reported the in vitro antioxidant activity of the soluble and bound phenolic extracts of foxtail millet and little millet cultivars. Cultivar-based differences in the antioxidant activity were observed in their study. Furthermore, Xiang et al. (2019) reported antioxidant activities in terms of DPPH radical scavenging, 2,2′-azino-bis(3-ethylbenzothiazoline-6-sulfonic acid) (ABTS) radical scavenging, and oxygen radical absorbance capacity (ORAC) in the soluble as well as bound phenolic extracts of dehulled foxtail millet. The antioxidant potential of foxtail millet extracts in terms of DPPH radical scavenging as well as ferric-reducing antioxidant power/ferric-reducing ability of plasma (FRAP) has been recently reported (Kuruburu et al. 2022). The antioxidant activities of phenolic extracts of proso millet have been demonstrated in terms of peroxyl radical scavenging capacity (Zhang et al. 2014), DPPH radical scavenging, ABTS assay, and FRAP assay (Shen et al. 2018; Yuan et al. 2022). In their study, Zhang et al. (2014) found that the peroxyl radical scavenging activity of free fraction ranged from 57.68 to 147.32 μmol of vitamin C equiv./100 g DW and that of the bound phenolic fraction ranged from 95.38 to 136.48 μmol of vitamin C equiv./100 g DW in the proso millet cultivars. Further, antioxidant activities (44% DPPH radical scavenging) of methanolic extracts of whole flour of kodo millet have been reported (Sharma et al. 2017). Similarly, Watanabe (1999) isolated phenolic compounds exhibiting antioxidant activity from Japanese barnyard millet. More recently, Ofosu et al. (2020) evaluated the in vitro antioxidant potential of barnyard millet cultivated in South Korea.

6.1.2 Antiproliferative Activity

Zhang et al. (2014) evaluated the antiproliferative activities of free and bound phenolic extracts of proso millet against human breast cancer and liver cancer cell lines. The authors reported that both free and bound extracts exhibited antiproliferative activity against both the tested cell lines with free extracts of proso millet exhibiting higher antiproliferative activity. The antiproliferative potential of methanolic extracts of whole proso millet flours was studied on MDA-MB-231 breast cell lines. The study showed that the proso millet extracts could inhibit proliferation in a dose-dependent mode. Also, the extracts have not exhibited any cytotoxicity at the tested doses (Shen et al. 2018). Further, the phytochemical-rich fractions of foxtail millet seeds extracted using 70% ethanol and alkali treatment exhibited antiproliferative potential against breast cancer lines (Kuruburu et al. 2022). The extracts were found to induce G2/M cell cycle arrest and DNA fragmentation leading to the inhibition of proliferation of the cancer cell lines.

6.1.3 Hypoglycaemic Activity

Most studies on hypoglycaemic activity of minor millet phenolics are based on the inhibitory potential of phenolic extracts on the carbohydrate digestive enzymes, viz., α-amylase and α-glucosidase, as these enzymes play an important role in regulating the blood glucose levels especially after a meal. Ethanolic extracts of foxtail and proso have been reported to exhibit α-amylase and α-glucosidase inhibitory activity (Kim et al. 2011). Also, the phenolic extracts of foxtail and barnyard millet among other millets were reported to possess hypoglycaemic potential in terms of α-amylase and α-glucosidase inhibition (Ofosu et al. 2020). The latter study also demonstrated the potential of small millet phenolics in inhibiting non-enzymatic glycation thus playing role in the amelioration of diabetes-associated complications (Ofosu et al. 2020).

6.2 Nutraceutical Properties of Minor Millet-Derived Bioactive Peptides

Bioactive peptides are low-molecular-weight molecules (<3 kDa) with specific amino acid sequences which are encrypted in the parent protein. These peptides, upon release from the parent protein, often through enzymatic proteolysis, become active and exhibit health-beneficial attributes. Studies have shown that peptides derived from foxtail millet by enzymatic hydrolysis could exhibit antioxidant activity as demonstrated by in vitro assays (Ji et al. 2019b). Further, the antioxidant potential (in terms of low ROS generation, enhanced glutathione (GSH), as well as superoxide dismutase) of peptides from germinated and heated foxtail millet were demonstrated in Caco-2 cell-based assays (Hu et al. 2020). In another study, Ji et al. (2020) reported the role of millet prolamin peptides derived from foxtail millet in reducing the oxidative damage. As observed by the studies in H_2O_2-induced HaCaT cells, the peptides reduced the generation of ROS and malondialdehyde, an important marker of oxidative stress. The peptides were also found to enhance the levels of GSH, suggestive of their antioxidant potential (Ji et al. 2020).

Xu et al. (2011) reported an antifungal peptide (MW 26.9 K Da) purified from foxtail millet seeds with activities against fungi including *Alternaria alternata*, *Trichoderma viride*, etc. As observed from the changes in morphology as well as ultrastructure in *A. alternata*, the study indicated that the foxtail millet-derived antifungal peptide acts by attacking the cell wall retraction of hyphal cytoplasm further leading to the mycelial death. Peptides with antioxidant and antimicrobial activity have been derived from *Lacticaseibacillus paracasei*-fermented foxtail millet meal (Amadou et al. 2014).

Bioactive peptides from the minor millets were found to show inhibitory activities against various key enzymes such as acetylcholinesterase (ACE), an important target for antihypertensive drugs. Karaś et al. (2019) studied the enzyme inhibitory properties of protein hydrolysates and peptides derived from heat-treated proso millet grains by in vitro gastrointestinal digestion. They reported that the peptides could show ACE inhibition activity with the peptide from prolamin fraction at 100 °C showing the highest activity. Also, peptide from globulin 11S at 65 °C and untreated prolamin (control) exhibited highest α-glucosidase and α-amylase activity, respectively.

Also, Jakubczyk et al. (2019) investigated the lipase inhibition as well as anti-inflammatory role of peptides derived from proso millet as mentioned above. They found that peptides from the prolamin fraction obtained by heat treatment at 65 °C exhibited highest pancreatic lipase inhibition. Highest inhibition of COX-1 and COX-2 was exhibited by peptides from the globulin 7S of grains subjected to heating at 65 °C. Peptides from the prolamin fraction of proso millet grains treated at 100 °C showed the highest LOX inhibition. The COX-1, COX-2, as well as LOX inhibition shows potential anti-inflammatory role of the peptides. The anti-inflammatory effect of peptides synthesized from germinated, heat-treated foxtail millet has been reported by Hu et al. (2020). As observed in their study, the foxtail millet-derived peptides exhibited inhibition against nitric oxide, tumour necrosis factor-α, as well as interleukin-6, demonstrating its anti-inflammatory potential. Furthermore, Ji et al. (2020) reported the anti-inflammatory role of millet prolamin peptide (peptides from foxtail millet prolamin) as demonstrated in the studies involving lipopolysaccharide (LPS)-stimulated RAW264.7 murine macrophages. The peptides were found to exhibit nitric oxide inhibition as well as inhibition of pro-inflammatory cytokines, viz., interferon-α (INF-α) and interleukin (IL)-6 and IL-1β.

7 Conclusion and Future Perspectives

Minor millets are heritage crops with superior agronomical and nutritional attributes. These grains are endowed with good amounts of protein, dietary fibre, micronutrients, viz., essential minerals and B-vitamins. Minor millets also house numerous phytochemicals, especially the phenolic compounds including phenolic acids and flavonoids. Owing to their excellent nutritional and phytochemical composition, minor millets exhibit excellent nutraceutical properties. These grains exert human health-beneficial effects through antioxidant, antiproliferative, hypolipidaemic, hypoglycaemic, as well as prebiotic properties. Also, bioactive peptides derived from minor millets were also found to exhibit health-beneficial effects to tackle some of the chronic diseases. With high prevalence of lifestyle-associated diseases as well as malnutrition including micronutrient deficiencies, it is imperative that crops such as minor millets would serve an important role in tackling these challenges. To reap the full benefits of these sustainable grains, it is important is conduct more in vivo studies on their health-beneficial roles and more studies on the bioavailability of the nutrients.

References

Amadou I, Gounga ME, Shi YH, Le GW (2014) Fermentation and heat-moisture treatment induced changes on the physicochemical properties of foxtail millet (Setaria italica) flour. Food Bioprod Process 92(1):38–45

Anitha S, Kane-Potaka J, Botha R, Givens DI, Sulaiman NL, Upadhyay S, Vetriventhan M, Tsusaka TW, Parasannanavar DJ, Longvah T, Rajendran A (2021) Millets can have a major impact on improving iron status, hemoglobin level, and in reducing iron deficiency anemia–a systematic review and meta-analysis. Front Nutr 8:725529

Annor GA, Marcone M, Bertoft E, Seetharaman K (2014) Physical and molecular characterization of millet starches. Cereal Chem 91(3):286–292

Bangoura ML, Nsor-Atindana J, Ming ZH (2013) Solvent optimization extraction of antioxidants from foxtail millet species' insoluble fibers and their free radical scavenging properties. Food Chem 141(2):736–744

Chandrasekara A, Shahidi F (2010) Content of insoluble bound phenolics in millets and their contribution to antioxidant capacity. J Agric Food Chem 58(11):6706–6714

Chandrasekara A, Shahidi F (2011a) Determination of antioxidant activity in free and hydrolyzed fractions of millet grains and characterization of their phenolic profiles by HPLC-DAD-ESIMS. J Funct Foods 3:144–158

Chandrasekara A, Shahidi F (2011b) Inhibitory activities of soluble and bound millet seed phenolics on free radicals and reactive oxygen species. J Agric Food Chem 59:428–436

Chandrasekara A, Shahidi F (2012) Bioaccessibility and antioxidant potential of millet grain phenolics as affected by simulated *in vitro* digestion and microbial fermentation. J Funct Foods 4(1):226–237

Devisetti R, Yadahally SN, Bhattacharya S (2014) Nutrients and antinutrients in foxtail and proso millet milled fractions: evaluation of their flour functionality. LWT-Food Sci Technol 9(2):889–895

Dykes L, Rooney LW (2006) Sorghum and millet phenols and antioxidants. J Cereal Sci 44(3):236–251

Farooq U, Mohsin M, Liu X, Zhang H (2013) Enhancement of short chain fatty acid production from millet fibres by pure cultures of probiotic fermentation. Trop J Pharm Res 12(2):189–194

Fujita S, Donghui L, Sugimoto Y, Inouchi N, Fuwa H (1989) Some properties of starch from foxtail millet (Setaria italica BEAUV.). Agric Biol Chem 53(4):1163–1165

Galanakis CM (2021) Functionality of food components and emerging technologies. Food Secur 10:128. https://doi.org/10.3390/foods10010128

Geervani P, Eggum BO (1989) Nutrient composition and protein quality of minor millets. Plant Foods Hum Nutr 39(2):201–208. https://doi.org/10.1007/bf01091900

Goswami D, Mandge HM, Rehal J, Rajput H (2023) Nutritional advancement in the ethnic and novel foods using the diverse minor millets. In: Pudake RN, Solanke AU, Kole C (eds) Nutriomics of millet crops, 1st edn. CRC Press, pp 141–156. https://doi.org/10.1201/b22809

Gulati P, Li A, Holding D, Santra D, Zhang Y, Rose DJ (2017) Heating reduces proso millet protein digestibility via formation of hydrophobic aggregates. J Agric Food Chem 65(9):1952–1959

Hedren E, Diaz V, Svanberg U (2002) Estimation of carotenoids accessibility from carrots determined by an in vitro digestion method. Eur J Clin Nutr 56(5):425–430

Hegde PS, Chandra TS (2005) ESR spectroscopic study reveals higher free radical quenching potential in kodo millet (*Paspalum scrobiculatum*) compared to other millets. Food Chem 92:177–182

Hithamani G, Srinivasan K (2014) Effect of domestic processing on the polyphenol content and bioaccessibility in finger millet (Eleusine coracana) and pearl millet (Pennisetum glaucum) Food Chemistry 16:455–462. https://doi.org/10.1016/j.foodchem.2014.04.107

Hithamani G, Srinivasan K (2017) Bioavailability of finger millet (*Eleusine coracana*) phenolic compounds in rat as influenced by co-administered piperine. Food Biosci 19:101–109

Hu S, Yuan J, Gao J, Wu Y, Meng X, Tong P, Chen H (2020) Antioxidant and anti-inflammatory potential of peptides derived from in vitro gastrointestinal digestion of germinated and heat-treated foxtail millet (Setaria italica) proteins. J Agric Food Chem 68(35):9415–9426

Ito Y, Suzuki A, Nasukawa H, Miyaki K, Yano A, Nagasawa T (2022) Ameliorative effects of Japanese barnyard millet (Echinochloa esculenta H. Scholz) bran supplementation in streptozotocin-induced diabetic rats. Food Sci Technol Res 28(5):431–439

Jakubczyk A, Szymanowska U, Karaś M, Złotek U, Kowalczyk D (2019) Potential anti-inflammatory and lipase inhibitory peptides generated by in vitro gastrointestinal hydrolysis of heat treated millet grains. CyTA-J Food 17(1):324–333

Ji Z, Feng R, Mao J (2019b) Separation and identification of antioxidant peptides from foxtail millet (Setaria italica) prolamins enzymatic hydrolysate. Cereal Chem 96(6):981–993

Ji J, Liu Y, Ge Z, Zhang Y, Wang X (2019a) Oleochemical properties for different fractions of foxtail millet bran. J Oleo Sci 68(8):709–718

Ji Z, Mao J, Chen S, Mao J (2020) Antioxidant and anti-inflammatory activity of peptides from foxtail millet (Setaria italica) prolamins in HaCaT cells and RAW264. 7 murine macrophages. Food Biosci 36:100636

Kaimal A, Mujumdar A, Thorat B (2021) Resistant starch from millets: recent developments and applications in food industries. Trends Food Sci Technol 111:563–580. https://doi.org/10.1016/j.tifs.2021.02.074

Kalsi R, Bhasin JK (2023) Nutritional exploration of foxtail millet (*Setaria italica*) in addressing food security and its utilization trends in food system. eFood 4(5):e111. https://doi.org/10.1002/efd2.111

Kamara MT, Huiming Z, Zhu K, Amadou I, Tarawalie F (2009) Comparative study of chemical composition and physicochemical properties of two varieties of defatted foxtail millet flour grown in China. Am J Food Technol 4:255–267. https://doi.org/10.3923/ajft.2009.255.267

Karaś M, Jakubczyk A, Szymanowska U, Jęderka K, Lewicki S, Złotek U (2019) Different temperature treatments of millet grains affect the biological activity of protein hydrolyzates and peptide fractions. Nutrients 11(3):550

Kim JS, Hyun TK, Kim MJ (2011) The inhibitory effects of ethanol extracts from sorghum, foxtail millet and proso millet on α-glucosidase and α-amylase activities. Food Chem 124(4):1647–1651

Krishna Kumari S, Thayumanavan B (1998) Characterization of starches of proso, foxtail, barnyard, kodo, and little millets. Plant Foods Hum Nutr 53:47–56

Kruger J, Taylor JR, Du X, De Moura FF, Lönnerdal B, Oelofse A (2013) Effect of phytate reduction of sorghum, through genetic modification, on iron and zinc availability as assessed by an in vitro dialysability bioaccessibility assay, Caco-2 cell uptake assay, and suckling rat pup absorption model. Food Chem 141(2):1019–1025

Kumar RR, Singh SP, Rai GK, Krishnan V, Berwal MK, Goswami S, Vinutha T, Mishra GP, Satyavathi CT, Singh B, Praveen S (2022) Iron and zinc at a cross-road: a trade-off between micronutrients and anti-nutritional factors in pearl millet flour for enhancing the bioavailability. J Food Compos Anal 111:104591

Kumari D, Madhujith T, Chandrasekara A (2017) Comparison of phenolic content and antioxidant activities of millet varieties grown in different locations in Sri Lanka. Food Sci Nutr 5(3):474–485

Kumari D, Chandrasekara A, Shahidi F (2019) Bioaccessibility and antioxidant activities of finger millet food phenolics. J Food Bioact https://doi.org/10.31665/JFB.2019.6187

Kuruburu MG, Bovilla VR, Leihang Z, Madhunapantula SV (2022) Phytochemical-rich fractions from foxtail millet (Setaria italica (L.) P. Beauv) seeds exhibited antioxidant activity and reduced the viability of breast cancer cells in vitro by inducing DNA fragmentation and promoting cell cycle arrest. Anticancer Agents Med Chem 22(13):2477–2493

Lee SH, Chung IM, Cha YS, Park Y (2010) Millet consumption decreased serum concentration of triglyceride and C-reactive protein but not oxidative status in hyperlipidemic rats. Nutr Res 30(4):290–296. https://doi.org/10.1016/j.nutres.2010.04.007

Longvah T, Ananthan R, Bhaskarachary K, Venkaiah K (2017) Indian food composition tables. National Institute of Nutrition, Indian Council of Medical Research, Hyderabad

Macierzanka A, Mackie AR, Krupa L (2019) Permeability of the small intestinal mucus for physiologically relevant studies: impact of mucus location and *ex vivo* treatment. Sci Rep 9:17516. https://doi.org/10.1038/s41598-019-53933-5

Malleshi NG, Agarwal A, Tiwari A, Sood S (2021) Nutritional quality and health benefits. In: Millets and pseudo cereals. Woodhead Publishing, pp 159–168

Mattila P, Pihlava JM, Hellström J (2005) Contents of phenolic acids, alkyl-and alkenylresorcinols, and avenanthramides in commercial grain products. J Agric Food Chem 53(21):8290–8295

Morah F, Etukudo UP (2017) Effect of sprouting on nutritional value of *Paniciummiliaceum* (proso millet). Edorium J Nutr Diet 4:1–4

Nakayama H, Afzal M, Okuno K (1998) Intraspecific differentiation and geographical distribution of Wx alleles for low amylose content in endosperm of foxtail millet, Setaria italica (L.) Beauv. Euphytica 102:289–293

Nishizawa N, Sato D, Ito Y, Nagasawa T, Hatakeyama Y, Choi MR, Choi YY, Wei YM (2002) Effects of dietary protein of proso millet on liver injury induced by D-galactosamine in rats. Biosci Biotechnol Biochem 66(1):92–96

Ofosu FK, Elahi F, Daliri EB, Chelliah R, Ham HJ, Kim JH, Han SI, Hur JH, Oh DH (2020) Phenolic profile, antioxidant, and antidiabetic potential exerted by millet grain varieties. Antioxidants 9(3):254

Pang M, He S, Wang L, Cao X, Cao L, Jiang S (2014) Physicochemical properties, antioxidant activities and protective effect against acute ethanol-induced hepatic injury in mice of foxtail millet (Setaria italica) bran oil. Food Funct 5(8):1763–1770

Panwar P, Dubey A, Verma AK (2016) Evaluation of nutraceutical and antinutritional properties in barnyard and finger millet varieties grown in Himalayan region. J Food Sci Technol 53(6):2779–2787. https://doi.org/10.1007/s13197-016-2250-8

Park KO, Ito Y, Nagasawa T, Choi MR, Nishizawa N (2008) Effects of dietary Korean proso-millet protein on plasma adiponectin, HDL cholesterol, insulin levels, and gene expression in obese type 2 diabetic mice. Biosci Biotechnol Biochem 72(11):2918–2925

Pattabiraman TN (1986) Trypsin/chymotrypsin inhibitors from millets. In: Friedman M (ed) Nutritional and toxicological significance of enzyme inhibitors in foods. Advances in experimental medicine and biology. Springer, Boston, MA. https://doi.org/10.1007/978-1-4757-0022-0_25

Pradeep PM, Sreerama YN (2017) Soluble and bound phenolics of two different millet genera and their milled fractions: comparative evaluation of antioxidant properties and inhibitory effects on starch hydrolysing enzyme activities. J Funct Foods 35:682–693

Pradeep PM, Sreerama YN (2018) Phenolic antioxidants of foxtail and little millet cultivars and their inhibitory effects on α-amylase and α-glucosidase activities. Food Chem 247:46–55

Raes K, Knockaert D, Struijs K, Van Camp J (2014) Role of processing on bioaccessibility of minerals: influence of localization of minerals and anti-nutritional factors in the plant. Trends Food Sci Technol 37(1):32–41

Ren X, Wang L, Chen Z, Hou D, Xue Y, Diao X, Shen Q (2021) Foxtail millet improves blood glucose metabolism in diabetic rats through pi3k/akt and nf-κb signaling pathways mediated by gut microbiota. Nutrients 13(6):1837

Rodríguez JP, Rahman H, Thushar S, Singh RK (2020) Healthy and resilient cereals and pseudo-cereals for marginal agriculture: molecular advances for improving nutrient bioavailability. Front Genet 11:49

Rooney LW (1996) Sorghum and millets. In: Henry RJ, Kettlewell PS (eds) Cereal grain quality. Springer, Dordrecht. https://doi.org/10.1007/978-94-009-1513-8_5

Samtiya M, Aluko RE, Dhewa T (2020) Plant food anti-nutritional factors and their reduction strategies: an overview. Food Prod Process Nutr 2:6. https://doi.org/10.1186/s43014-020-0020-5

Seo MC, Ko JY, Song SB, Lee JS, Kang JR, Kwak DY, Oh BG, Yoon YN, Nam MH, Jeong HS, Woo KS (2011) Antioxidant compounds and activities of foxtail millet, proso millet and sorghum with different pulverizing methods. J Korean Soc Food Sci Nutr 40(6):790–797

Serna-Saldivar SO, Espinosa-Ramírez J (2019) Grain structure and grain chemical composition. In: Sorghum and millets. AACC International Press, pp 85–129. https://doi.org/10.1016/B978-0-12-811527-5.00005-8

Serna-Saldivar S, Rooney LW (1995) Structure and chemistry of sorghum and millets. In: Dendy DAV (ed) Sorghum and millets: chemistry and technology. American Association of Cereal Chemists, St. Paul, MN, pp 69–124

Shahidi F, Chandrasekara A (2013) Millet grain phenolics and their role in disease risk reduction and health promotion: a review. J Funct Foods 5:570–581

Sharma B, Gujral HS (2019) Influence of nutritional and antinutritional components on dough rheology and in vitro protein & starch digestibility of minor millets. Food Chem 299:125115. https://doi.org/10.1016/j.foodchem.2019.12511

Sharma S, Saxena DC, Riar CS (2016) Analysing the effect of germination on phenolics, dietary fibres, minerals and γ-amino butyric acid contents of barnyard millet (*Echinochloafrumentaceae*). Food Biosci 13:60–68. https://doi.org/10.1016/j.fbio.2015.12.007

Sharma S, Sharma N, Handa S, Pathania S (2017) Evaluation of health potential of nutritionally enriched Kodo millet (Eleusine coracana) grown in Himachal Pradesh, India. Food Chem 214:162–168

Sharma R, Sharma S, Dar BN, Singh B (2021) Millets as potential nutri-cereals: a review of nutrient composition, phytochemical profile and techno-functionality. Int J Food Sci 56(8):3703–3718. https://doi.org/10.1111/ijfs.15044

Sheethal HV, Baruah C, Subhash K, Ananthan R, Longvah T (2022) Insights of nutritional and anti-nutritional retention in traditionally processed millets. Front Sustain Food Syst 5:735356

Shen R, Ma Y, Jiang L, Dong J, Zhu Y, Ren G (2018) Chemical composition, antioxidant, and antiproliferative activities of nine Chinese proso millet varieties. Food Agric Immunol 29(1):625–637

Shyam R, Singh RP (2018) Evalution of nutritional value and anti-nutritional factors of kodo millet (*Paspalum scrobiculatum* L.) germplasm grown in Eastern (U. P.). Plant Arch 18:247–250

Sireesha Y, Kasetti RB, Nabi SA, Swapna S, Apparao C (2011) Antihyperglycemic and hypolipidemic activities of *Setaria italica* seeds in STZ diabetic rats. Pathophysiology 18(2):159–164. https://doi.org/10.1016/j.pathophys.2010.08.003

Sudharshana L, Monteiro PV, Ramachandra G (1988) Studies on the proteins of kodo millet (Paspalum scrobiculatum). J Sci Food Agric 42(4):315–323

Tatala S, Ndossi G, Ash D, Mamiro P (2007) Effect of germination of finger millet on nutritional value of foods and effect of food supplement on nutrition and anaemia status in Tanzania children. Tanzan J Health Res 9(2):77–86

Taylor J (2017) Millets: their unique nutritional and health-promoting attributes. In: Gluten-free ancient grains, Woodhead publishing series in food science, technology and nutrition, pp 55–103. https://doi.org/10.1016/B978-0-08-100866-9.00004-2

Taylor JRN, Duodu KG (2015) Effects of processing sorghum and millets on their phenolic phytochemicals and the implications of this to the health-enhancing properties of sorghum and millet food and beverage products. J Sci Food Agric 95:225–237

Tharifkhan SA, Perumal AB, Elumalai A, Moses JA, Anandharamakrishnan C (2021) Improvement of nutrient bioavailability in millets: emphasis on the application of enzymes. J Sci Food Agric 101(12):4869–4878

Ugare R, Chimmad B, Naik R, Bharati P, Itagi S (2014) Glycemic index and significance of barnyard millet (Echinochloa frumentacae) in type II diabetics. J Food Sci Technol 51:392–395

Veena B, Chimmad BV, Naik RK, Shantakumar G (2005) Physicochemical and nutritional studies in barnyard millet. Karnataka J Agric Sci 18:101–105

Vetriventhan M, Azevedo VCR, Upadhyaya HD et al (2020) Genetic and genomic resources, and breeding for accelerating improvement of small millets: current status and future interventions. Nucleus 63:217–239. https://doi.org/10.1007/s13237-020-00322-3

Watanabe M (1999) Antioxidative phenolic compounds from Japanese barnyard millet (*Echinochloa utilis*) grains. J Agric Food Chem 47:4500–4505. https://doi.org/10.1021/jf990498s

Wen Y, Liu J, Meng X, Zhang D, Zhao G (2014) Characterization of proso millet starches from different geographical origins of China. Food Sci Biotechnol 23:1371–1377

Xiang J, Zhang M, Apea Bah FB, Beta T (2019) Hydroxycinnamic acid amide (HCAA) derivatives, flavonoid C-glycosides, phenolic acids and antioxidant properties of foxtail millet. Food Chem 295:214–223. https://doi.org/10.1016/j.foodchem.2019.05.058

Xiang J, Yuan Y, Du L, Zhang Y, Li C, Beta T (2023) Modification on phenolic profiles and enhancement of antioxidant activity of proso millets during germination. Food Chem: X 18:100628

Xu W, Wei L, Qu W, Liang Z et al (2011) A novel antifungal peptide from foxtail millet seeds. J Sci Food Agric 91(9):1630–1637. https://doi.org/10.1002/jsfa.4359

Yin SY, Kuo SM, Chen YR, Tsai YC, Wu YP, Lin YR (2019) Genetic variation of physicochemical properties and digestibility of foxtail millet (Setaria italica) landraces of Taiwan. Molecules 24(23):4323

Yuan Y, Xiang J, Zheng B, Sun J, Luo D, Li P, Fan J (2022) Diversity of phenolics including hydroxycinnamic acid amide derivatives, phenolic acids contribute to antioxidant properties of proso millet. Lwt 154:112611

Zhang LZ, Liu RH (2015) Phenolic and carotenoid profiles and antiproliferative activity of foxtail millet. Food Chem 174:495–501. https://doi.org/10.1016/j.foodchem.2014.09.089

Zhang L, Liu R, Niu W (2014) Phytochemical and Antiproliferative activity of Proso Millet. PLoS One 9(8):e104058

Agronomic and Cultivation Practices in Minor Millets

Rajendra Prasad Meena, Mahendar S. Bhinda, Kamal Garg, Arjun Singh, and Dinesh Chandra Joshi

Abstract

Minor millets are nutrient-dense and climate-resilient crops that play an important role in the food and nutritional security of developing world. Globally, millets are mostly produced and consumed by the developing world. India remains the world's largest producer of millet, with over 37% of global production. The last 5-year mean data of millets in India revealed that, out of total millet growing area (~129 lakh ha), only 12.1% area (15.7 lakh ha) belongs to minor millets. Similarly, out of total millet production (167.9 lakh tons), minor millets are contributing only 12.4% (20.8 lakh tons). Other than area expansion, the productivity of minor millets is still very low compared to major crops, mainly due to poor emphasis on research and the adoption of improved agro-technologies. These crops are mainly grown by resource-poor farmers on marginal land with poor management. Minor millets are grassy in nature; the initial slow growth, the resemblance of seedlings with weeds, and the lack of availability of selective herbicides make them more susceptible to weed infestation. The weeds that cause yield reduction can reach up to 75% in minor millets. Similarly, poor nutrient management and monocropping in minor millets are responsible for yield reduction and soil health degradation. So, apart from the varietal improvement, market and value chain development. Agronomic interventions such as good crop husbandry, nutrient management, weed management, mechanization, and postharvest management are the utmost needed for enhancing the productivity of

R. P. Meena (✉) · M. S. Bhinda · D. C. Joshi
Division of Crop Improvement, ICAR-Vivekananda Parvatiya Krishi Anusandhan Sansthan, Almora, India

K. Garg · A. Singh
ICAR-Vivekananda Parvatiya Krishi Anusandhan Sansthan, Almora, India

ICAR-Indian Agricultural Research Institute, New Delhi, India

D. C. Joshi et al. (eds.), *Minor Millets*,
https://doi.org/10.1007/978-981-96-4265-6_3

minor millets. This chapter explores crucial agronomic practices essential for enhancing the productivity and sustainability of minor millets. Key areas addressed include sowing methods, crop diversification, integrated nutrient management, effective weed control measures, mechanization options, and postharvest management techniques. These interventions are pivotal not only for boosting yields but also for improving soil health and promoting the sustainable cultivation of minor millets.

Keywords
Minor millets · Agronomic and cultivation practices · Global production · Millet conservation

1 Introduction

Millets refer to a group of small-seeded crops having resilience to adverse climates and are mostly preferred for resource-poor farming. They are nutrient-rich (Farnandez et al. 2013) and play a pivotal role in addressing malnutrition and curing various diseases in the tribal population (Joshi et al. 2018, 2020; Bhinda et al. 2023). Therefore, they are referred to as nutria-cereals (Gull et al. 2014; Kumar et al. 2021). Minor millets are becoming popular in urban areas among health-conscious people (Fig. 1). These crops are superior to other major cereals in terms of adaptability and resilience to unfavorable conditions; therefore, they are referred to as crops of famine (Devi et al. 2014; Shukla et al. 2015). These days, millets are getting more attention on account of their short growing period and their distinctive capacity to thrive even on marginal and poor soils, where other crops typically fail to turn a profit (Devi et al. 2014; Trivedi et al. 2015). Particularly in tribal agriculture, they provide superior harvests in arid regions, enhancing the security of food, feed, and nutrition (Doggett 1989; Pradhan et al. 2019). In emerging nations across the continents of Asia and Africa, millets also have a significant impact on the economy and food security. Though small millets are essential for achieving food and nutrition security, there has been a noticeable decrease in their production and cultivation. Based on cultivated area, they are broadly classified as major and minor millets (Joshi et al. 2021; Meena et al. 2021).

Pearl millet and sorghum come under major millets, whereas finger millet, little millet, kodo millet, barnyard millet, foxtail millet, proso millet, and browntop millet come under minor or small millets in the order of decreasing cultivated area in India (Meena et al. 2021). Millets are known by a single name; however, their habit, climate, and input requirements vary from crop to crop (Meena et al. 2021; Parihar et al. 2021). Pearl millet is the most widely grown millet; it alone contributes ~56% to India's millet production (Meena et al. 2021). It has a fast-growing habit, and the plant can attain 6 feet in 60 days. It is preferred for grain as well as for cattle fodder in the drier regions of Rajasthan, Gujrat, Haryana, etc. Similarly, sorghum is another major millet, mostly grown for grain and green fodder purposes. Winter-sown sorghum gave a higher yield than kharif sown due to the less pressure of weed

Fig. 1 Field photographs of minor millets

infestation and efficient management of water and nutrients. Among the minor millets, finger millet has the largest acreage and production in India. It is mostly grown in Karnataka, Uttarakhand, Tamil Nadu, and some other parts of the country. It is rich in calcium, iron, and protein and almost free of gluten. It is a healthy substitute for major food grain crops (Gupta et al. 2017). High calcium and protein content makes it a perfect food for the brain development of growing kids, and its high fiber content makes it ideal for weight loss in fatty people. It is a warm-season crop that is mainly grown in the *kharif* season, but it is also grown in the winter season in some parts of southern India. Barnyard millet is grown in a medium- to high-altitude irrigation-assured region, especially where rice is unfit for cultivation due to the low temperature. The crop is known for its nutrient-rich grain and quality fodder for animals in hills (Gupta et al. 2009). Foxtail millet is the second most grown minor millet in India; however, it is first in the world. With ancient evidence of its cultivation (Zohary and Hopf 2000), China holds the first position in foxtail millet production. The crops have high iron, almost free from pests and disease, and are recommended for lowering blood sugar and cholesterol. It is also grown for hay, silage, and birdseed in some parts of Europe and America. Its early maturity (80–90 days) and efficient use of residual moisture make it a good catch crop in rainfed areas (Jijau 1989). Kodo millet is the coarsest grain millet among minor millets. It is a warmer-season crop grown with a low water requirement. It has high levels of polyphenols, antioxidants, and lecithin compounds and is very good for the nervous system. Proso millet is a short-duration, warm-season crop that can grow with low water where other millets would fail (Sahib 1997; Seetharam 1998). Its shallow root system can use residual moisture more efficiently and conserve moisture in a deeper soil profile for succeeding crops. The common names and special characteristics of minor millets are depicted in Table 1.

Table 1 Special characteristics of small millets

Millet	Common name	Botanical name	Special characteristics
Finger millet	Ragi mandua, nachni, kapai, nagli, marua	*Eleusine coracana* (L.) Gaertn	Rich source of calcium and micronutrients and have diverse adaptability (Seetharam 1998)
Foxtail millet	Kauni, kangni, korra, tenai, rala	*Setaria italica* (L.) Beauv.	Short duration, grow in low soil fertility and drought (Jijau 1989)
Kodo millet	Kodo, varagu, haraka, arikalu	*Paspalum scrobiculatum* L.	Long duration, grown well in shallow and deep soil (Hegde and Gowda et al. 1989)
Little millet	Kutki, samai, samalu, same	*Panicum sumatrense* Roth ex Roem. and Schult	Short duration, withstand both drought and water logging (Doggett 1989)
Proso millet	Cheena, panivaragu, variga, baragu	*Panicum miliaceum* L.	Short duration, tolerant to heat and drought (Sahib 1997)
Barnyard millet	Sawan, jhingora, kuthiraivali, oodalu	*Echinochloa frumentacea* Link.	Fastest growing, voluminous fodder (Gupta et al. 2009)

Despite the many superiority over major cereals, minor millets are lagging far behind in production per unit area. The lack of biotic and abiotic resistance, high-yielding cultivars, improper agronomical management, and lack of policy support were the major off-putting factors associated with shrinking cultivated area and production per unit area. In the consequently national year of millet (2018) and the international year of millet (2023), the government policies favor research, public sector investment, and expansion in the cultivated area of millets in India. Most millet growers are resource-poor farmers; they grow crops without external inputs or agronomic management (Kumar et al. 2021). Therefore, best agronomic practices are needed for uniform germination, better crop stand, and to harness the maximum solar radiation and available nutrient and water resources for enhancing production per unit area. In the present chapter, we summarize the improved agronomic and cultivation practices of minor millets for enhancing the production per unit area at farmers' fields. In this chapter, we also discussed about modern concepts and frontier areas for boosting minor millet productivity.

2 Cropping Area, Production, and Productivity

Developing countries produce and consume the majority of the small millets. Globally during 1960s millets (except sorghum) were grown in 43.94 M ha area which was reduced to 30.93 M ha in 2021 (Table 2). The data revealed that Africa (18.6 M ha) and Asia (11.6 M ha) together contribute >97% in millet growing area of entire globe (Table 2). Similarly, both continents contribute >96% of millet production. However, highest productivity was recorded in Europe (1.60 t ha) followed by Asia (1.40) and lowest in Africa (0.65 t ha). India remains the world's largest producer of millet, with over 37% of global production. The last 5-year mean data of millets cultivated area in India revealed that out of total millet growing area (~129 lakh ha.), only 12.1% area (15.7 lakh ha) belongs to minor millets (Table 3; Fig. 2). Similarly, out of the total millet production (167.9 lakh tons), minor millets are contributing only 12.4% (20.8 lakh tons). The productivity of minor millets (1187 kg/ha) remained almost the same as that of major millets, largely due to the higher productivity of finger millet. According to state-wise production statistics in India, Rajasthan contributes the highest share in total millet production (~50 lakh tons), mainly due to pearl millet production. Karnataka shared the highest in finger millet production (~11 lakh tons), followed by Tamil Nadu and Uttarakhand. The lower production of minor millets mainly attributed to shrinking cultivated area and lack of improved production technologies compared to major cereals.

3 Climatic and Soil Requirement

The minor millets are grown in diverse soils and climatic condition (Powar et al. 2020). They were grown in tropical and subtropical condition; they flourish well when temperature is ~25 ± 5 °C. These crops are grown in the areas where rainfall

Table 2 Global millets (except sorghum) harvested area, production and productivity (FAOSTAT 2023)

	1961–1970	1971–1980	1981–1990	1991–2000	2001–2010	2011–2020	2021
	Millet harvested area (lakh ha)						
Africa	123.1	127.6	131.7	188.3	204.0	194.3	186.0
America	2.73	2.84	2.07	2.07	2.00	1.74	2.74
Asia	277.7	250.1	208.1	161.9	139.5	114.0	116.6
Europe	35.60	28.53	27.46	14.79	7.06	5.23	3.68
Oceania	0.28	0.32	0.34	0.32	0.35	0.36	0.36
World	439.4	409.4	369.7	367.3	352.8	315.6	309.3
	Millet production (lakh tons)						
Africa	71.8	78.2	91.7	121.8	158.0	128.9	121.1
America	3.23	3.44	2.74	3.03	3.02	2.93	3.59
Asia	171.7	168.9	159.0	138.2	137.4	145.4	170.0
Europe	26.02	21.84	26.69	12.01	8.34	7.28	5.92
Oceania	0.28	0.30	0.35	0.35	0.34	0.36	0.37
World	273.0	272.7	280.5	275.5	307.1	284.8	300.9
	Millet grain productivity (kg/ha)						
Africa	582.3	613.7	697.7	647.4	773.0	664.0	651.0
America	1179.9	1205.5	1340.2	1454.8	1516.8	1658.5	1307.7
Asia	616.7	675.5	762.8	854.1	981.0	1279.5	1457.6
Europe	739.4	763.7	969.8	850.4	1174.3	1380.2	1607.2
Oceania	999.8	935.8	1015.6	1070.6	970.8	1019.7	1020.9
World	620.3	665.8	758.4	750.0	868.4	903.7	972.7

ranges from 250 to 1250 mm or more. Climatic and soil requirement of these crops varied crop to crop; the detailed information is given in Table 4.

4 Improved varieties

Various research institutions and universities have developed and introduced numerous high-yielding varieties across all small millet types, tailored to different states. Detailed information can be found in Table 5.

5 Land Preparation and Sowing

Minor millets are mostly grown as rainfed crops; therefore, two plowings at different depths are needed for moisture conservation. First, deep plowing should be done when pre-monsoon showers occur, and second, light plowing should be done at the onset of the monsoon for final seedbed preparation. A gentle slope should be provided to ensure drainage. Even water stagnation in small patches drastically reduces the germination percentage. The seeds of small millets are very small (except kodo

Table 3 Harvested area, production, and productivity of millet in India (APEDA 2024)

	Area (Lakh ha)					
	2019–2020	**2020–2021**	**2021–2022**	**2022–2023**	**2023–2024**	**05 years average**
Jowar	48.24	43.78	38.01	35.35	36.47	40.37
Bajra	75.43	76.52	68.41	75.72	70.08	73.23
Minor millet	14.62	16.03	16.47	15.91	15.33	15.67
Total	138.29	136.33	122.89	126.98	121.88	129.27
	Production (Lakh tone)					
	2019–2020	**2020–2021**	**2021–2022**	**2022–2023**	**2023–2024**	**05 years average**
Jowar	47.72	48.12	41.51	38.14	40.34	43.17
Bajra	103.63	108.63	97.81	114.31	95.31	103.94
Minor millet	21.26	23.45	20.68	20.75	18.15	20.86
Total	172.61	180.20	160.00	173.20	153.80	167.96
	Productivity (kg/ha)					
	2019–2020	**2020–2021**	**2021–2022**	**2022–2023**	**2023–2024**	**05 years average**
Jowar	989	1099	1092	1079	1106	1073
Bajra	1374	1420	1430	1510	1360	1419
Minor millet	1278	1253	1127	1176	1100	1187
Mean	1214	1257	1216	1255	1189	1226

millet), and they take 5–7 days for germination. Therefore, good seedbed preparation helps in soil moisture conservation for uniform germination and minimizes crop-weed competition at the initial stage. Millets are grown by tribal people or resource-poor farmers. Hence, most of the minor millets are sown by the broadcasting method, which is highly cumbersome for agronomic management. So, seed or seedlings should be sown or transplanted by following proper crop geometry. The detailed information is given in the below Table 6.

For transplanted finger millet, the nursery should plant it in the first fortnight of June. For the same reason, an area of ~750 m^2 is ideal for raising seedlings in a 1-hectare area. On a well-leveled, well-drained field, apply ~100 kg of well-decomposed FYM with 1 kg of SSP and 0.5 kg of MOP. Seed should be sown at 10 cm row-to-row spacing with a 3 cm seed depth. Slightly cover with fine soil, and sprinkle the water on the bed. Apply urea just after 1 week of seed emergence, and allow the seedlings to grow for up to ~25 days when the plant reaches the near-four leaf stage. Transplant the 02 seedlings in each hill in 22.5 or 25 cm row-to-row and 10 cm plant-to-plant spacing. In the system of ragi intensification, the finger millet seedlings can be transplanted at a distance of 25 cm in a square pattern. That facilitates the interculture operation and weed management.

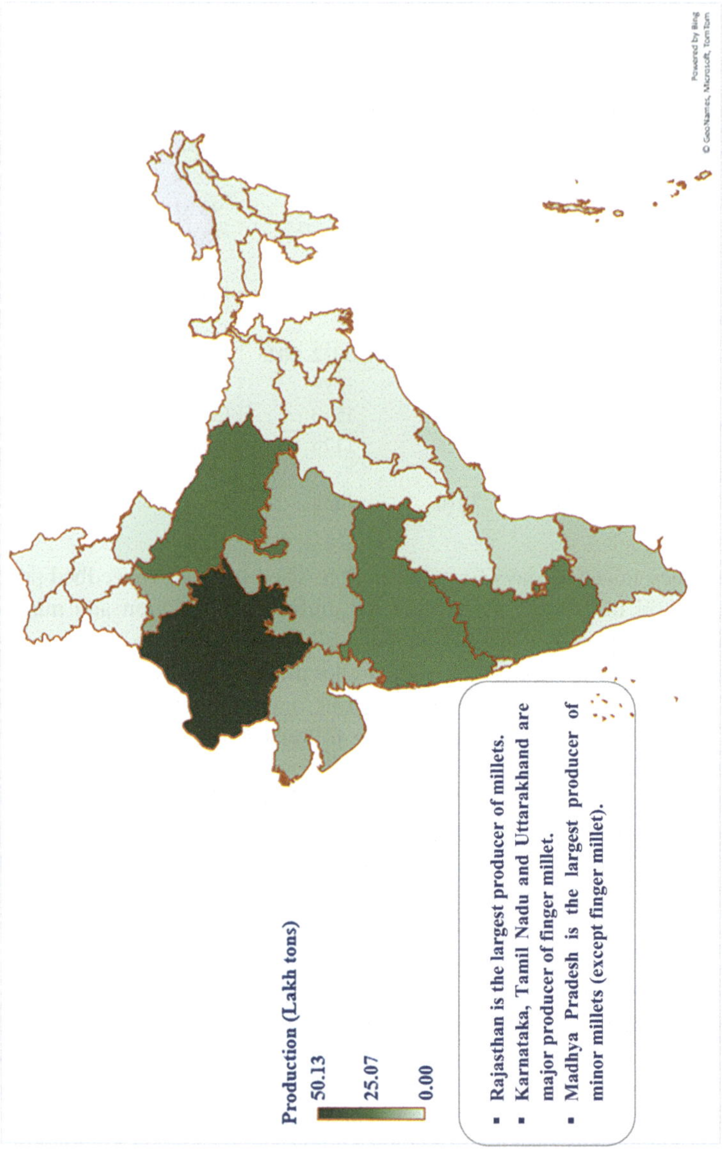

Fig. 2 State wise production of millets in India (Average of 2019 to 2023) (Data Source: APEDA 2024)

Table 4 Climatic and soil requirements of minor millets

Crop	Climatic requirement	Soil requirement
Finger millet	Mainly grown in tropical and subtropical climate up to an altitude of 2100 m. For better growth and development, 25–30 °C temperature range is needed	Well-drained alluvial, loamy, and sandy loam soils
Foxtail millet	Same as finger millet	Well-drained loamy soil is ideal for better crop growth. Susceptible to high and low moisture
Little millet	It can be cultivated up to 2000 m above mean sea level. Tolerant to drought and water logging	Well-drained sandy loam soil with fine seedbed is required for better crop stand
Proso millet	Crop can be grown both in summer and kharif season. Short duration, resistant drought, and water-logging make it perfect crop contingent planning	Well-drained sandy loam soils with high organic matter are ideal for proso millet cultivation
Kodo millet	Grown mostly in warm and dry climate, highly drought-tolerant Therefore, can be grown in areas where rainfall is scanty and erratic	Grown in wide range of soil from gravelly and stony upland poor soils to loam soils
Barnyard millet	Crop can grow in well-distributed rainfall with absence of prolonged dry spell. It can cultivate up to 2700 m above sea level, where rice is unfit for cultivation	Well-drained sandy loam soil is ideal for its cultivation
Browntop millet	Annual warm season crop, grown on sea level up to 2500 mabs.	Easily grown in rocky and shallow soils but susceptible to less soil moisture

6 Nutrient Management

Minor millets are mainly grown under low nutrient inputs on marginal lands. So, that was also a reason for low productivity. They respond well to organic and inorganic nutrient management. Among the nutrients, they respond very well to the application of nitrogen and phosphorus but not to potassium. The nutrient demand of minor millets is mainly influenced by the indigenous nutrient-supplying capacity of the soil and the yield potential of crops or varieties. Therefore, the recommended dose of fertilizer is varied from crop to crop and region to region (Table 7). Details are given in Table 4. When compared to nutrient management that relies exclusively on chemicals, minor millets react favorably to integrated nutrient management (INM). The main aim of the INM is to enhance crop yield and nutrient use efficiency while synchronizing crop nutrient demand and supply. The efficiency of nitrogenous fertilizers can be enhanced by scheduling, i.e., 50% N should be applied as a basal and 50% N at the maximum tillering stage. The basal dose of N can be avoided in heavily weed-infested areas to check for early weed growth. The two-split application of N can be done after first weeding (first at 20 DAS and second at ~50 DAS). The foliar application of micronutrients, i.e., Zn, Cu, Fe, B, etc., also

Table 5 Improved varieties of small millets

Varieties	Released year	Recommended area for cultivation	Special characteristics
Finger millet			
GPU 67	2010	All-millet-growing area	Long duration, non-lodging (semidwarf)
VL Mandua 347	2012	All-finger millet-growing states except Tamil Nadu and Maharashtra	Neck and finger blast resistance Suitable for contingent crop planning
VL Mandua 352	2014	Andhra Pradesh, Bihar, Gujarat, Jharkhand, Karnataka, Madhya Pradesh, Odisha, Uttarakhand, Maharashtra, and Tamil Nadu	High calcium content Finger blast resistance
VL Mandua 376	2016	All-finger millet-growing areas of country	Short duration, fertilizer-responsive and suitable for organic cultivation
VL Mandua379	2018	Uttarakhand hills	Short duration, blast-resistant
VL Mandua 380	2019	Uttarakhand, Bihar, Jharkhand and Madhya Pradesh Uttarakhand hills	High protein content. Blast-resistant.
VL Mandua382	2020	Uttarakhand hills	White grains
Barnyard millet			
VL Madira 172	2000	Uttar Pradesh, Gujarat, and Karnataka	Tolerant to grain smut
VL Madira 181	2001	Bihar, Karnataka, Madhya Pradesh, and Tamil Nadu	High grain yield
VL Madira 207	2008	All-barnyard millet-growing states except Tamil Nadu and Gujarat	Tolerant to grain smut
PRJ 1	2009	Uttarakhand hills	Small awn in spike. Stiff straw, Immune to grain smut
DHBM 93-3	2016	National	Fertilizer-responsive
DHB-93-2	2018	Karnataka	Ideal for contingent planning
MDU-1	2018	Tamil Nadu, Karnataka	Non-shattering, high milling percentage (70%)
Foxtail millet			
TNAU 43	1994	Tamil Nadu	High straw yield
Meera (SR 16)	2001	Rajasthan	Stay green character, resistant to downy mildew
SiA 3085	2011	National	Resistant to downy mildew and blast

(continued)

Table 5 (continued)

Varieties	Released year	Recommended area for cultivation	Special characteristics
SiA 3156	2012	Andhra Pradesh, Bihar, Gujarat, Karnataka, Madhya Pradesh, Tamil Nadu, and Uttarakhand	Fertilizer-responsive
RAU (Rajendra Kauni 1–2)	2018	Central India	~81 days maturity, resistant to leaf blast, rust, smut, brown spot, downy mildew, and leaf blight
Kodo millet			
JK 98	2010	All states of India	Resistant to shoot fly and moderately resistant to head smut
DPS 9-1	2010	All states of India	Tolerant to shoot fly
RK 390-25	2016	All states of India	Responsive to NPK, resistant to drought, lodging, and key pest shoot fly and moderately resistant to head smut
TNAU-86	2017	All states of India	Moderately resistant to shoot fly and head smut and has better nutritional qualities
Little millet			
OLM 208	2009	Odisha, Chhattisgarh, and Gujarat	
BL 6	2016	All states of India	High iron content (28.3 mg/100 g grain). Tolerant to major pests
DHLM-14-1	2017	Tamil Nadu, Karnataka, Gujarat, Maharashtra, and Orissa	Late maturing, bold-seeded
Proso millet			
TNAU 164	2010	Andhra Pradesh, Bihar, Karnataka, Maharashtra, Tamil Nadu	Long panicle and high-yielding
TNAU 151	2010	All-proso millet-growing states	Bold grains, tolerant to shoot fly
TNPm-230	2017	All-proso millet-growing states	Short duration, drought-tolerant variety
DHP-2769	2018	Karnataka	Suitable for contingency planting

benefits crops in nutrient-deficient regions (Kumawat et al. 2019). Nutrients can also be supplied through biofertilizers, which are cost-effective and environmentally friendly (Fig. 3). But using various strains or plant growth-promoting rhizobacteria (PGPRs) for solubilizing native-fixed nutrients is the most effective approach for sustainable agriculture (Subramanian et al. 2009; Ahmed et al. 2011). Their diversity in the root rhizosphere promotes root surface areas, resulting in higher nutrient and water uptake (Lugtenberg 2015).

Table 6 Sowing time, seed rate, and sowing method of minor millets

Crop	Sowing time	Seed rate and method of sowing	Spacing (cm) (row-to-row and plant-to-plant)
Finger millet	Sowing time differs region to region because of environmental conditions. Ideal time for *kharif* sown is June to July, for *rabi*, September to October	8–10 kg/ha seed for line sowing and 4–5 kg/ha for transplanting. Seedlings of ~25 days old (four-leaf stage) are ideal for transplanting. Broadcasting sown should be avoided	22.5 cm × 10 cm
Foxtail millet	*Kharif*-irrigated crop can be sown in the first week of June. However, June last to July last ideal for rainfed crop. August to September is ideal for *rabi* season crop in South India	Recommended seed rate is 8–10 kg/ha for line sowing and 15 kg/ha for broadcasting	25 cm × 10 cm
Little millet	For *kharif* sown crop June–July and for *rabi* season September to October is ideal	For line sowing 6–8 kg/ha and for broadcasting 10–12 kg/ha	22.5 cm × 10 cm
Proso millet	First fortnight of July is ideal for *kharif* sown crop	For line sowing 10 kg/ha and for broadcasting 15 kg/ha	22.5 cm × 10 cm
Kodo millet	Mainly grown in Central India, June–July is ideal time for sowing	For line sowing 10 kg/ha and for broadcasting 15 kg/ha	22.5 cm × 10 cm
Barnyard millet	For *kharif* sown crop June–July and for *rabi* season September to October is ideal. Dry seeding before onset of monsoon is practiced in NW Himalayas	For line sowing 08 kg/ha and for broadcasting 10–12 kg/ha	25 cm × 10 cm
Browntop millet	Crop can be sowed from April to August	For line sowing 05 kg/ha seed is sufficient. Seed should not be sown >2.5 cm depth	20 cm × 08 cm

These microorganisms also enhance the production of siderophores, organic acids, and solubilizing native nutrients, which improve tolerance to salinity, drought, and other stresses. The study highlights the capacity of *Bacillus* sp. AZ6 to produce various organic acids, including ferulic acid, syringic acid, chlorogenic acid, cinnamic acid, and caffeic acid, when cultivated in a liquid medium (Hussain et al. 2015). Wu et al. (2006) noticed a decrease in soil pH by up to 0.47 units following bacterial inoculation, attributed to the secretion of organic acids and H^+ ions, consequently promoting the solubility and uptake of zinc and iron in plants. The use of biofertilizers, viz., *Azotobacter*/*Azospirillum*, phosphate solubilizing bacteria (PSB), potassium solubilizing microorganisms (KSM), mycorrhiza, etc., to fix or enhance the solubility of native nutrients is also recommended to enhance grain

Table 7 The recommended dose of N, P_2O_5, and K_2O for different minor millets of various regions

State name	Finger millet	Foxtail millet	Kodo millet	Little millet	Proso millet	Barnyard millet
Tamil Nadu	40:20:20	40:20:0	40:20:0	40:20:0	40:20:0	40:20:0
Karnataka	50:40:37.5	30:15:0	20:20:0	20:20:0	20:20:0	20:20:0
Maharashtra	60:30:30	20:20:0	20:20:0	20:20:0	20:20:0	20:20:0
Andhra Pradesh	40:20:20	40:30:0	20:20:0	20:20:20	20:20:20	20:20:20
Uttar Pradesh	–	–	20:20:0		40:20:0	40:20:0
Bihar	0:20:20	20:20:0	–	20:10:0	40:20:0	40:20:0
Uttarakhand	50:40:25	20:20:0	–	–	20:20:0	40:20:0
Odisha	40:20:20	–	40:20:0	–	–	–
Madhya Pradesh	60:40:20	–	40:20:0	–	–	–
Chhattisgarh	60:40:20	20:20:0		–	20:20:0	–

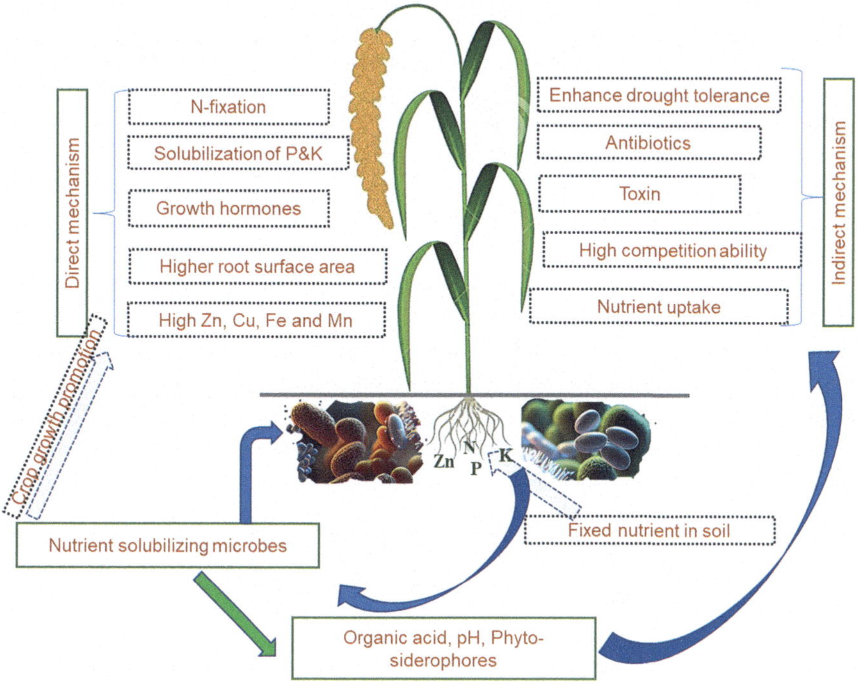

Fig. 3 Direct and indirect impacts of biofertilizers on crop growth

yield in marginal lands (Hussain et al. 2015; Pradhan et al. 2021). Thus, biofertilizers offer a cost-effective, efficient, and environmentally friendly alternative to conventional fertilizers. Their use not only addresses nutrient deficiency in modern agriculture but also reduces reliance on chemical fertilizers, showcasing their significant role in nutrient solubilization.

7 Organic Farming and Minor Millets

Organic farming is an agricultural system that uses ecologically based pest controls, and biological fertilizers derived largely from animal and plant wastes and nitrogen-fixing cover crops. Modern organic farming has many ecological advantages and was created in reaction to the harm that synthetic fertilizers and chemical pesticides in conventional agriculture were causing to the environment (Eyhorn et al. 2019). Minor are mainly grown by resource-poor farmers in tribal region with little no user of fertilizer and other agrochemicals under rainfed conditions. Such farming is by default organic farming, which can be certified as organic farming so farmers can get a good price for their product. Various varieties of minor millets have been tested for organic farming their special characteristics are given in Table 5. Minor millets are suitable for organic farming since they are less susceptible to insect-pest and disease. Moreover, these crops can grow with limited nutrient and water. Gowda et al. (2004) noted that the application of poultry manure, farmyard manure, and vermicompost, at rates of 1.32, 3.0, and 1.5 tons/ha, respectively, equivalent to 20 kg/ha of nitrogen, resulted in finger millet yields of 2211, 1740, and 1942 kg/ha, respectively. Similarly, significant higher grain yield (2279 kg/ha) of barnyard millet was recorded with the application of enriched vermicompost @ 1 t ha^{-1}than unfertilized plot (Kumar et al. 2022).

8 Cropping Systems

Cropping systems play a pivotal role in the cultivation of minor millets. These underutilized crops exhibit resilience and possess substantial nutritional value, making them well-suited for cultivation across diverse agroecological settings. The implementation of appropriate cropping systems tailored to the unique requirements of minor millets is imperative to harness their full potential. Such systems can effectively enhance soil health and fertility through strategic crop rotation, intercropping, and the incorporation of organic amendments. Additionally, they can facilitate integrated pest and disease management by disrupting pest life cycles and promoting natural biological control mechanisms (Li and Wu 2018). Moreover, cropping systems can optimize water use efficiency, improve yield stability, and promote sustainable agricultural practices (Yang et al. 2018). The integration of minor millets into diversified cropping systems holds promise for enhancing food security, addressing nutritional deficiencies, and fostering the cultivation of climate-resilient crops.

Minor millet can be grown as a sole crop or as part of various legume crop-based cropping systems. Here are some prominent examples of minor millet-based cropping systems:

Intercropping: Minor millet is often intercropped with legumes to maximize land utilization and enhance overall productivity (Maitra 2020). Intercropping offers numerous benefits, including enhancing resource utilization, reducing the presence of harmful organisms, soil and water conservation, and ultimately boosting productivity and sustainability within the system (Launay et al. 2009; Maitra 2020). This practice also provides multiple benefits such as improved soil fertility through nitrogen fixation (Giller and Wilson 1991; Maingi et al. 2001), reduced pest and disease incidence, and diversified food production. Common intercropping systems of minor millets with their salient finding are given in below Table 8.

Mixed cropping: Minor millet is also grown in mixed cropping systems where it is cultivated alongside other crops such as soybean, horse gram, amaranths, buckwheat, etc. This practice helps in minimizing the risk of crop failure and optimizing resource utilization (Franco et al. 2015). Different crops in the mixture can have complementary growth habits and nutrient requirements, leading to increased overall yield and stability (Opole 2019).

Agroforestry systems: Millets are also integrated into agroforestry systems, especially in regions with limited arable land. In such systems, they grew in combination with trees or shrubs such as peach, plum, pear, pecan nut, etc., providing shade and protection to the crop while utilizing the vertical space. This approach helps in soil conservation, microclimate regulation, and diversification of farm products.

Crop rotation: Minor millet can be included in crop rotation systems with legumes to break pest and disease cycles and enhance soil fertility (Maitra 2020). Crop

Table 8 Minor millets based intercropping systems with their salient findings

Minor millets	Prominent cropping systems/rotations with salient findings
Finger millet	• Intercropping of finger millet+ horse gram (2:1) recorded higher system productivity, LER, and low weed infestation compared to sole cropping (Pradhan et al. 2018) • Finger millet followed by winter horse gram during residual fertility recorded higher-yield attributes and grain yield (Aliveni et al. 2021) • Finger millet intercropped (4:2) with soybean, groundnut, and black gram recorded higher equivalent yield of finger millet (3521, 4013, and 4233 kg/ha) than sole finger millet cropping (2017 kg/ha) (Kumar and Ray 2020) • Intercropping red gram with finger millet has shown to enhance the yield of finger millet while simultaneously reducing both weed population and dry weight (Sidar and Thakur 2017) • The higher grain yield of finger millet was recorded when it grown with black gram at 6:2 and 8:2 row proportion (Nigade et al. 2012; Milenkovic et al. 2019)
Barnyard millet	• Barnyard millet and rice bean (4:1) intercropping recorded higher yield than sole cropping (Chapke et al. 2018)
Foxtail millet	• The higher system equivalent yield (52.7 q/ha) was recorded in foxtail millet + pigeon pea followed by foxtail millet + mesta (30.5 q/ha) (Aliveni et al. 2021)
Kodo millet	• Intercropping of various legumes with kodo millet (2:1) recorded was given higher economical returns than sole cropping (Chapke et al. 2018)
Little millet	• The higher productive tiller and test weight of little millet were recorded when it was intercropped with pigeon pea at 6:1 ratio (Sharmili and Manoharan, 2018) • Intercropping of little millet + pigeon pea followed by blackgram (8:2) recorded higher system yield than horse gram was sown as sequence crop after little millet + pigeon pea (Sivagamy et al. 2020)

rotation involves alternating the cultivation of different crops in a specific sequence over time. All minor millet fits well in diversified rotation systems due to their short duration. It's important to note that the choice of cropping system may vary depending on local agroecological conditions, farmer preferences, and market demands. Farmers often adopt specific cropping systems based on their goals, available resources, and the specific agroclimatic conditions of their region.

9 Weed Management

Weeds are the major cause of lower productivity in the minor millets (Fig. 4); losses in grain yield have been recorded up to 75% (Vinothini and Arthanari 2017; Mishra et al. 2018; Rao 2021). Weeds compete with crops for nutrients, water, space, sunlight, and CO_2, which leads to poor crops (Rao 2021). The degree of yield reduction varies depending on factors such as weed flora, density, the timing of their infestation, soil characteristics, rainfall patterns, and agricultural management techniques

Fig. 4 Major weed flora found in minor millets field

employed. At the early growth stage, tiny seedlings and slow growth make minor millets very susceptible to heavy weed infestation (Silva et al. 2014). It's crucial to control weeds during the critical period of crop-weed competition, typically occurring between 15 and 42 days after sowing (Dubey et al. 2023). The critical period of crop-weed competition refers to the maximum duration during which weeds can not be tolerated without negatively impacting the final yields of the crop (Zimdahl 1980). During this critical period, it's imperative to remove weeds to minimize crop losses. Because selective herbicides for millets are limited, weed management for these crops mainly relies on cultural and mechanical methods. To achieve optimal weed management and maximize millet crop yields, a combination of methods is necessary. This includes using weed-competitive crop varieties, adjusting spacing, applying the right amount and placement of fertilizers, mulching with crop residues, practicing intercropping, employing cultural and mechanical methods, and selectively using herbicides where appropriate. Broadcasted sown crops invite weeds; therefore, line-sown or transplanted sown crops should be promoted (Shamina et al. 2019). The line-sown crop with narrow row spacing and basal nitrogen dose enhances early crop cover and reduces impacts from weeds (Nelson 1977; Agdag 1995; Kumar et al. 2019). The major weed flora found in minor millets is described in Table 9. However, weed flora varies with the growing condition, soil type, and location (Mashingaidze et al. 2012).

Table 9 Dominant weed flora found in minor millets

Minor millets	Dominant weed flora	References
Finger millet	*Paspalum dilatatum*, *Oxalis latifolia*, *Cyperus rotundus* L. *Cynodon dactylon* (L.), *Galinsoga parviflora*, *Persicaria capitatum*, *Dactyloctenium aegyptium*, *Echinochloa colona*, *E. indica*, *Acanthospermum hispidum*, *Cyperus difformis*, *Cyperus iria*, and *Leucas aspera*	Rao (2021)
Barnyard millet	*Echinochloa colona*, *Cynodon dactylon*, *Phyllanthus niruri*, *Solanum nigrum*, *Echinochloa crus-galli*, *Dactyloctenium aegyptium*, *Eleusine indica*, *Setaria glauca*, *Phragmites karka*, *Cyperus rotundus*, *Sorghum halepense*, *Celosia argentea*, and *Commelina benghalensis*	Chapke et al. (2020)
Foxtail millet	*Cyperus rotundus*, *Echinochloa colona*, *Phyllanthus niruri*, *Cynodon dactylon*, *Echinochloa crus-galli*, *Sorghum halepense*, *Eleusine indica*, *Dactyloctenium aegyptium*, *Commelina benghalensis*, *Celosia argentea*, *Amaranthus viridis*, and *Solanum nigrum*	Prabhakar et al. (2017)
Kodo millet	*Trianthema portulacastrum*, *Boerhaavia diffusa*, *Parthenium hysterophorus*, *Digera arvensis*, *Brachiaria reptans*, *Acrachne racemosa*, *Dactyloctenium aegyptium*, *Panicum repens* under grasses, *Cyperus rotundus*	Vinothini and Arthanari (2017)
Little millet	*Setaria glauca*, *Cynodon dactylon*, *Echinochloa colona*, *Echinochloa crus-galli*, *Celosia argentea*, *Commelina benghalensis*	Chapke et al. (2020)

In the absence of suitable selective pre-emergence herbicides, two manual weedings at 15–20 days and 30–35 days after sowing are sufficient to control the weeds. Despite their phytotoxicity, some herbicides are recommended for minor millets. In finger millet, among the herbicidal treatments, pre-emergence application of isoproturon at 0.50 kg a.i./ha was found to be superior to reducing the weed density up to 25 DAS (Ungata et al. 2020). Compared to various herbicides (isoproturon, oxadiargyl, bensulfuron methyl + pretilachlor, bispyribac sodium, ethoxysulfuron, tembotrione, and 2,4-D sodium salt), the application of metsulfuron methyl + chlorimuron ethyl 20 WP @ 20 g a.i./ha (15–20 DAS) recorded a higher finger millet grain yield (1540 kg/ha); however, it was significantly lower in two hand-weeded plots (AICSMIP 2021).

Isoproturon @ 0.5–0.75 kg a.i./ha applied as pre-emergence spray or mixed with sand or soil along with one-hand weeding at 30–45 days after sowing has been found effective in controlling weed problems and gives a higher yield in kodo, foxtail, and barnyard millet (AICSMIP 2002, 2005). Thambi et al. (2021) found that a combination of bensulfuron-methyl 60 and pretilachlor 495 g/ha (RM) PE applied 3 days after sowing (DAS) was effective in controlling weeds in barnyard millet. Isoproturon at 0.5 kg/ha applied pre-emergence, followed by 2,4-D (Na salt) at 0.5 kg/ha postemergence (Kujur 2016), and oxadiargyl at 0.08 kg/ha pre-emergence at 3 DAS, followed by ethoxysulfuron at 0.012 kg/ha post-emergence at 30 DAS, were reported to provide effective broad-spectrum weed control in finger millet (Shubhashree and Sowmyalatha 2019). Similarly, in little millet, the application of metsulfuron methyl + chlorimuron ethyl 20 WP @ 20 g a.i./ha (15–20 DAS) recorded a higher grain yield (2449 kg/ha), which was almost the same as in two hand weeded plots (2449 kg/ha) (AICSMIP 2021). Field trials were conducted on proso millet tolerance to saflufenacil applied pre-emergence at 36, 50, and 100 g a.i./ha. Despite foliar injury, the grain yields of proso were not reduced by any rate of saflufenacil compared to untreated controls (Reddy et al. 2014).

10 Mechanization for Drudgery Reduction

Mechanization has undoubtedly been instrumental in enhancing the production of food crops. However, minor millets, often cultivated by resource-poor farmers, have not seen the same level of mechanization. These farmers typically employ traditional broadcasting methods for sowing seeds, leading to higher cultivation costs for weed management and subsequently lower productivity. By embracing mechanized techniques such as precision seeding, mechanical weeding, and efficient irrigation systems, farmers can optimize their cultivation processes, mitigate weed-related challenges, and enhance overall productivity. Moreover, mechanization can streamline labor-intensive tasks, reducing drudgery and freeing up valuable time for farmers to focus on other aspects of farm management. For minor millets especially for the Indian Himalayan region, several crucial agricultural tools and machinery are designed. Details are given in Table 10.

Table 10 Tools for minor millets for drudgery reduction

S. no.	Tools and their uses	Photographs
1.	*VL Syahi Hal (VL iron plow)*: used for plowing and planking; operated by pair of bullocks and a person. It's an environmentally friendly save trees from deforestation (Singh et al. 2015; Shankar and Haokip 2020)	
2.	*VL small tool kit*: manually operated tools used for sowing, intercultural activities, and harvesting of the crops. (a) *VL garden rake*: used for weed removal and cleaning of crop field; (b) *VL line maker*: used for line sowing of finger millet and barnyard millet; (c) *VL hand hoe*: used for intercultural operations; (d) *VL Kutla*: very effective tool for weeding in millets; (e) *VL Darati:* used for harvesting of millets (Parihar et al. 2021)	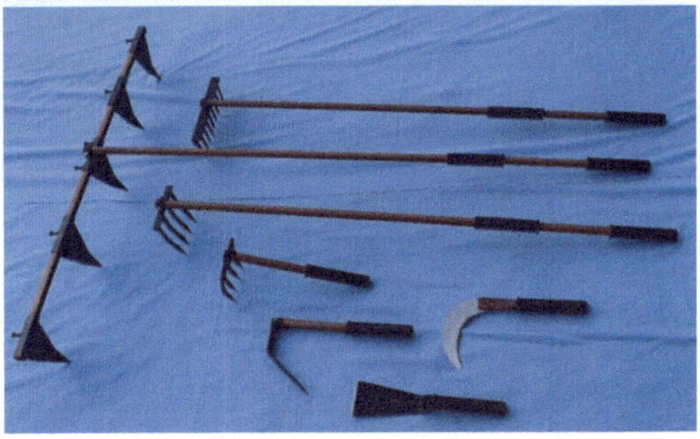
3.	*VL seed*-cum-*ferti drill*: used for line sowing of seeds along with fertilizer. It can be operated by two persons or pair of bullocks. It promotes uniform germination; better plant stands and reduces early weed competition which results in higher grain yield	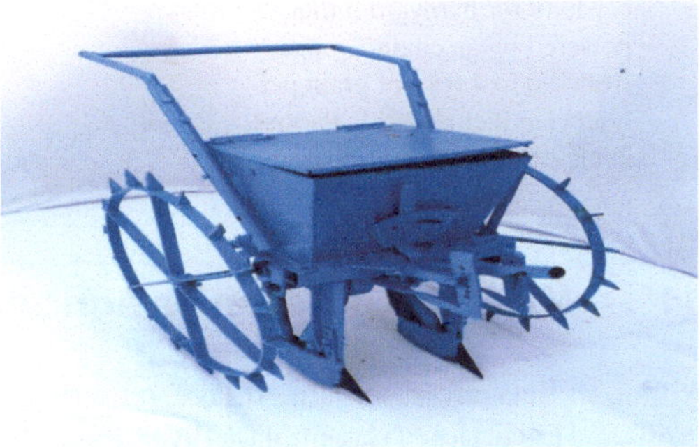

(continued)

Table 10 (continued)

S. no.	Tools and their uses	Photographs
4.	*Vivek millet thresher*–cum–*pearler*: the threshing of the minor millets (finger millet, barnyard millet, and foxtail millet, etc.) is laborious and time-consuming due to multilayer seed coating. In hills and tribal areas, threshing of millets is done by beating the ear-head with a stick or by revolving a bullock upon the ear-head on the floor, and in some places, threshing is done by pressing it with legs. Therefore, a multitasked electric-operated thresher is developed for the threshing, dehusking, pearling, and polishing of the grains. The threshing capacity for grains ranges from 40 to 60 kg per hour, while the pearling capacity for finger millet, operated by an electric motor, is between 60 and 80 kg per hour. Additionally, the dehulling capacity for barnyard millet, powered by an engine, ranges from 2.0 to 4.0 kg of grain per hour (Dixit et al. 2011; Parihar et al. 2021)	

11 Millets in Conservation Agriculture

Minor millets are minimum tillage requiring shallow-rooted and short-duration crops and have a lot of potential for diversified conservation agriculture. Conservation tillage practices, such as no-till or minimum tillage, can play a crucial role in soil water conservation and yield enhancement in minor millet cultivation (Acharya et al. 2019). By minimizing soil disturbance and preserving crop residue on the soil surface, conservation tillage methods contribute to the maintenance of soil structure, reduction of soil erosion, and enhancement of organic matter level. This results in improved water infiltration, retention, and availability for minor millet crops, particularly in regions with erratic rainfall or limited irrigation (Chalise et al. 2019). Additionally, conservation tillage helps to mitigate soil compaction, promoting deeper root penetration and more efficient water and nutrient uptake by minor millet plants. Consequently, these practices create a favorable environment for crop growth and development, ultimately leading to higher yields and increased agricultural sustainability. Study revealed that diversification of cereals with legumes enhances soil organic carbon and biological activities (Tiemann et al. 2015).

12 Production Constraints

Minor millets are mainly grown by resource-poor farmers on marginal land with a lack of access to improved technologies and agronomic practices. Furthermore, poor management of abiotic and biotic stress causes lower average yields compared to major cereals (Sood et al. 2020). Climatic factors, viz., rainfall patterns, distribution, and mean seasonal temperatures, as well as topography, alongside edaphic factors such as soil type, fertility, and drainage, play crucial roles in the performance of minor millet production systems (Meena et al. 2021). Additionally, agronomic management factors such as weed, water, and nutrient management, coupled with the socio-economic status of the farming community, collectively contribute to the overall success of minor millet cultivation. Weed infestation is indeed a significant constraint in global millet production (Burkill 1985; Rao 2021). The slow initial growth of small millets often promotes weed growth, leading to increased competition for sunlight, nutrients, space, and water during the early stages of growth. This competition ultimately diminishes productivity. Bird damage also poses a considerable threat to millet growers, with yield reductions potentially reaching 100% in isolated crop fields. The extent of damage varies depending on the size of millet grains, with small-grain millets being more susceptible to bird damage compared to large-grain millets. In small millets, small and multi-layered grains, unsynchronized maturity, and seed shattering cause problems for mechanization and post-harvest management. Furthermore, the traditional method for dehulling and pearling of minor millet grains is very cumbersome and time-consuming (Sood et al. 2015).

The socio-economic conditions of the farming community, particularly in Uttarakhand, represent a significant constraint on low millet productivity. Factors such as the limited availability of improved cultivars seed (Rakshit and Wang 2016), inadequate access to agricultural inputs, deficient market infrastructure, overreliance on natural resources, and a lack of scientific vision all contribute to lower millet productivity in the region (Sood et al. 2015; Meena et al. 2021).

13 Conclusion

Minor millets are crucial for food and nutritional security in climate change-hit regions of developing world. With their distinct nutritional makeup, immense health advantages, C_4 photosynthetic pathway, and ephemeral nature, they are excellent crops to diversify cropping systems for agriculture that is adaptable to climate change. Minor millets inherently possess tolerance to withstand biotic and abiotic stresses. With their ability to address challenges in the current agricultural system, they should be recognized as the "crop of the future." Ever since, minor millets are mostly produced and consumed by the resource-poor farmers of tribal communities under fragile ecosystems. Globally, millet cultivated area and production are declining constantly. India remains the world's largest producer of millet, with over 37% of global production. However, out of total millet growing area (~129 lakh ha) and production (167.9 lakh tons), only 12.1% area (15.7 lakh ha) and 12.4% production (20.8 lakh tons) belongs to minor millets. The productivity of minor millets is still very low compared to major crops, mainly due to poor emphasis on research and the adoption of improved agronomic management. Minor millets are grassy in nature; the initial slow growth, the resemblance of seedlings with weeds, and the lack of availability of selective herbicides make them more susceptible to weed infestation. In-depth molecular research is essential for developing herbicide-tolerant millet cultivars. Additionally, there is a need to develop fertilizer-responsive, nutrient-dense cultivars that can achieve productivity levels comparable to major cereals. Agronomic interventions such as good crop husbandry, nutrient management, weed management, mechanization, and postharvest management are the utmost needed for enhancing the productivity of minor millets. These interventions are pivotal not only for boosting yields but also for improving soil health and promoting the sustainable cultivation of minor millets.

References

Acharya BS, Dodla S, Gaston LA, Darapuneni M, Wang JJ, Sepat S, Bohara H (2019) Winter cover crops effect on soil moisture and soybean growth and yield under different tillage systems. Soil Tillage Res 195:104430

Agdag MI (1995) Row spacing in proso millet. [Dissertation]. University of Nebraska, Lincoln

Ahmed MA, Ahmed AG, Magda MH, Tawfik MM (2011) Effect of organic and biofertilizers on wheat productivity in new reclaimed soils. Res J Agric Biol Sci 7(1):105–114

AICSMIP (2002) Annual report. All India Co-ordinated Small Millets Improvement Project (ICAR), Bangalore

AICSMIP (2005) Annual report. All India Co-ordinated Small Millets Improvement Project (ICAR), Bangalore

AICSMIP (2021) Annual report. All India Co-ordinated Small Millets Improvement Project (ICAR), Bangalore

Aliveni A, Venkateswarlu B, Sree Rekha M, Prasad PRK, Jayalalitha K (2021) Performance of rabi Horsegram as influenced by kharif Finger Millet Crop Geometry and Nutrient Management Practices. Biol Forum Int J 13(4):852–856

APEDA (2024) Harvested area, production, and millet productivity in India 2019-24. https://apeda.gov.in/

Bhinda MS, Joshi DC, Parihar M, Meena RP, Joshi P, Gupta A, Kant L, Sood S (2023) Genetics, breeding, and genomics of Indian barnyard millet (*Echinochloa frumentacea*): status and perspectives. In book: Neglected and Underutilized Crops Future Smart Food.ACADEMIC PRESS, p 115–135

Burkill HM (1985) The useful plants of west tropical Africa, vol 2. Royal Botanic Gardens, Kew

Chalise KS, Singh S, Wegner BR, Kumar S, Pérez-Gutiérrez JD, Osborne SL, Nleya T, Guzman J, Rohila JS (2019) Cover crops and returning residue impact on soil organic carbon, bulk density, penetration resistance, water retention, infiltration, and soybean yield. Agron J 111:99–108

Chapke RR, Prabhakar Shyamprasad G, Das IK, Tonapi VA (2018) Improved millets production technologies and their impact. Technology Bulletin, ICAR-Indian Institute of Millets Research, Hyderabad, India. pp. 84

Chapke RR, Shyam Prasad G, Das IK, Hariprasanna K, Singode A, Kanthi SBS, Tonapi VA (2020) Latest Millet production and processing technologies. Booklet, ICAR–Indian Institute of Millets Research, Hyderabad, India: 82. (ISBN: 81–89335–90–X)

Devi PB, Vijayabharathi R, Sathyabama S, Malleshi NG, Priyadarisini VB (2014) Health benefits of finger millet (*Eleusine coracana* L.) polyphenols and dietary fiber: a review. J Food Sci Technol 51(6):1021–1040

Dixit AK, Nanda SK, Singh KP, Aleksha Kudos SK (2011) Economic benefits of Vivek millet thresher-cum-pearler and agro processing centre in hilly region of Uttarakhand. J Hill Agric 2:177–182

Doggett H (1989) Small millets—a selective overview. In: Seetharam A et al (eds) Small millets in global agriculture. Oxford and IBH Publishing Company, Delhi, India, pp 59–70

Dubey RP, Chethan CR, Choudhary VK, Mishra JS (2023) A review on weed management in millets. Indian J Weed Sci 55(2):141–148

Eyhorn F, Muller A, Reganold JP et al (2019) Sustainability in global agriculture driven by organic farming. Nat Sustain 2:253–255

FAOSTAT (2023) Production-yield quantities of millets in world + (total) 1961–2021. https://www.fao.org/faostat/en/#data/QV Accessed December 2023

Farnandez DR, Vanderjagt DJ, Millson M et al (2013) Fatty acid, amino acid and trace mineral composition of *Eleusine coracana* (Pwana) seeds from northern Nigeria. Plant Foods Hum Nutr 58:1–10

Franco JG, King SR, Masabni JG, Volder A (2015) Plant functional diversity improves short-term yields in a low input intercropping system. Agric Ecosyst Environ 203:1–10

Giller KE, Wilson KJ (1991) Nitrogen Fixation and Tropical Cropping Systems. CAB International, Wallingford, pp 10–120

Gowda KTK, Jena BK, Ramamoorthy K, Dubey OP, Rao TV, Shankarlingappa BC, Ashok EG (2004) Augmenting legumes production in small millets based cropping systems. AICRP on Small Millets, Bangalore, p 144

Gull A, Jan R, Nayik GA, Prasad K, Kumar P (2014) Significanceof finger millet in nutrition, health and value-added products: a review. J Environ Sci Comput Sci Eng Technol 3:1601–1608

Gupta A, Mahajan V, Kumar M, Gupta HS (2009) Biodiversity in the barnyard millet (*Echinochloa frumentacea* Link: Poaceae) germplasm in India. Genetic Resources and Crop Evolution 56:883–889

Gupta SM, Arora S, Mirza N, Pande A, Lata C, Puranik S, Kumar J, Kumar A (2017) Finger millet: a "certain" crop for an "uncertain" future and a solution to food insecurity and hidden hunger under stressful environments. Front Plant Sci 8:643

Hegde BR, Gowda BKL (1989) Cropping systems and production technology for small millets in India. In: Seetharam A et al (eds) Small millets in global agriculture. Oxford & IBH Publishing Co, Delhi, India, pp 209–235

Hussain A, Arshad M, Zahir ZA, Asghar M (2015) Prospects of zinc solubilizing bacteria for enhancing growth of maize. Pak J Agric Sci 52:915–922

Jijau C (1989) Importance and genetic resources of small millets with emphasis on foxtail millet (Setaria italica) in China. In: Seetharam A et al (eds) Small millets in global agriculture. Oxford and IBH Publishing Company, Delhi, India, pp 93–100
Joshi DC, Sood S, Hosahatti R, Kant L, Pattanayak A, Kumar A, Yadav D, Stetter MG (2018) From zero tohero: the past, present and future of grain amaranth breeding. Theor Appl Genet 131(9):1807–1823
Joshi DC, Zhang K, Wang C, Chandora R, Khurshid M, Li J, He M, Georgiev MI, Zhou M (2020) Strategic enhancement of genetic gain for nutraceutical development in buck wheat: a genomics–driven perspective. Biotechnol Adv 39:107479
Joshi DC, Meena RP, Chandora R (2021) Genetic resources: collection, characterization, conservation, and documentation. In Millets and pseudo cereals. Genetic resources and breeding advancements series in food science, Technology and Nutrition, Woodhead Publishing, 19–31
Kujur S (2016) Weed management in finger millet (*Eleusine coracana* L. Gaertn). [Dissertation]. Raipur: Indira Gandhi Krishi Vishwavidyalaya
Kumar B, Ray PK (2020) Performance of intercropping of legumes with finger millet (*Eleusine coracana*) for enhancing productivity, sustainability and economics in Koshi region of Bihar. J Pharmacogn Phytochem 9(3):1568–1571
Kumar A, Paliwal A, Rawat L, Kumar P, Paliwal A, Chaudhary S (2019) Barnyard millet (Echinochloa frumentacea) productivity enhancement through establishment methods and weed management practices under hilly rain fed conditions. Int J Chem Stud 7:1360–1362
Kumar A, Tripathi MK, Joshi D, Kumar V (2021) Millets and millet technology. Springer, Singapore, p 438
Kumar E, Hari M, Mohamed Yassin E, Somasundaram MR, Latha RA (2022) Effect of organic manures and organic foliar sprays on yield attributes and yield of barnyard millet (*Echinochloa Frumentacea* L.). Int J Environ Clim Change 12(11):886–892
Kumawat N, Kumar R, Khandkar UR, Yadav RK, Saurabh K, Mishra JS, Dotaniya ML, Hans H (2019) Silicon (Si)- and zinc (Zn)-solubilizing microorganisms: role in sustainable agriculture. Biofertilizers for sustainable agriculture and environment. Soil Biol 55
Launay M, Brisson N, Satger S, Hauggaard-Nielsen H, Corre-Hellou G, Kasynova E, Ruske R, Jensen ES, Gooding MJ (2009) Exploring options for managing strategies for Pea-Barley intercropping using a modelling approach. Eur J Agron 31:85–98
Li S, Wu F (2018) Diversity and cooccurrence patterns of soil bacterial and fungal communities in seven intercropping systems. Front Microbiol 9:15–21
Lugtenberg B (2015) Life of microbes in the rhizosphere. In: Lugtenberg B (ed.) Principles of plant-microbe interactions, microbes for sustainable agriculture; Springer, Cham
Maingi MJ, Shisanya AC, Gitonga MN, Hornetz B (2001) Nitrogen fixation by common bean (Phaseolus vulgaris L.) in pure and mixed stands in semi-arid South east Kenya. Eur J Agron 14:1–12
Maitra S (2020) Potential horizon of brown-top millet cultivation in drylands: a review. Crop Res 55:57–63
Mashingaidze N, Madakadze IC, Twomlow S (2012) Response of weed flora to conservation agriculture systems and weeding intensity in semi–arid Zimbabwe. J Agric Res 36:5069–5082
Meena RP, Joshi D, Bisht JK, Kant L (2021) Global scenario of millets cultivation. In: Kumar A, Tripathi MK, Joshi D, Kumar V (eds) Millets and millet technology. Springer, Singapore, pp 33–50
Milenkovic M, Simic M, Brankov M, Milojkovic-Opsenica D, Kresovic B, Dragicevic V (2019) Intercropping of soybean and proso millet for biomass production. J Process Energy Agric 23:38–40
Mishra JS, Kumar R, Upadhyay PK, Hans H (2018) Weed management in millets. Indian Farm 68(11):77–79
Nelson LA (1977) Influence of various row widths on yields and agronomic characteristics of proso millet. Agron J 69:351–353
Nigade RD, Karad SR, More SM (2012) Agronomic manipulations for enhancing productivity of finger millet based on intercropping system. Adv Res J Crop Improve 3(1):8–10

Opole RA (2019) Opportunities for enhancing production, utilization and marketing of finger millet in Africa. Afr J Food Agric Nutr Dev 19(13863–82):284

Parihar M, Kumar A, Bisht JK, Bhinda MS, Nath N, Meena RP, Mondal T, Joshi DC, Bijarniya H, Singh S, Kant L (2021) Reviving the forgotten food network of potential crops to strengthen nutritional and livelihood security in North Western Himalayas. Indian J Agron 66(5th IAC Special issue):S44–S59

Powar RV, Aware VV, Patil SB et al (2020) Finger Millet threshing and pearling technologies in India: a comprehensive review. J Biosyst Eng 45:71–84

Prabhakar PC, Ganiger BB, Bhat S, Nandini C, Kiran, Tippeswamy V, Manjunath HA (2017) Improved production technology for foxtail millet. Technical Bulletin– 3/2017–18. ICAR–AICRP on Small Millets p. 22

Pradhan A, Patil SK, Sao A, Nag SK, Mukherjee SC (2018) Finger millet and horse gram intercropping systems in replacement series for advantages and weed smothering. Agric Sci Dig 38(2):73–80

Pradhan A, Panda AK, Bhavani RV (2019) Finger millet in tribal farming systems contributes to increased availability of nutritious food at household level: insights from India. Agric Res 8:540–547

Pradhan S, Meena RP et al (2021) Zn-solubilizing microorganism: a novel perspective for sustainable agriculture. In: Biofertilizers. Volume 1: advances in bio-inoculants. Woodhead Publishing, pp 69–81

Rakshit S, Wang Y-H (2016) The Sorghum genome. Springer International Publishing, Singapore

Rao AN (2021) Weed management in finger millet in India—an overview. Indian J Weed Sci 53(4):324–335

Reddy SS, Stahlman PW, Geier PW, Charvat LD, Wilson RG, Moechnig MJ (2014) Tolerance of foxtail, proso and pearl millets to saflufenacil. Crop Prot 57:57–62

Sahib KH (1997) Importance of Proso millet in Indian agriculture. pp 11–12. National seminar on small millets. 23–24 April 1997. Coimbatore, India

Seetharam A (1998) Small millets research: achievement during 1947–97. Indian J Agric Sci 68:431–438

Shamina C, Annadurai K, Hemalatha M, Suresh S (2019) Effect of spacing and weed management practices on Barnyard millet (*Echinochloa frumentaceae*) under rainfed condition. Int J Curr Microbiol Appl Sci 8:330–337

Shankar K, Haokip SW (2020) Recent scenario of agricultural research for tribal and hill regions of India. Agric Food e-Newsl 2(2):170–172

Sharmili K, Manoharan S (2018). Studies on Intercropping in Rainfed Little Millet (Panicum sumatrense). Int. J. Curr. Microbiol. App. Sci 7(2): 323–327

Shubhashree KS, Sowmyalatha BS (2019) Integrated weed management approach for direct seeded finger millet (*Eleusine coracana* L). Int J Agric Sci 11:8193–8195

Shukla A, Lalit A, Sharma V, Vats S, Alam A (2015) Pearl and finger millets: the hope of food security. Appl Res J 1:59–66

Sidar S, Thakur AK (2017) Effect of tillage and conservation farming on weed population and yield of finger millet (Eleusine coracana L.)under rain fed ecosystem. Int J Curr Microbiol Appl Sci 6:3650–3664

Silva C, Silva A, Vale W, Galon L, Petter F, May KD (2014) Weed interference in the sweet sorghum crop. Bragantia 73(4):1

Singh S, Sahoo DC, Singh NK, Bisht JK (2015) Operator physiological response and bullock draught ability during primary tillage. Agric Eng Int CIGR J Open Access 14(04):115–120

Sivagamy K, Ananthi K, Kannan P, Vijayakumar M, Sharmili K, Rajesh M, Nirmalakumari A, Parasuraman P (2020) Studies on agro techniques to improve the productivity and profitability of Samai +red gram intercropping system under rainfed conditions. Int J Curr Microbiol App Sci 9(6):4126–4130

Sood S, Khulbe RK, Gupta AK, Agrawal PK, Upadhyaya HD, Bhatt JC (2015) Barnyard millet—a potential food and feed crop of future. Plant Breed 134:135–147

Sood S, Joshi DC, Pattanayak A (2020) Breeding advancements in barnyard millet. Gosal S, Wani S, (eds) Accelerated Plant Breeding, Volume 1. Springer, Cham

Subramanian KS, Tenshia V, Jayalakshmi K, Ramachandran V (2009) Role of arbuscular mycorrhizal fungus (Glomus intraradices) (fungus aided). Agric Biotechnol Sustain Dev 1:29–38

Thambi B, Latha KR, Arthanari PM, Djanaguiraman M (2021) Integrated weed management practices in barnyard millet (Echinochloa frumentacea) under irrigated condition. Pharma Innov J 10(10):1404–1408

Tiemann LK, Grandy AS, Atkinson EE, Marin-Spiotta E, McDaniel MD (2015) Crop rotational diversity enhances belowground communities and functions in an agroecosystem. Ecol Lett 18:761–771

Trivedi AK, Arya L, Verma M, Verma SK, Tyagi RK, Hemantranjan A (2015) Genetic variability in proso millet (Panicum miliaceum) germplasm of central Himalayan region based on morphophysiological traits and molecular markers. Acta Physiol Plant 37:23. UK, p 648

Ungata T, Dorasila N, Yasarapu SR, Patro TSSK, Narasupalli A (2020) Impact of diverse weed management practices on direct seeded finger millet (*Eleusine coracana*). Indian J Agric Sci 90:1135–1139

Vinothini G, Arthanari PM (2017) Pre-emergence herbicide application and hand weeding for effective weed management in irrigated kodo millet (Paspalum scrobiculatum L.). Int J Chem Stud 5:366–369

Wu SC, Cheung KC, Luo YM (2006) Wong effects of inoculation of plant growth promoting rhizobacteria on metal uptake by Brassica juncea. Environ Pollut 140:124–135

Yang C, Fan Z, Chai Q (2018) Agronomic and economic benefits of pea/maize intercropping systems in relation to N fertilizer and maize density. Agronomy 8:52. https://doi.org/10.3390/agronomy8040052

Zimdahl RL (1980) Crop–weed competition: a review. International Plant Protection Centre, Oregon State University, Corvallis, Oregon, USA

Zohary D, Hopf M (2000) Domestication of plants in the old world, 3rd edn. Oxford University Press, Oxford

Genetic Resources of Small Millets: Collection, Characterization, and Utilization

K. Venkatesh, K. N. Ganapathy, B. Amasiddha, and C. Tara Satyavathi

Abstract

Genetic resources management encompasses a scientific field primarily involving the collection, characterization, evaluation, conservation, documentation, distribution, and utilization of the diverse elements present within a crop species. This genetic diversity manifests in various forms, including cultivated varieties, wild and weedy progenitor species, and any line resulting from plant breeding activities. Among the small millets, such as finger millet, foxtail millet, proso millet, barnyard millet, kodo millet, and little millet, there is a growing recognition of their significance due to their adaptability to harsh and arid climates, as well as their nutritional richness.

Millets, including finger millet and proso millet with higher protein content; barnyard millet, little millet, and kodo millet with increased dietary fiber; finger millet with elevated levels of calcium; barnyard millet with iron; and foxtail millet with zinc, contribute to the nutritional value of these crops. Recognizing these attributes, the management of genetic resources becomes crucial for conserving the existing diversity in these crops, ensuring their availability for future utilization and development.

The Indian Council of Agricultural Research (ICAR)-Indian Institute of Millets Research (IIMR) serves as a National Active Germplasm Site (NAGS), tasked with collecting, conserving, evaluating, documenting, and distributing millet genetic resources to authorized users within the country. ICAR-IIMR-Hyderabad, between 2015 and 2021, collected a total of 559 small millet germplasms, which are now preserved in the Millets Genebank. Additionally, around 15,000 accessions of small millets were acquired from various national and international centers, with over 9300 small millet accessions characterized at ICAR-

K. Venkatesh (✉) · K. N. Ganapathy · B. Amasiddha · C. T. Satyavathi
ICAR-Indian Institute of Millets Research, Hyderabad, India
e-mail: karnam.venkatesh@icar.gov.in

D. C. Joshi et al. (eds.), *Minor Millets*,
https://doi.org/10.1007/978-981-96-4265-6_4

IIMR. Currently, approximately 16,500 small millet accessions are conserved at the Millets Genebank.

Keywords

Small millets · Genetic resources · Protein · Nutrient · Collection · Characterization · Genebank · ICAR-Indian Institute of Millets Research (IIMR)

1 Introduction

In recent decades, both global and Indian agriculture have experienced remarkable growth in food production. Despite this progress, the agricultural sector faces significant challenges, including climate change and malnutrition (Sharma et al. 2015; Kumar et al. 2018). Historically, there has been an overexploitation of irrigated agricultural lands in the nation. However, there is a critical need to shift the focus toward dry and marginal lands to address the adverse effects of current and future climate change scenarios.

The inherent challenges of achieving higher production in dry and marginal lands, characterized by lower fertility, pose hurdles for conventional crops like rice and wheat. Millets emerge as a promising solution due to their climate-resilient and robust nature, making them suitable alternatives for conventional crops in such challenging environments. Beyond their climate resilience, millets are nutritionally rich, offering a wealth of essential micronutrients and vitamins crucial for addressing malnutrition (Hariprasanna et al. 2014; Maruthamuthu et al. 2022). Recognized as nutricereals, millets serve as a valuable source of vitamins, minerals, essential fatty acids, phytochemicals, and antioxidants, providing a means to combat hidden hunger.

The nutritional abundance of millets, especially in polyphenols and other bioactive composites, is associated with potential health benefits. Millets are reported to play a role in reducing the rate of fat absorption, promoting the slow release of sugars (low glycemic index), and thereby mitigating the risks of cardiac diseases, diabetes, and high blood pressure (BP) (Kumar et al. 2018).

Primary millets include pearl millet (*Pennisetum glaucum*, also known by as *Pennisetum americanum*, *Pennisetum typhoides*, and *Pennisetum typhoideum*), foxtail millet (*Setaria italica*), proso millet or common millet (*Panicum miliaceum*), and finger millet (*Eleusine coracana*). Additionally, the category of minor millets encompasses barnyard millet (*Echinochloa* spp.), kodo millet (*Paspalum scrobiculatum*), little millet (*Panicum sumatrense*), and browntop millet (*Brachiaria ramosa*) (Upadhyaya et al. 2006).

Genetic resources management is a scientific field primarily involving the collection, characterization, and conservation of diversity within a crop species. This management aims to ensure the availability and utilization of trait-specific germplasm in national crop improvement programs for the future (Singh et al. 2019). The genetic diversity in question may take the form of cultivated varieties, wild and weedy progenitor species, and any lines resulting from plant breeding activities. Plant Genetic Resources encompass materials with actual or potential use for altering the genetic constitution of a plant species, contributing to the development of improved cultivars. These resources comprise valuable natural variants, including exotic and indigenous collections such as advanced cultivars, released varieties, recombinant inbred lines (RIL), near isogenic lines (NIL), mutants, genetic stocks, landraces, and wild and weedy relatives (Singh et al. 2019).

The management of genetic resources is crucial to conserve the existing diversity in these crops for future use and development, addressing the needs of both present and future generations of researchers (Singh et al. 2019).

2 Global Status of Small Millets Conservation

The International Crops Research Institute for the Semi-Arid Tropics (ICRISAT), National Bureau of Plant Genetic Resources (NBPGR), New Delhi, and Millet Genebank (MGB) at ICAR-Indian Institute of Millets Research (IIMR) collectively preserve an extensive pool of millet germplasm from 92 nations. The national genebank (NGB) at ICAR-NBPGR in New Delhi safeguards a total of 22,971 accessions of small millet germplasm. Among these, 11,587 accessions belong to finger millet, 4244 to foxtail millet, 2362 to kodo millet, 1885 to little millet, 1005 to proso millet, and 1888 to barnyard millet. Furthermore, the Millets Genebank at ICAR-IIMR in Hyderabad preserves an additional 17,330 accessions of small millets. For detailed information on the status of other small millet genetic resources in these genebanks, please refer to Table 1.

Table 1 Status of millet genetic resources in the Genebank

Crop name	ICRISAT	NGB-ICAR-NBPGR	MGB-ICAR-IIMR
Finger millet (IE)	7519	11,587	7806
Foxtail millet (ISe)	1542	4244	4653
Kodo millet (IPs)	665	2362	344
Little millet (IPmr)	473	1885	694
Proso millet (IPm)	849	1005	2128
Barnyard millet (IEc)	749	1888	1705
Total	**11,797**	**22,971**	**17,330**

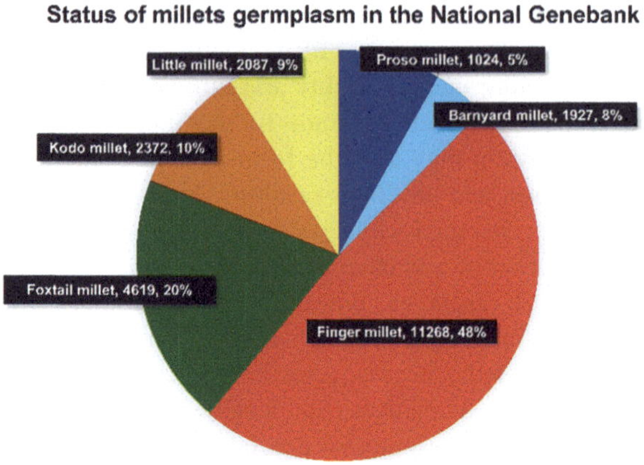

3 Species Diversity of Small Millets

3.1 Finger Millet (*Eleusine coracana*)

Finger millet's closest wild relative is *Eleusine coracana* subsp. *africana* (Kennedy-O'Byrne) Hilu and de Wet. Originating in Africa, *Eleusine coracana* subsp. *africana* is the ancestral form of finger millet (Upadhyaya et al. 2008). Finger millet and subsp. *africana* both exhibit tetraploid characteristics, featuring a chromosome count of $2n = 36$ and a basic chromosome number of $x = 9$. These subspecies exhibit hybridization where they coexist in Africa, and the resulting crosses often give rise to derivatives that function as weeds in cultivated fields (Upadhyaya et al. 2008).

3.2 Foxtail Millet (*Setaria italica*)

Foxtail millet, with its origins in the highlands of central China, holds a historical significance as it was first domesticated around 5000 years ago. The process of domestication marked a crucial turning point, leading to the cultivation and utilization of this versatile grain. Over time, foxtail millet expanded its geographical reach, making its way to Europe and India, where it played a significant role in local agricultural practices and diets.

The likely progenitor species of foxtail millet is identified as *Setaria viridis*, sharing the same somatic chromosome number of 18 ($2n = 18$) as the cultivated foxtail millet. This genetic similarity underscores the evolutionary connection between the wild and domesticated forms of this grain.

Foxtail millet exhibits diversity in its cultivation, leading to the classification into three distinct races. The moharia race is prevalent in regions spanning Europe, Russia, Afghanistan, and Pakistan. The maxima race is cultivated in East China,

Japan, Korea, and the United States. The indica race, on the other hand, is primarily found in India and Sri Lanka. This classification highlights the adaptability and spread of foxtail millet across diverse geographical and climatic conditions, contributing to its global significance as a staple crop. The intricate details of its genetic lineage and regional variations provide valuable insights into the agricultural history and biodiversity associated with foxtail millet.

3.3 Proso Millet (*Panicum miliaceum*)

Proso millet, alternatively known as broomcorn and common millet, has a rich history dating back to its domestication in Neolithic China, occurring as early as 10,000 years ago (Lu et al. 2009). This ancient cultivation marked a significant milestone in the development of human agriculture and the establishment of proso millet as a staple crop.

The cultivated proso millet exhibits a diverse array of characteristics, leading to its classification into five races based on the structure of the panicle. The first two races, namely, miliaceum (characterized by open inflorescences and suberect branches with few subdivisions) and patentissimum (featuring a narrow and diffuse panicle), are prevalent in Eurasia. In contrast, the latter three races—contractum, compactum, and ovatum—display more compact inflorescences with distinctive traits such as drooping, cylindrical, and curved structures, respectively (Reddy et al. 2007).

3.4 Little Millet (*Panicum sumatrense*)

Little millet, scientifically known as *Panicum sumatrense*, holds a common presence in the agricultural practices of India, Nepal, and Western Myanmar. This small-grained cereal has roots in the ancestral species *Panicum psilopodium*, reflecting its historical cultivation and utilization in these regions. Little millet, with its adaptability to diverse agroecological conditions, has become a staple in local farming systems, contributing to both food security and dietary diversity.

Within the cultivated species of little millet, two distinct races have been identified: robusta and nana. The robusta race predominantly thrives in Northwestern Andhra Pradesh and specific areas of Orissa, showcasing the crop's regional variations and adaptations. On the other hand, the nana race adds to the diversity of little millet cultivation, with its unique characteristics influencing its growth patterns and nutritional content.

3.5 Kodo Millet (*Paspalum scrobiculatum*)

Kodo millet, scientifically referred to as *Paspalum scrobiculatum*, exhibits a noteworthy level of diversity that is systematically classified into three distinct races.

This categorization is based on the morphology of the panicle, a pivotal part of the plant's reproductive structure (de Wet et al. 1983).

The three races identified within the cultivation of Kodo millet are regularis, irregularis, and variabilis. Each race possesses unique traits related to the structure and arrangement of the panicle. This panicle-centric classification system provides a comprehensive understanding of the variations existing within kodo millet, offering insights into the plant's genetic diversity and adaptability.

The regularis race of Kodo millet is characterized by consistent and standardized panicle morphology. In contrast, the irregularis race exhibits variations and irregularities in the structure of its panicles. The variabilis race showcases a diverse range of characteristics, reflecting the adaptability and variability inherent in this particular millet species.

3.6 Barnyard Millet (*Echinochloa crus-galli* and *Echinochloa colona*)

Barnyard millet, classified under two main species, *Echinochloa crus-galli* (native to Eurasia and domesticated in Japan) and *E. colona* (domesticated in India), presents an intriguing botanical landscape (Upadhyaya et al. 2008). Both species share a common hexaploid chromosome number of $2n = 54$, underscoring their genetic similarity and evolutionary connections.

However, a nuanced perspective emerges when considering alternate reports suggesting that two distinct species contribute to the barnyard millet complex. According to this viewpoint, *Echinochloa esculenta* (synonymous with *Echinochloa utilis* and *Echinochloa crus-galli*) finds cultivation in Japan, Korea, and the north-eastern part of China. Concurrently, *Echinochloa frumentacea* (synonymous with *Echinochloa colona*) is prevalent in regions such as Pakistan, India, Nepal, and Central Africa (Wanous et al. 1990).

This dual-species perspective enriches our understanding of the geographical distribution and cultivation practices associated with barnyard millet. The divergent species demarcation aligns with distinct agroecological regions, each contributing to the cultivation and utilization of barnyard millet in unique ways. The hexaploid chromosome number shared by both species highlights the genetic complexities and evolutionary adaptations that have shaped barnyard millet over time.

4 Small Millet Genetic Resources Management

The ICAR-Indian Institute of Millets Research (IIMR) holds the designation of being one of the National Active Germplasm Sites (NAGS). It shoulders the responsibility of collecting, conserving, evaluating, documenting, and distributing genetic resources of millets to bona fide users within the country.

5 Collection

A total of 2273 accessions were collected through 45 explorations conducted by NRCS/DSR/IIMR, with an additional 423 accessions collected by various individuals. The Indigenous Collection numbers were assigned to 1928 accessions obtained from the National Genebank, NBPGR-New Delhi. Notably, 10 out of the 45 explorations were conducted in collaboration with NBPGR (regional stations), resulting in the collection of all eight millets.

The millet varieties collected during these explorations include finger millet (244 accessions), foxtail millet (102 accessions), proso millet (19 accessions), barnyard millet (7 accessions), little millet (146 accessions), and kodo millet (41 accessions).

Geographically, the maximum number of accessions was collected from Andhra Pradesh (474 accessions), followed by Maharashtra (447 accessions), Tamil Nadu (431 accessions), Madhya Pradesh (266 accessions), Karnataka (190 accessions), Gujarat (183 accessions), Uttar Pradesh (172 accessions), Odisha (141 accessions), Rajasthan (135 accessions), Chhattisgarh (60 accessions), Jharkhand (58 accessions), Uttarakhand (30 accessions), Telangana (16 accessions), Bihar (13 accessions), and West Bengal (1 accession). This comprehensive collection effort provides a valuable repository of millet genetic resources from diverse regions, contributing to research, conservation, and agricultural development.

6 Augmentation

A total of 20,000 accessions have been augmented from diverse national and international centers, encompassing a variety of millets such as finger millet (10,704 accessions), foxtail millet (5096 accessions), proso millet (1666 accessions), barnyard millet (1661 accessions), and more. Notably, ICAR-NBPGR-New Delhi and its regional stations have significantly contributed to this collection, augmenting a total of 58,572 millet accessions. Additionally, CGIAR-ICRISAT-Patancheru-Hyderabad has contributed 7887 accessions, further enriching the diverse genetic resources available for research and agricultural development.

7 Characterization

Characterizing small millet germplasm holds paramount significance for several reasons, contributing to sustainable cultivation, agricultural resilience, and nutritional security. Key benefits of small millet germplasm characterization include preserving genetic diversity within small millets, which is vital for breeding programs aimed at developing new varieties with improved traits such as disease resistance, drought tolerance, and nutritional richness. Understanding the genetic makeup aids in the development of improved varieties, while identifying well-adapted varieties to diverse climatic conditions enhances agricultural resilience.

Nutritional characterization of different small millet germplasms allows for the development of varieties with enhanced nutritional profiles. Additionally, characterization helps identify germplasms resistant to common pests and diseases. Understanding growth habits, maturity periods, and other agronomic traits assists farmers in selecting varieties aligned with their specific agricultural practices and goals.

Furthermore, well-characterized germplasm facilitates the exchange of genetic resources between research institutions, ensuring a broader pool of diversity for global agricultural development. Overall, the comprehensive characterization of small millet germplasm plays a pivotal role in advancing sustainable agriculture and ensuring food security.

Characterization of small millet germplasm was carried out at ICAR-IIMR. The highlights of characterization of different small millets have been presented in detail in the following paragraphs:

Finger Millet A comprehensive assessment was conducted at IIMR, characterizing a total of 5400 accessions of finger millet germplasm for 14 quantitative and 18 qualitative traits. Among these traits, the grain yield per plant exhibited the highest variability, with days to 50% flowering and leaf blade length closely following suit.

The plant growth habit was predominantly erect (3521 accessions), followed by decumbent (1628 accessions). Dark green leaf color was the most prevalent (1541 accessions), with light green following closely (1002 accessions). Absence of plant pigmentation at the leaf juncture was the most common (3973 accessions), compared to its presence (959 accessions). Leaf sheath pubescence was predominantly absent (3004 accessions), with presence noted in 2382 accessions.

Culm branching was most frequently observed as present (2645 accessions), followed by absent (2200 accessions). Semi-compact ear head shape had the highest occurrence (1313 accessions), while fist-shaped ear heads followed closely (1259 accessions). Intermediate-sized ear heads were the most common (1065 accessions), with large-sized ear heads trailing (280 accessions). Absence of finger branching was widespread (3901 accessions), as opposed to its presence (878 accessions). Branching predominantly occurred at the thumb finger position (336 accessions), followed by branching in all fingers (287 accessions). The presence of multiple whorls on fingers was mostly absent (2874 accessions) compared to its presence (1914 accessions). Gaps on fingers were more commonly absent (1157 accessions) than present (338 accessions).

Spikelet shattering was observed more frequently as present (1789 accessions) than absent (1023 accessions). Glume color tended to be light brown (823 accessions), followed by brown (507 accessions). Pericarp persistence on seeds after threshing was mainly persistent (724 accessions), with partially persistent observed in 489 accessions. Copper brown seed color was the most prevalent (1569 accessions), followed by light brown (1318 accessions) and brown (24 accessions). Round seed shape dominated (3161 accessions), with reniform shape following (1514 accessions). Smooth seed surface was more common (2622 accessions) than rough (651 accessions).

Furthermore, specific characteristics were identified in select accessions, such as those with more tillers (>12.00, 95 accessions), early flowering (<50.00 days, 13 accessions), more leaves (>20.00, 157 accessions), longer leaves (>70.00 cm, 61 accessions), wider leaves (>2.00 cm, 10 accessions), shorter plants (<50.00 cm, 24 accessions), taller plants (>160.00 cm, 34 accessions), longer finger length (>15.00 cm, 35 accessions), wider ear head (>6.00 cm, 9 accessions), more fingers on the main axis of the ear head (>13.00, 34 accessions), more productive tillers (>13.00, 47 accessions), higher grain yield (>90.00 g/plant, 64 accessions), and more significant 100-seed weight (>1.20 g, 11 accessions).

Foxtail Millet At ICAR-IIMR, a thorough characterization activity was conducted on 2580 accessions of foxtail millet germplasm, focusing on 11 quantitative and 19 qualitative traits. Notably, plant height demonstrated the highest variability, with grain yield, days to 50% flowering, and other traits following suit in the order of variability.

The predominant frequency in flag leaf shape, specifically the leaf tip shape, was observed to be pointed (2384 accessions), followed by pointed to rounded (154 accessions). The basal leaf sheath exhibited maximum instances of absence or weak pigmentation in 2089 accessions, compared to strong pigmentation (383 accessions). Dark green foliage intensity was most frequently observed in 1859 accessions, followed by medium intensity (567 accessions). An upright plant growth habit was observed in 1791 accessions, with spreading observed in 743 accessions. The absence or weak anthocyanin coloration of the leaf collar was noted in 2115 accessions, while strong coloration was observed in 363 accessions. A semi-erect leaf attitude of the blade was predominant in 1813 accessions, with a slightly drooping attitude in 552 accessions.

The presence of inflorescence bristles was most common in 2383 accessions, compared to absence (29 accessions). Long panicle length of bristles was observed in 1495 accessions, while short length was noted in 478 accessions. The presence of inflorescence apical sterility was more frequent in 1964 accessions, compared to absence (447 accessions). Absence of anthocyanin coloration of the bristles was observed in 2081 accessions, with presence noted in 320 accessions. The absence or weak anthocyanin coloration of the flag leaf was identified in 2374 accessions, while medium coloration was observed in 25 accessions. Absence of glume anthocyanin coloration was more prevalent in 2010 accessions, compared to its presence (387 accessions). A semi-erect panicle attitude to the stem was most frequent in 1336 accessions, with a drooping attitude observed in 564 accessions. The conical panicle type was predominant in 1135 accessions, followed by the spindle type in 636 accessions. The oblong inflorescence shape was observed more frequently in 1090 accessions, compared to the pyramidal shape (808 accessions). Absence of inflorescence lobes was more common in 1259 accessions, while presence was noted in 1059 accessions. Circular grain shape was most frequent in 1025 accessions, followed by medium ovate shape in 777 accessions. Brown grain color was observed more frequently in 1489 accessions, compared to yellow color (332 accessions).

Additionally, specific characteristics were identified in select accessions, including those with early flowering (<40.00 days, 97 accessions), longer flag leaf (>40.00 cm, 46 accessions), wider flag leaf (>2.90 cm, 23 accessions), longer peduncle (>40.00 cm, 67 accessions), thicker stem (>1.00 cm, 5 accessions), more basal tillers (>20.00, 24 accessions), longer panicle (>27.00 cm, 60 accessions), wider panicle (>3.00 cm, 28 accessions), taller plant (>150.00 cm, 27 accessions), shorter plant (<40.00 cm, 34 accessions), higher grain yield (>70.00 g/plant, 41 accessions), and more 100-seed weight (>0.35 g, 79 accessions).

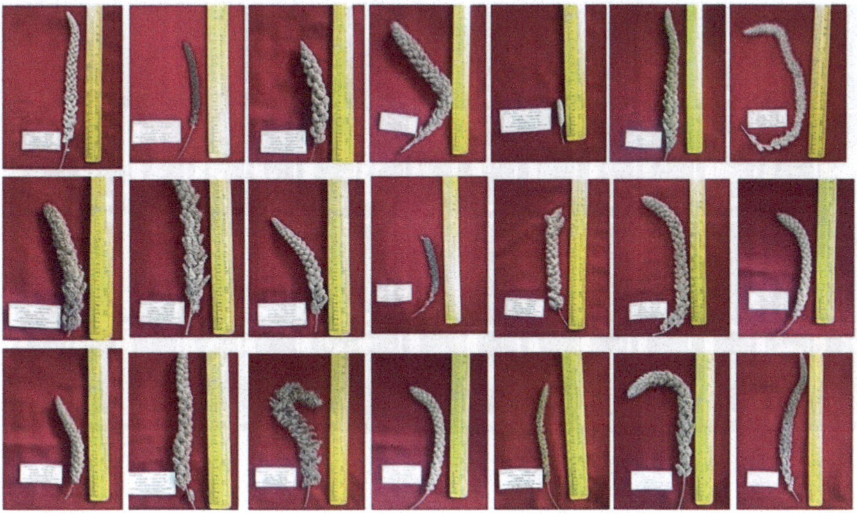

Proso Millet A total of 645 accessions of proso millet germplasm, along with four checks (GPUP 8, TNAU 145, TNAU 164, and TNAU 202), were comprehensively characterized at ICAR-IIMR-Hyderabad for 23 agro-morphological traits, encompassing 11 quantitative and 12 qualitative traits. Among the quantitative traits, plant height exhibited the highest variability, followed by date of maturing and days to 50% flowering.

In the qualitative dataset, the most frequent occurrences were observed for erect growth habit in 444 accessions, absence of leaf sheath pigmentation in 630 accessions, sparse leaf sheath pubescence in 333 accessions, absence of ligule pubescence in 639 accessions, glabrous leaf blade pubescence in 559 accessions, arched inflorescence shape in 443 accessions, presence of culm branching in 378 accessions, absence of lodging in 677 accessions, absence of seed shattering in 661 accessions, intermediate panicle compactness in 303 accessions, golden yellow grain color in 470 accessions, and oval grain shape in 379 accessions.

Moreover, a selection of 442 accessions from trait-specific proso millet germplasm was discerned, with 82 of them showcasing specificity for multiple traits, ranging from 2 to 6. This extensive characterization offers valuable insights into the diverse traits inherent in the proso millet germplasm collection. Such insights prove instrumental in the targeted selection and development of varieties endowed with desirable agricultural attributes.

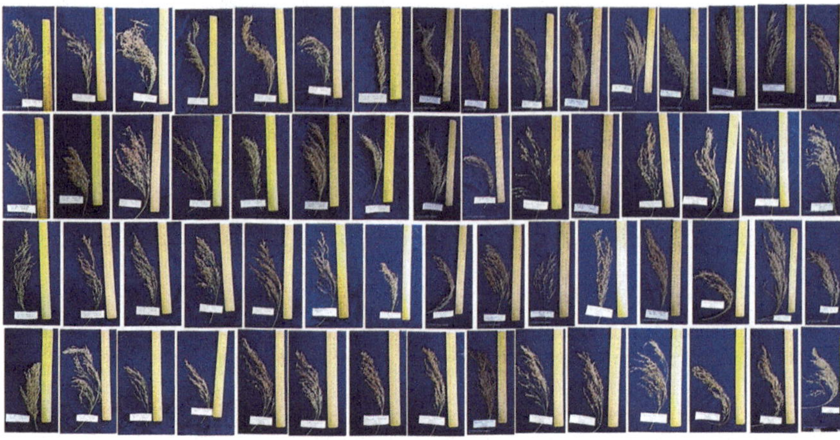

Barnyard Millet A comprehensive characterization was undertaken at ICAR-IIMR-Hyderabad, covering a total of 553 accessions of Barnyard millet germplasm, alongside 4 reference checks (DHBM 93–3, VL 207, VL 172, and PRJ 1). The assessment included 25 agro-morphological traits, comprising 12 quantitative and 13 qualitative traits. Notably, plant height emerged as the trait with the highest variability, closely followed by days to maturity, days to 50% flowering, and grain yield per plant.

Within the qualitative dataset, the most prevalent occurrences were observed for erect plant growth habit in 349 accessions, absence of leaf sheath pigmentation in 310 accessions, arched inflorescence shape in 359 accessions, green inflorescence color in 376 accessions, intermediate panicle compactness in 340 accessions, onside spikelet arrangement on the rachis in 383 accessions, curved lowest raceme shape in 340 accessions, thick lowest raceme thickness in 317 accessions, absence of lowest raceme branching in 494 accessions, presence of culm branching in 443 accessions, absence of lodging in 541 accessions, light grey grain color in 267 accessions, and oval grain shape in 473 accessions.

Furthermore, a collection of 411 trait-specific accessions of barnyard millet germplasm were identified, of which 92 accessions demonstrated multi-trait specificity for two to five traits. This detailed characterization provides valuable insights into the diversity of traits present in the barnyard millet germplasm collection, facilitating the selection and development of varieties with desired agricultural attributes.

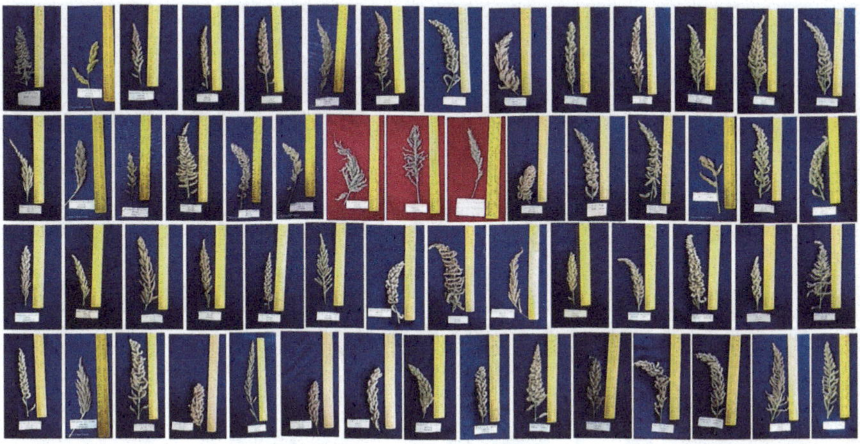

8 Kodo Millet

A comprehensive characterization was conducted on 69 accessions from the core collections of kodo millet germplasm. This evaluation encompassed 13 quantitative and 15 qualitative traits, with days to 50% flowering emerging as the most variable trait, closely followed by plant height and others.

In terms of growth habit, the predominant frequency was observed in the erect category (27 accessions), closely followed by decumbent (23 accessions). When examining leaf erectness, the highest frequency was noted for the droopy type (31 accessions). Leaf sheath pigmentation manifested in 51 accessions, while leaf juncture pigmentation was absent in 46 accessions. Internode pigmentation was lacking in 27 accessions, and leaf blade pigmentation was present in 42 accessions.

Concerning panicle compactness, the highest frequency was in the open category (25 accessions), followed by semi-compact (22 accessions). Complete panicle exertion was observed in 30 accessions. Regular spikelet arrangement on the rachis was noted in 45 accessions, and spike branching was present in 33 accessions. Lodging was observed in 40 accessions. In terms of grain color, brown was the most prevalent (28 accessions), followed by dark brown (13 accessions). The grain shape was predominantly oval in 45 accessions.

Moreover, specific characteristics were identified in select accessions, including those with more basal tillers (>18.00, 11 accessions), early flowering (<56.00 days, 10 accessions), longer flag leaf blade (>35.00 cm, 9 accessions), wider flag leaf blade (>1.50 cm, 11 accessions), longer peduncle (>3.00 cm, 3 accessions), longer panicle (>9.00 cm, 6 accessions), wider panicle (>2.50 cm, 7 accessions), longer thumb raceme (>7.00 cm, 4 accessions), longer raceme (>10.00 cm, 9 accessions), more number of racemes (>6.00, 4 accessions), taller plant (>90.00 cm, 8 accessions), shorter plant (<60.00 cm, 12 accessions), higher grain yield (>20.00 g/plant, 4 accessions), and more 100-seed weight (>0.40 g, 17 accessions).

9 Little Millet

At ICAR-IIMR, 66 accessions of recently discovered little millet germplasm underwent characterization for 9 quantitative and 12 qualitative traits. Plant height emerged as the most variable trait, closely followed by days to 50% flowering and the length of the flag leaf blade, among others.

The highest frequency in plant growth habit was noted in the erect category (46 accessions). Along with the absence of leaf sheath pigmentation (52 accessions), a glabrous leaf sheath pubescence (58 accessions), lack of ligule pubescence (56 accessions), and a glabrous leaf blade pubescence (56 accessions) were also observed. Additionally, a diffused inflorescence shape was observed in 11 accessions, culm branching in 12 accessions, and intermediate panicle compactness in 8 accessions, lodging was found in 9 accessions, and seed shattering was absent in 14 accessions. Furthermore, light brown grain color was predominant (19 accessions), followed by brown (17 accessions), and elliptical grain shape in 30 accessions.

Furthermore, four accessions were identified for early flowering (<55.00 days), eight accessions were identified with longer flag leaf blade (>30.00 cm), five accessions were identified with a longer peduncle (>17.00 cm), eight accessions were identified with taller plants (>100.00 cm), six accessions with longer panicles (>25.00 cm), seven accessions with wide panicles (>1.00 cm), and four accessions with higher grain yield (>10.00 g/main plant).

10 Conservation

The Millet Genebank (MGB) is conserving a total of 48,462 accessions of millets in bulk. Sorghum has the highest representation with 27,366 accessions, followed by finger millet (8057), foxtail millet (4573), pearl millet (4094), proso millet (1463), barnyard millet (1159), little millet (670), kodo millet (333), tef (36), browntop millet (25), quinoa (12), and Job's tears (1).

11 Utilization

A total of 48 finalized products have been successfully developed through the dedicated selection and breeding efforts employing ICAR-IIMR germplasm in trials conducted by the All India Coordinated Research Projects. Among these, the highest number comprises 22 rabi sorghum varieties, followed by 6 sweet sorghum varieties, 7 kharif sorghum varieties, 1 kharif sorghum hybrid, 5 single-cut forage varieties, and 1 each of sweet sorghum variety, sweet sorghum hybrid, and dual-purpose varieties. In addition to these, three specialty sorghum varieties have made significant contributions to the trials.

12 Distribution

A total of 121,077 accessions have been distributed by scientists, researchers, and students across the country for research use. A total of 1079 Material Transfer Agreements (MTAs) have been signed to facilitate the supply of germplasm. The highest distribution occurred in 2012–13, with 24,827 accessions reaching bona fide users, followed by 17,981 in 2021–2022. A maximum frequency of 17,000 acc. of millet germplasm was distributed and utilized for salinity and drought-tolerant screening at Gujarat/Rajasthan followed by 7500 acc. Utilized for screening leaf blast, neck blast, and finger blast, 6160 acc. of germplasm evaluated for yield attributing traits, etc.,

13 Genetic Stocks Registration with ICAR-NBPGR

A cumulative total of 112 millet genetic stocks have been officially registered with ICAR-NBPGR. The highest registration count is attributed to sorghum genetic stocks, with a notable 84 accessions. Following this, there are registrations for finger millet (19 accessions), barnyard millet (4 accessions), foxtail millet (3 accessions), and little millet (2 accessions).

14 Documentation

A total of 3585 sorghum genotypes have been identified for their unique traits, utilizing information compiled from published annual reports spanning from 2000–2001 to 2013–2014. Additionally, 1775 millet genotypes have been recognized for their distinctive traits, drawing from compiled information found in the published ICAR-IIMR Annual Reports spanning from 2014–2015 to 2018–2019. Furthermore, a comprehensive count of 17,794 sorghum genotypes has been established for their unique traits in the AICSIP trials, collating information from published AICSIP reports covering the period from 2004–2005 to 2013–2014. 19,710 accessions of pearl millet genotypes were identified based on its uniqueness of trait in the All India Coordinated Research Project (AICRP) on pearl millet trials from the compiled information published from the AICRP on pearl millet reports from 2001–2002 to 2018–2019. Small millet genotypes identified based on its uniqueness of trait in the AICRP on small millet trials from the compiled information published from the AICRP on small millet reports from 2007–2008 to 2017–2018, viz., 3828 accessions of finger millet, 1000 accessions of foxtail millet genotypes, 775 accessions of barnyard millet genotypes, 1032 accessions of kodo millet genotypes, 483 accessions of proso millet genotypes, and 827 accessions of little millet genotypes. Sorghum genotypes were identified based on its uniqueness of trait in the AICRP on Sorghum from the compiled information published from the AICRP on Sorghum report from 2014 to 2020. The Pedigree databases for Sorghum (Elite Breeding Stocks: AICRP on Sorghum, 1975–2022), small millets (Elite Breeding Stocks: AICRP on Small Millets, 1986–2022), and pearl millet (Elite Breeding Stocks: AICRP on Pearl Millet, 1981–2018) serve as comprehensive records detailing the germplasm used as parental lines in their respective crop improvement programs. The documentation includes the distribution of millet germplasm to bona fide users, and these records have been published in four volumes covering the periods 2000–2007, 2008–2012, 2012–2015, and 2015–2021.

15 Challenges to Small Millet Germplasm Resources Management

- Cross-pollinated nature of few of the small millets.
- The data in the form of passport data and characterization data is in huge volume and requires special efforts and funding to handle and use such huge quantum of data.
- Mechanisms to use the characterization data for use in crop improvement need to be developed.
- Often it is difficult to track the utilization of germplasm accessions supplied by the GenBank due to poor feedback mechanism.
- Creating platforms for the integration and analysis of data from diverse domains is crucial. These platforms can be tailored to suit the specific requirements of stakeholders (Weise et al. 2020).
- More emphasis is required to characterize the germplasm resources at molecular level and special provisions are necessary to handle the generated data.
- Application of molecular tools in curation of GenBank collections with respect to minimizing potential duplications.

References

de Wet JM, Brink D, Rao K, Mengesha M (1983) Diversity in kodo millet, Paspalum scrobiculatum. Econ Bot 37:159–163

Hariprasanna K, Agte V, Elangovan M, Patil JV (2014) Genetic variability for grain iron and zinc content in cultivars, breeding lines and selected germplasm accessions of sorghum [Sorghum bicolor (L.) Moench]. Indian J Genet Plant Breed 74(01):42–49

Kumar A, Tomer V, Kaur A, Kumar V, Gupta K (2018) Millets: a solution to agrarian and nutritional challenges. Agric Food Secur 7:31

Lu H, Zhang J, Liu KB, Wu N, Li Y, Zhou K, Ye M, Zhang T, Zhang H, Yang X, Shen L (2009) Earliest domestication of common millet (Panicum miliaceum) in East Asia extended to 10,000 years ago. Proc Natl Acad Sci 106(18):7367–7372

Maruthamuthu E, Venkatesh K, Bellundagi A, Pandey S, Pandey CD (2022) Assessment of variability in sorghum [Sorghum bicolor (L.) Moench] germplasm for agro-morphological traits. Electron J Plant Breed 13(2):488–497

Reddy VG, Upadhyaya HD, Gowda CL (2007) Morphological characterization of world's proso millet germplasm collection. J SAT Agric Res 3:4

Sharma I, Tyagi B, Singh G, Venkatesh K, Gupta O (2015) Enhancing wheat production—a global perspective. Indian J Agric Sci 85:3–13

Singh K, Kumar S, Kumar SR, Singh M, Gupta K (2019) Plant genetic resources management and pre-breeding in genomics era. Indian J Genet Plant Breed 79(Sup-01):117–130

Upadhyaya HD, Gowda CL, Buhariwalla HK, Crouch JH (2006) Efficient use of crop germplasm resources: identifying useful germplasm for crop improvement through core and mini-core collections and molecular marker approaches. Plant Genet Resour 4(1):25–35

Upadhyaya, HD, Gowda, CLL, Reddy, VG and Singh, S (2008) Diversity of small millets germplasm in genebank at ICRISAT. In: 5th International Symposium on New Crops and Uses: their role in a rapidly changing world, 3-4 September, 2007, University of Southampton, Southampton, UK.

Wanous M et al (1990) Origin, taxonomy and ploidy of the millets and minor cereals. Plant Varieties Seeds 3:99–112

Weise S, Lohwasser U, Oppermann M (2020) Document or lose it—on the importance of information management for genetic resources conservation in genebanks. Plan Theory 9:1050

Advances in Hybridization Techniques in Small Millets

C. Nandini, Dinesh Chandra Joshi, B. Nandini, T. E. Nagaraja, Sujata Bhat, S. Gazala Parveen, Suresh D. Ekbote, and B. M. Dushyanthakumar

Abstract

Small millets are important crops for future food supplies. They withstand drought, are nutrient-rich, and are also used for fodder purposes. Millet crops include finger millet, barnyard millet, kodo millet, foxtail millet, little millet, browntop millet, and proso millet. These crops belong to the family Poaceae, and they are predominantly self-pollinated in nature. Crossing is necessary to create genetic variation and to make use of the genetic resources that are accessible. Because the flowers of these crops are tiny, physical emasculation and pollination are almost impossible. Although the contact method of hybridization is mostly used in all millets, only 2% to 3% of them successfully produce true F_1. Hot water treatment is followed for emasculation other than contact method in the case of foxtail millet, finger millet, and barnyard millet. The success rate is more in hot water treatment compared to contact method in all the crops. Utilization of male sterile lines is also in practice in foxtail millet. In finger millet partial male sterile (PS) line is used for crossing, which sets 27.4–49.4% of seeds

C. Nandini (✉) · S. D. Ekbote
Zonal Agricultural and Horticultural Research Station, Babbur Farm, Hiriyur, KSNUAHS, Shivamogga, Karnataka, India
e-mail: nandinic@uahs.edu.in

D. C. Joshi
Division of Crop Improvement, ICAR-Vivekananda Parvatiya Krishi Anusandhan Sansthan, Almora, Uttarakhand, India

B. Nandini
College of Horticulture, Kolar, University of Horticultural Sciences, Bagalkot, India

T. E. Nagaraja · S. Bhat · S. Gazala Parveen
AICRP on Small millets, University of Agricultural Sciences, GKVK, Bengaluru, India

B. M. Dushyanthakumar
KSNUAHS, Shivamogga, Karnataka, India

D. C. Joshi et al. (eds.), *Minor Millets*,
https://doi.org/10.1007/978-981-96-4265-6_5

upon crossing and ~10 percent of seeds upon selfing. This PS line can be effectively utilized in finger millet crossing program. Treatment with hot water, manual emasculation-pollination, and USSR methods of hybridization are frequently used in proso, little, and browntop millets in addition to the contact method for effective crossing. To generate genetic variability among proso and little millet genotypes, a novel modified crossing method called Small millets, University of Agricultural Sciences, Bengaluru (SMUASB) is used, which has a 56%–60% success rate in producing true hybrids. A 75% success rate was observed in seed establishment during manual emasculation and hand pollination of foxtail millet under growth chamber and greenhouse conditions. Kodo millet has cleistogamous flowering behavior; therefore mutation breeding and pure line selection is practiced more. In each different millet crop, different hybridization techniques are followed to create variability depending upon the breeding behavior of the crops.

Keywords

Small millets · Hybridization · Partial sterile line · Emasculation · Crossing · Pollination · Cleistogamous

1 Introduction

Small millets commonly referred to as nutri-cereals and are novel diet source and forage crops in arid and semiarid areas. They are less susceptible to environmental pressures, have an array of adaptations, are responsive to climate change, need less water for growth, and show high resistance toward pests and diseases (Bandyopadhyay et al. 2017; Saxena et al. 2018; Vetriventhan et al. 2020). Millets are C4 plants that have a significant tolerance toward drought and heat stress, a short-growing season, a greater capacity to produce dry matter, and an extremely high photosynthetic efficiency (Amit et al. 2021).

Ecologically, these millet crops are short duration and have flexibility to adjust to wide class of temperature, moisture regime, and varied management practices (Tonapi et al. 2022). Small millets are, viz., foxtail millet (*Setaria italica*), finger millet (*Eleusine coracana* (L.) Gaertn.), little millet (*Panicum sumatrense*), kodo millet (*Paspalum scrobiculatum*), proso millet (*Panicum miliaceum*), browntop millet (*Brachiaria ramosa* (L.), and barnyard millet (*Echinochloa frumentacea*). Millets are gaining relevance in recent years all over the world because of its ability to fight hunger and malnourishment as well as offer food and nutritive security. India proclaimed 2018 to be the national year of millets. Recently, the General Assembly of the United Nations has approved more than 70 countries supported and welcomed India's declaration of 2023 as the International Year of Millets (Anonymous 2022).

Globally, small millets are grown in nearly 93 countries; roughly 97% are grown and consumed in developing nations, particularly those in Africa and Southeast Asia. With 83% of Asia's millet-growing area and 26.6% of the global total, India is the world's largest millet grower. Indian states', viz., Madhya Pradesh, Uttarakhand, Odisha, Karnataka, and Jharkhand, millets are an essential component of tribal food. However, due to their immense potential as a nutraceutical, they have also become more popular in cities recently.

Vetriventhan et al. (2020) report that 248 different improved varieties of small millets are released in India. This includes 121 finger millet, 24 proso millet, 20 little millet, 33 kodo millet, 32 foxtail millet, and 18 barnyard millet. Of these, 65% were released as a result of germplasm selection, pedigree selection accounts 30%, and mutation breeding contributed 5% to improved varieties' development.

Hybridization is the initial step in crop improvement, which combines desired traits to create a variable population for selection. Hybridization in cross-pollinated crops is simple due to the biology and morphology of their flowers. It's challenging in self-pollinated crops. These crops are members of the Poaceae family. These crops have cleistogamous flowering, where pollination occurs before flower opening. Therefore, in these crops, self-pollination is predominant. So, hybridization is necessary in order to create variability in these crops. Breeders have to find easy-to-use methods for emasculation and pollination of these crops because of the difficulties in creating artificial crosses. Hybridization followed by the Pedigree method of selection, has been a critical breeding approach to develop varieties in small millets., Although there is a lot of variability in small millet crops accessible in current collections of germplasm, it hasn't yet been used to its full potential. Selection and hybridization in segregating populations enable the utilization of available diversity to generate new, superior cultivars. Nandini et al. (2019) remarked that understanding floral biology and having a candid, practical hybridization method and an appropriate gene marker to identify real F_1 are prerequisites for artificial hybridization.

2 *Eleusine coracana* (L.) Gaertn. (Finger Millet/Ragi)

It's a millet crop grown in Asia and Africa. Being a hardy crop, it has wide adaptability to adverse environmental conditions, and cultural conditions make it a potential food crop (Seetharam 1998). Ragi is a nutraceutical-valued annual herbaceous cereal crop. Expanding as a primary crop, especially in Asia and Africa, this ancient crop is extremely significant historically, culturally, and nutritionally. It is a self-pollinated C4 allotetraploid ($2n = 4x = 36$, AABB) (Sood et al. 2023a) crop, specifically in famine-prone areas and to support developing nations (Gupta et al. 2017; Mgonja et al. 2007). The ragi grain is extremely resistant to insect infestations and is long-lasting when stored.

3 Finger Millet Floral Morphology and Its Anthesis

Ragi plants have multiple fingers, or inflorescences, each finger having several spikelets, which in turn have five florets with three anthers encircling the stigma of each floret. Two large barren leaves that are each encased in two palea, or scales, cover the florets. The lower blooming glumes, called lemma, have a tiny appendage on the axial side that contains the florets. Little scaly two lodicules are located close

Fig. 1 Floral biology of finger millet: (**A**) ear head/inflorescence, (**B**) individual spikelet, (**C**) outer glume, (**D**) ovary, (**E**) Lemma, (**F**) palea, (**G**) spikelet at maturity, (**H**) Lemma and palea covering the grain, (**I**) Lemma and palea covering matured grain. (Source: Gupta et al. 2011)

to the base of the ovary (Dodake and Dhonukshe 1998). Gynoecium is bicarpellary and unilocular and has two plumose stigmatized styles in the superior ovary. Stigma is nearly encircled by the androecium and makes sure the self-pollination. Filament measures between 0.48 and 0.85 mm in length, and the anthers are bigger, when compared to filaments (Dodake and Dhonukshe 1998; Emrey 2022).

One floret per day opens from the bottom to the top of a spikelet in an inflorescence, whereas an individual spike open from top to bottom occurs in a spikelet. From third day, following the start of flowering, more number of flowers opens. It takes 5–7 days for blooming to complete. From 1:00 a.m. to 5:00 a.m. is when the anthesis happens (Gowda 1997). Almost simultaneously, both stigmas and anthers appear simultaneously when the lemma and palea open. Prior to the flowers opening, the anthers undergo a longitudinal dehiscence (Seetharam et al. 2003). During the dehiscence stage, the wet stigma and anthers reached identical height within the flower. Through their dehiscence, the anthers fertilize their own stigmatic surfaces. Pollen has a 20-minute viable lifespan; in contrast stigma has 5-h receptive period (Dodake and Dhonukshe 1998). Consequently, estimated spontaneous outbreeding in ragi does not exceed 1% (Fig. 1).

4 Hybridization Techniques Used in Finger Millet

4.1 Contact Method

It's a simplest crossing method, according to Ayyangar (1934). Hybridization includes identification of right male and female panicles for crossing, followed by male and female fingers tied together by intertwining the fingers. Butter paper bags are used to cover the knotted panicles to prevent pollination by outside pollen. One of the most widely used emasculation and pollination techniques is contact approach. It has a 2–3% success rate in obtaining true hybrids. Genotypes with dominant characteristics, such as pigment on nodes and fingers, have been utilized as male parents in the contact technique (Gupta 2006). This helps to identify real hybrid in the F_1 generation. This procedure recovers very few hybrid seeds; all female panicle seeds must be raised in order to identify the hybrids in the field. With this strategy, growing all the seeds for the purpose of identifying a small number of hybrid plants takes more time, space, labor, and resources. Thus, the hot water emasculation method (Raj et al. 1964) is an improved option for hybridization in finger millet. To circumvent the difficulty of physically extracting the anthers from the florets, many researchers have employed a hot water treatment for emasculation (Keller 1952).

4.2 The Pollination and Emasculation Techniques Using Hot Water Treatment

At the right stage, the female panicles are immersed in 48–52 °C warm water for 5 minutes during the process of hot water emasculation procedure. Female air-dried panicles are tied to designated male panicle in a way similar to contact method. As a result of this technique, in female panicle few seed set can be observed and majority of which are true hybrids. This results in a reduction in resources as well as time needed to evaluate an extensive number of $F_1{}^s$. This method's drawback, if raising the hot water's temperature, has the deleterious effect of drying up of the ears (Fig. 2).

In finger millet it's observed that flower opening is induced by temperature and surrounded humidity. The panicle is covered at the proper stage using a 7.5 cm × 10 cm polythene bag lined with humid filter paper and stuffed with absorbing cotton wool. The anthers emerge from the florets without spilling pollen because of the extreme humidity. By tapping the bag, pollen grains from male genotype are collected and then dusted on the female panicle that has been emasculated (Nagaraja et al. 2022).

In finger millet, there have also been reports on genetic male sterility (GMS). One of the best ways to induce male sterility is to apply gametocide gibberellic acid (GA3) at 1600 ppm at the fifth leaf stage. The GMS line, called "INFM 95001," was created in collaboration with the Division of Agronomy at the University of Nebraska in the United States and ICRISAT in Matapos, Zimbabwe, and Kano, Nigeria. However, a maintenance issue with the GMS prevented it from being used for hybridization in finger millet. In the genetic background of GPU 28, identified partial GMS lines (INGR14015PS; 1-IC0598201) limit its application because of significant fluctuations in fertility and sterility rates indifferent genetic backgrounds and locations (Gowda et al. 2014).

Fig. 2 Treating with hot water for effective emasculation and pollination in finger millet

5 Utilization of *Partial Male Sterile 1* (PS 1) Mutant in Hybridization

Partial male sterile mutant PS 1 (Gowda et al. 2014), identified and developed at PC Unit, Small Millets, Bengaluru, has made it very easy to maintain sterility through selfing or growing in isolation. This mutant sets about 10% of seeds upon selfing and about 20% during open pollination (Annual Report 2012–2013). When PS 1's seed set was assessed over several seasons, it was discovered to be consistent. More options for hybridized breeding in finger millet are provided by PS 1. Crossed seeds formed from PS 1 are a mixture of self-fertilized and cross-fertilized seeds since PS 1 is partially sterile (Manjappa et al. 2022).

The hybrids produced with a total of 46 genotype (28 improved cultivars from various Indian states and 18 superior cultivars of African and Asian nations) were evaluated over the seasons for PS 1 heterosis level (Manjappa 2017). An analysis was conducted on the correlation among heterosis and parental divergence using 16 SSR markers and 18 morphological characteristics. Considering an average of 40.1%, the percentage of seeds set in crosses originating from PS1 varied from 27.4% to 49.4% (Manjappa et al. 2022). According to a pollen germination study, only a small percentage of PS1 grains of pollen had the ability to grow their pollen tube to the ovule and set seeds, and the bulk of pollens were unable to germinate on their stigma. Some pollen also showed defective pollen tube expansion in the finger millet. This PS1 line can be used for crossing with effectiveness.

Fig. 3 Partial sterile line. Virescent mutant line

6 Utilization of *Virescence 1 (v1)* Mutant in Hybridization

PC Unit, Small Millets, Bengaluru, had isolated a virescent mutant "*v1*" (Acc. No. GE 1), which expresses yellowish leaf color at seedling stage and later turns to green color during growing stages (Manjappa et al. 2019). Identification of F_1s developed from PS 1 requires more land resources because all the plants (including both sterile and fertile plants) have to be grown till seed set stage. Identification of stable visual marker expressed at seedling stage would be an appropriate solution for F_1 identification at seedling stage itself; this will restrict breeder to handle only few breeding populations. True F_1 identification under the nursery itself is benefited by the development of partial male sterile lines with early seedling markers.

Virescent mutant has similar performance as the control genotype for many agronomic traits except delayed flowering and reduced plant height. Mutant expresses its virescent yellow leaf color at first leaf stage to pre-anthesis stage; however, it expresses complete yellow leaf color up to tillering and at later stages begins green leaf reversion. Leaf chlorophyll content in v1 was half of the control genotype, and during green reversion stage, it started increasing. Virescent expression at seedling stage is environmentally insensitive, whereas green reversion is environmentally influenced (Manjappa et al. 2022) (Fig. 3).

7 Foxtail Millet (*Setaria italica*)

Italian millet (*Setaria italica* (L.) P. Beauv) is another name for foxtail millet. It is among the oldest crops grown for grazing, hay, and food grain purpose. The seeds are good source of minerals (3.3%) and protein (12.3%). The millet's panicles resemble the tail of a fox; that is why this millet is called foxtail millet. In the central Chinese highlands, foxtail millet was thought to have been domesticated (Yang et al. 2012). This crop is diploid and self-pollinating and has $2n = 18$ chromosomes.

Fig. 4 Floral biology of foxtail millet: (**a**) spike/inflorescence, (**b**) clusters of spikelets, (**c**) spikelet which is subtended, (**d**) opened spikelet, (**e**) outer glume, (**f**) Lemma and palea enclosed grain, (**g**) individual grain. (Source: Gupta et al. 2011)

8 Foxtail Millet Floral Morphology and Anthesis

A main stalk with short side branches covered in bristles and spikes covers the foxtail millet flower. The terminal spike is known as inflorescence. There are two glumes on each spikelet, which enfold 2-min flowers. According to Nirmalakumari and Vetriventhan (2010), the upper spikelet is bisexual and has two long feathery styles, a long oval and smooth ovary, and three stamens. The bottom spikelet is sterile and possesses yellow/white-colored anther and ovary surrounded by two long style and lodicules with feathery stigma (Jayaraman et al. 1997). Spikelet opens from upward to downward direction on main axis of the spike (Sundararaj and Thulasidas 1976), and 8–16 days are needed for complete flowering. It takes 80 min for a single floret to fully open, and it stays open for around 30 min (Malm and Rachie 1971). According to Siles et al. (2001), in foxtail millet anthesis typically occurs shortly after midnight and in the early hours between 8 and 10 a.m., although it can vary depending on environmental conditions. An ear head needs 10–15 days to finish complete flowering. From sixth day of emergence, the highest number of florets opens (Sundararaj and Thulasidas 1976). The two main factors affecting pollination are humidity and temperature (Fig. 4).

9 Methods of Hybridization in Foxtail Millet

Earlier, pure line selection was primarily utilized for foxtail millet improvement because of its predominant nature of self-pollination. Development of new hybridization technologies led to the utilization of pedigree method for cultivar improvement. Considerable amount of heterosis observed in foxtail millet for grain yield and midparent heterosis shows nearly 60% higher yield in F_1 compared to respective parental genotypes (Siles et al. 2004). Hybridization occurs through contact mode of crossing; however, the likelihood of obtaining true hybrids is minimal. A hot water-treated emasculation method was later adopted and extensively utilized for hybridization. To identify true-to-type F_1, a dominating marker character was typically utilized. Rapid generation advancement scheme and advancement of selection were widely followed for foxtail improvement (Jijau 1989).

Methods for achieving artificial hybridization in foxtail millet were detailed by Siles et al. (2001). This process involves emasculation and then hybridization. The florets opened on the first day is removed with a fine forceps during early hours (8–9 a.m.) of the morning; then on the second day of anthesis, female parent is isolated from other plants and flowers. The male and female parents exercise their bristles with scissors as their anthers period comes closer. Emasculation is performed as soon as the first anther appears and before the pollen sac bursts, and anthers are then extracted with a fine pair of forceps. After labeling each emasculated spikelet, the unlabeled spikelets are removed, and the entire panicles are covered with a glassine bag. Immediately pollination effected by placing the emasculated female spike slightly below the male spike which sheds the pollen later cover both the spikes with a glassine bag. After pollination; during the third day, take out the male panicle carefully, and inspect each female panicle for any possible later-developing florets. This crossing technique yields successful results with an average seed set of 75%, more than 90% true hybrid seed.

It can be challenging to complete emasculation in the short window of time between the flower's blooming and pollen shed in the pollination and emasculation procedure. A single flower's opening and closing times can vary based on the accession and surrounding conditions. In *S. italica*, anthers must be removed rapidly and carefully because they shed pollen as they emerge from the floret. Furthermore, the number of crossings that can be completed by one individual each day is restricted because pollinations take place right after emasculation.

Another technique that has the benefit of allowing for the cross-pollination of a huge quantity of mature panicles is the immersion of mature panicles in hot water to heat-kill maturing grains of pollen (Darmency and Pernes 1985; Chang 1958; Wang et al. 1998; Miyaji and Samura 1954). However, there are substantial differences observed across trials on the heat treatment's temperature and duration (e.g., 47 °C for 10 min and 42 °C for 20 min). Moreover, no systematic publications are available in foxtail millet on evaluations of the effectiveness of hot water-treated emasculation. Therefore, the development of genetic resources and the adoption of *S. viridis* as an ideal system in the scientific community would be considerably accelerated by the adoption of a standardized and straightforward procedure for carrying out genetic crosses in *S. viridis*.

In 2013, Hui Jiang and colleagues devised an optimal standard crossing methodology for *S. viridis*. In this technique plants are grown in controlled environmental conditions in order to obtain synchronized flowering and increase reproducibility. To help identify successful controlled genetic crosses, transgenic *S. viridis* lines with reporter GUS gene is used as male parental line in cross experiments. The GUS transgene facilitates the rapid scoring of F_1 seed subsequent to harvesting, hence offering a simple qualitative assay for assessing pollination and emasculation effectiveness. In an effort to eliminate seeds from newly grown flowers following emasculation, hot water treatment at 48 °C for 3–6 min was followed. The flowers that have been emasculated are also labeled. Other Setaria accessions were also crossed using this technique.

10 The Use of Male Sterile Lines

Male sterile lines, viz., recessive and dominant-based genic male sterile lines, photoperiod-sensitive nuclear lines, male sterile lines based on gene interaction, cytoplasmic-based male sterile lines, and cytoplasmic genetic male sterility lines, were also developed in China for the purpose of exploiting heterosis (Cui et al. 1979; Hao et al. 2009; Zhao et al. 1996; Wang et al. 1993). Partial genetic male sterile line (PAGMS) is employed successfully in hybrid production even after several male sterile lines were developed (Diao and Jia 2017). In China higher yields are largely attributed due to the use of genetic male sterile lines to take advantage of heterosis. Jigu 16 and Suanxi 28 × Zhangnong 10 are hybrid cultivars of foxtail millets that are derived from genetic male sterility (Cui et al. 1979). The first foxtail millet PTGMS line at Baxia Institute of Agricultural Sciences, "292" (Cai 5 Ce35-1), produced significant male fertility when treated with a short photoperiod (13 h or less) and rendered male sterility when treated with a long photoperiod (14 h or more) (Zhao et al. 1996). Treating with warm water is advised to create sterility in foxtail millets male lines, according to research by Nugroho et al. (2020). This is because the treatment kept the stigma receptive efficiently destroyed pollen and induced sterile rachis in the upper panicle and pollination sections. Therefore, we may conclude that further efforts are required to take advantage of male sterility in the direction to encourage the hybrid progenies, which will help foxtail millet overall crop improvement through new segregants and genetic recombination.

11 Proso Millet (*Panicum miliaceum* L.)

Panicum miliaceum L., or proso millet, is a cereal that is extensively grown in Southeast Asian nations such as Sri Lanka, Pakistan, Nepal, India, and Western Burma. Proso millet with a diploid chromosomal number of $2n = 36$, self-pollinating in nature, while outcrossing may occur in excess of 10% (Popov 1946). In affluent nations, it is fed to birds, used as animal feed, and considered as food in few regions in Asia (Rajput et al. 2014). In terms of nutrition, grains are on par with or even better than the major cereals, and amount of grain protein is high (Gomeshe 2017). It is a C4 plant that is extremely tolerant to heat and drought and is a short-life cycle millet and demands less water for cultivation. Crop is efficient to produce a reasonable crop yield even in damaged soils, under adverse weather circumstances, highly suitable for extreme profiles of soil and adverse climatical situations (Changmei and Dorothy 2014).

12 Floral Biology of Proso Millet

Inflorescence of proso millet is drooping-type panicle that bears several spikelets which are either open-type or compact in nature (Hulse et al. 1980). There are two lemmas and two glumes in each spikelet. The glume's length varies; the length of inner glumes is the same as the spikelet length, while the outside glumes are shorter. There is one floret in per lemma. The upper region lemma is shorter and more fertile compared to lower lemma, while the floret of the lower lemma lacks stamen and sterile in nature. The palea of the lower lemma is sterile, and length is also reduced in comparison to the palea of the upper lemma, which is fertile (Seetharam et al. 2003). Individual floret consists of three anthers, which are tan, dark brown, blackish, or amber in color. The shape of the ovary is bifid and possesses plumose stigmas (Nanda and Agrawal 2008).

The panicle's blooming progresses from top to bottom (Sundararaj and Thulasidas 1976). During 10:00 a.m. to 12:00 noon, anthesis occurs (Jayaraman et al. 1997), and it requires 12–15 days to complete the anthesis of first and last floret. In general synchronism is observed between stigma receptivity and pollen shedding. It was noted that the anthers were sticky, and the pollen did not shed, while the florets were open (Nelson 1984). The anthers were dried out and started to release pollen within a minute after the florets opened. For 10–15 min, the florets stay open (Zhuravel 1956). Bright sunshine, low humidity, and high temperatures all encourage flowering. Cloudy days reduce the amount of flowering. By warming a panicle with a lens, it can be stimulated (Fig. 5).

13 Little Millet (*Panicum sumatrense*)

It is one among very few millets that are grown in limited quantities throughout India. Although little and proso millet are related, little millet's seeds are substantially smaller. This crop is tetraploid and self-pollinating and has chromosomal number $2n = 36$. Cultivation of little millet demands very less water and highly suitable for subtropical and tropical regions. Little millet is very well-known for its drought resistance (Doggett 1989). It has a high protein, carbohydrate, vitamin, and mineral content. The protein is enriched with lysine, cystine, and methionine and has a balanced profile of amino acids.

14 Anthesis and Floral Morphology of Little Millet

The panicle-shaped, constricted or thyrsiform inflorescence measures 15–45 cm in length and 1.0–5.0 cm in width (Seetharam et al. 2003). Spikelet length ranges from 2 to 3.5 mm, and it is persistent (Bor 1960). At the time of maturity, panicle branches droop and become scabrous. At the terminal of the branches, spikelets are solitary and generate on uneven pedicels. Every spikelet has 2-min florets in it. The upper one lacks rachilla extension and is either bisexual or fertile, while the bottom is sterile in nature. Staminate floret is encircled by palea and lemma I, whereas fertile floret is encircled by palea and lemma II (Sundararaj and Thulasidas 1976). The shape of the florets are compressed dorsally, elliptical and acute in appearance. It consists of three anthers and roughly measures about 1.5 mm length. The lower glume is oblong (0.7 to 1.2-mm-long) and

Fig. 5 Floral biology of proso millet: (**A**) spike/inflorescence, (**B**) opened floret, (**C**) outer glume, (**D**) inner glume, (**E**) inner lemma, (**F**) palea, (**G**) inner glume, (**H**) outer glume, (**I**) upper lemma, (**J**) pollen grain, (**K**) lemma and palea covered the grain, (**L**) individual grain (Source: Gupta et al. 2011)

membrane-like, keels are absent, and has one to three veins. Compared to the fertile lemma, the florets' apex reaching the glume is thinner (Fig. 6). Lower glumes are absent in lateral vein, and its tip is sharp. In addition to being ovate and keelless, lower glume is shorter compared to the upper glume, and 11–15 veins are seen (Nanda and Agrawal 2008).

Starting from second or third day following the panicle's appearance, the spikelets start to open. The flowering takes place from upward to downward direction of the panicle. On the sixth or seventh day, a highest number of flowers were opened, and nearly 15 days is required for complete flowering (Sundararaj and Thulasidas 1976). Anthesis happens between 9.30 and 10.30 a.m. (Jayaraman et al. 1997). According to Seetharam et al. (2003), autogamy is prevailing, and the glumes open for a short period. The anthesis process goes fairly quickly; it takes 2–5 min to complete.

15 Crossing Techniques Employed in Proso Millet and Little Millet

1. Contact method.
2. Pollination after manual emasculation.
3. Treatment with hot water.
4. USSR technique.
5. SMUASB method.

Fig. 6 Floral biologyof little millet: (**A**) panicle/inflorescence, (**B**) individual floret, (**C**) side view of individual floret, (**D**) opened spikelet, (**E**) outer view of glume, (**F**) first lemma, (**G**) sterile spikelet, (**H**) fertile spikelet, (**I**) upper glum, (**J**) palea and lemma enclosed grain (Source: Gupta et al. 2011)

15.1 Contact Method

The crossing block for the pollination will be prepared in advance by planting the selected desired male and female parents close to each other. The panicles of the chosen male and female remain tied loosely or freely. Both the parental panicles will be separated after pollination and fertilization; and care should be taken while choosing the male parent which possesses the phenotypic markers to support in distinguishing the true-to-type F_1 (http://agritech.tnau.ac.in). In self-pollinated crops, it is the most effective crossing technique; though there is very little chance of getting a true-to-type F_1; only 2–3% of the true F_1 is visible. Further, large F_1 generation needs to be evaluated to find the true-to-type hybrids which demands high resource for F_1 screening.

15.2 Pollination After Hand Emasculation

This process involved choosing panicle that will be opened in the following day. Evening hours of the day emasculation are carried out followed by pollination during the subsequent morning. The emasculated female flower is pollinated by the male flowers that will open on that date in order to be pollinated. After that, a butter paper bag is used to cover them and tie them together. In 2–5 days, cross-pollination occurs naturally inside the butter paper bag. To identify the true-to-type F1 hybrid, genotypic marker will be utilized (http://agritech.tnau.ac.in).

For crops that self-pollinate, this is the conventional technique predominantly used for crossing. The primary problem with this strategy is that lemma and palea are extremely tight; these damage the ovary and limit on seed set when tried to open the floret before the natural anthesis. Furthermore, the amount of time between the anthesis of the first and last florets is quite long. Because of this, it is challenging to ascertain right time for emasculation before the initiation of anthesis naturally (Jasovskij 1960).

15.3 Treatment With Hot Water

Many researchers exploited treatment with hot water for emasculation to minimize the difficult task of physical exertion of anthers from individual floret (Keller 1952). Ear head expected to flower in next 2 or 3 days are selected, and they are dipped in 52^0 C warm water for 2 min; this ideal temperature and time was concluded based on the hybrid seed set percentage (http://agritech.tnau.ac.in). In 1972, Srivastav and Yadav reported little millet panicles effectively emasculated with hot water treatment for 8–10 min at 49 °C or 5 min for 50 °C. In a comparable manner, Primak and Jakovlev (1964) found that treating proso millet panicle in 50 °C hot water for 5 min improves seed set. After female parent emasculation, probable next day opening male panicle is tied with emasculated panicle and covered with butter paper bag.

The major limitation of this technique is it requires precise equipment to maintain the right temperature. Otherwise, stigma will be affected by higher/lower temperature, which could lead to a comparatively low seed set rate.

15.4 USSR Method

In order to address the issue with pollen removal in manual emasculation and treatment with hot water, this modified crossing approach was suggested (Seetharam et al. 1986).

15.4.1 Induced Flower Opening (USSR Method)

New lines and genotypes developed successfully using the USSR approach; the specific details are given below:

1. Gently hand-massage the panicle to stimulate the flowers mechanically to open.
2. Well ahead of actual flowering, flowers open 2–3 min in advance.
3. Dip in ambient temperature water to prevent the anther from bursting.
4. Using the forefingers, thrash anthers from the opened florets.
5. Keep the flowers that have opened and clip off the unopened florets.

15.4.2 Pollination Method

For the pollination, the desired male panicle is placed above the emasculated female panicle, followed by both male and female spikes covered in glassine bag to ensure effective pollination. For 2 days, panicles were shaking together during regular anthesis time. On the third day after pollination, every male panicle was taken out with care and female panicle inspected closely for any possible later-developing florets. When such florets are not entirely removed, they frequently continue to grow and develop into seed that could be mistaken with crossed seed. Re-bagging and maintenance were taken care for the female spike until it's harvested.

The drawback of this method is that mechanically massaging the panicles to encourage the florets to open and removing the florets' anthers with the forefinger may cause damage to the stigma.

15.5 SMUASB Method (Modified Crossing Method)

This method of modified crossing is also called as SMUASB method; this method of crossing in little and proso millet which is developed at PC Unit, Small Millets, UAS, Bengaluru, during 2014. Lemma and palea in little millet and proso millet are extremely tight in nature, and any efforts to open in advance may prevent it from seed set due to floral damage. In the SMUASB method, as a mechanical stimulator, ice-cold water is used for floret opening.

16 Emasculation

The florets of little millet and proso millet begin to open from top to bottom in the spike. To facilitate crossing, the desired male and female parents are sown in alternative rows at 8–9 a.m. is ideal for effective crossing morning.

Details about emasculation are listed below:

- Effective emasculation required to select the female panicle in which the first floret is opened.
- The female spike is sprayed with cold water (5 and 8 °C) that aid in opening the female floret naturally an hour before the actual flower opening.
- Once all of the florets have opened spontaneously, emasculation is carried out by putting the spike in cold water and wash with cool water to remove all of the anthers.
- The non-opened flowers were clipped off which include underdeveloped florets nearbase and fertilized florets at the top (Fig. 7).

Fig. 7 In proso and little millets, the modified crossing (PCUASB) approach is used (source: Nandini et al. 2019)

17 Pollination

- The selection of the male parent ensures that the first floret has opened.
- Similar to the female panicle, the male panicle is sprayed with cold water (5–8 °C) to trigger flower opening and to maintain the anthers wetness.
- The male spikes connect/tied loosely around the female parent panicle as soon as they open to facilitate adequate pollination and aeration.
- To maintain the anthers and stigma moist, water is then gushed on tied panicles.
- To prevent cross-pollination, butter cover is placed over tied female and male spikes.

Little amount of cross-pollination, i.e., 3–5%, may be observed in these crops, since they are predominantly self-pollinated. To identify and harvest true F_1 seeds from the female parent, label the panicles of the crossed parent, and mention parental combination and crossing date.

18 Barnyard Millet (*Echinochloa frumentacea*)

This is a member of the *Echinochloa* genus, which includes two cultivated species that are primarily farmed for food and fodder: *Echinochloa frumentacea* (Indian barnyard millet) and *Echinochloa esculenta* (Japanese barnyard millet) (Sood et al. 2023b). Barnyard millet is autogamous, hexaploid, with a chromosome number of 54 and belongs to Poaceae family. The crop is ready for harvest in 80–90 days, which is why it's known as fastest-growing millet (Sood et al. 2020). Typically, this cultivated in China, Japan, India, and Korea for human and animal consumption (Upadhyaya et al. 2014). This millet is highly treasured for its fast growth, exceptional nutritional content, and resistance to drought (Dwivedi et al. 2012; Saleh et al. 2013), due to its increasing significance barnyard millet taking forefront as a food and feed.

19 Floral Biology and Anthesis

Barnyard millet has upright terminal panicles that rarely droop. Recemes are few to abundant, tightly packed at top, and arranged into four irregular rows onto the triquetrous rachis.

Oblong to elliptical florets, with awnless lemma but sharp pointy tips, subsessile and arranged on rough, short pedicels with two glumes supporting them (De Wet et al. 1983). About one-third of the spikelet's length is made up of the lower glume, while the top glume is a little bit shorter. The lower lemma and glumes appear pubescent. The lower floret has a lemma and a tiny palea, which is sterile, whereas it contains bisexual upper florets (Gupta et al. 2010). The fertile lemma is plano-convex, elliptic, smooth, shiny, abruptly sharp, pointy, or cuspidate, with edges enrolling below over the palea but not enclosing the palea's tip. The sterile lemma has five veins. According to Napper (1965), the palea is flat and has a surface roughness similar to a fertile lemma. There are three stamens, the superior ovary has two distinct styles and has plumose stigma (Sundararaj and Thulasidas 1976). The grain is securely encased with hardened white palea and lemma.

First, the upper receme of the flower opens, and then the inflorescence blooms from the top to downward. It takes 10–14 days for the panicle to fully open, and an additional 10–15 days for complete flowering. The majority of the florets open between 6–8 days after flower initiation (Sundararaj and Thulasidas 1976). Flowering starts from 5 am to 10 am, and peak flowering occurs between 6 a.m. to 7 a.m. (Sundararaj and Thulasidas 1976; Jayaraman et al. 1997). Within a solitary raceme, blooming begins at both peripheral ends and then progresses to the raceme's center. The stigma branches spread, and the flower opens prior to the anther's dehiscence (Seetharam et al. 2003). The flower closes in 30 min (Fig. 8).

Fig. 8 Floral biology of barnyard millet; (**a**) inflorescence/spike (**b**) Spikelet arrangement on raceme, (**c**) individual raceme, (**d**) individual floret, (**e**) lemma(lower) (**f**), glume(upper), (**g**) Cross section of fertile lemma enclosing grain; (**h**) Spikelet view from lower glume region (**i**) Lower glume (**j**) lemma which is fertile (**k**) Individual grain (**l**) Lemma and Palea enclosed grain (Source: Gupta et al. 2011)

20 Hybridization Techniques in Barnyard Millet

The method of contact crossing is frequently used, additionally treating with hot water (48 °C) for 4–5 min has been shown to be helpful (Gupta et al. 2011).

20.1 Kodo Millet (*Paspalum scrobiculatum* L.)

Kodo millet originated in India and believed that domesticated nearly three thousand year ago (House et al. 1995). Paspalum is the genus with a chromosome number of $2n = 4x = 40$ (Hiremath and Dandin 1975). Compared to other millets, it is well-known for its severe drought tolerance and produces high yield in a short duration thus is of great economic value. Kodo millet has the highest free radical quenching potential, suggests that it may have some beneficial antioxidant activity (Hegde and Chandra 2005).

21 Anthesis and Floral Morphology of Kodo Millet

Inflorescence of kodo millet consists of two to six recemes that are extensively scattered along a short axis or subdigitate (Seetharam et al. 2003). Typically, the florets are sessile or rest on pedicel. On a compressed rachis, they often solitary and organized in two rows (Ramakrishna et al. 2002). Few florets are paired in the receme middle. According to Nanda and Agrawal (2008), the spikelets are arranged in two series that alternate between being long and short pedicelled. Glume II is equal to the spikelet's length and glume I is absent. Lemma II encloses the two florets, whereas lemma I is nearly identical to glume II. The spikelet's lower floret is reduced to a valve and is sterile, whereas the upper floret has a hermaphrodite flower (Sundararaj and Thulasidas 1976). Hard, horny, persistent husks enclose the grain (Seetharam et al. 2003).

Kodo millet shows cleistogamous nature of flower opening (Yadava 1997), with a flower opening percentage of not greater than 15–20%; therefore, self-pollination is the normal. Early in the morning, spikelet's open (Jayaraman et al. 1997) between 2.30 and 6 am, spikelet opening occurs (Jayaraman et al. 1997). It is more difficult to force the flower open artificially when the lemma is highly intact, which damages the flower (Fig. 9).

There is limited recombination breeding due of the cleistogamous flowering pattern. However, many improved varieties are available for commercial cultivation. The majority of these were developed from landrace or superior germplasm accessions selections. To create cultivars with higher yields, pure line selection is

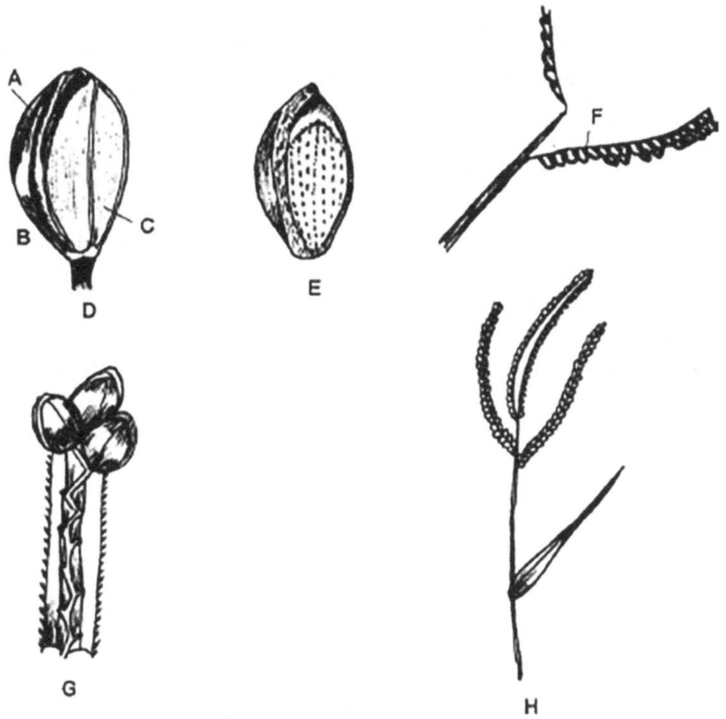

Fig. 9 Floral biology of kodo millet: (**A**) upper floret, (**B**) second glume, (**C**) lemma of lower floret, (**D**) spikelet, (**E**) floret, (**F**) rachis; (**G**) Arrangement of rachis in spikelet (**H**) Inflorescence (Source; **Nanda JS, Agrawal** PK (2008) *Botany of Field crops* (vol 1), Kalyani Publishers, India, 381 pp.)

generally used. APK 1, KMV 20, and GK 2 are three varieties that were obtained by selection using germplasm that was introduced into various agroecosystems. Furthermore, a significant amount of induced mutagenesis is also used to create variety in kodo millet (Yadava 1997).

21.1 Browntop Millet (*Brachiaria ramosa* (L)

Among all small millet crops, browntop millet fetches higher price in the market. Unique characteristic nature of browntop millets are drought tolerance and resistance to climate change and contain high levels of proteins, carbohydrates, vitamins and minerals. Browntop millet locally called as korale in kannada and ukorralu in telgu. It is adaptable to many cropping systems and cultivated in varied soil profiles and temperatures. According to Basappa et al. (1987), browntop millet is a tetraploid and has four basic chromosome numbers; because of its propensity to thrive as a mimic weed in tiny millet fields, it is also known as the "illegal wife of little millet."

22 Floral Morphology, Anthesis, and Hybridization Techniques in Browntop Millet

The inflorescence is a branching panicle with two types, viz., open type and compact type. With three anthers and thick petals, the plants are bisexual. A noticeable vein can be seen in the center of the leaf in certain species. Blooming takes place from the panicle's top to bottom. Most flowers open on the fourth or fifth day of flower initiation. Research on anthesis and floral biology in this crop has been minimal.

In browntop millet, crossing is done using contact method; however, the success rate is low. There are no known attempts to cross browntop millet using any crossing procedures. Although there is a lower success rate in creating true hybrids, in browntop millet SMUASB approach is employed for hybridization program (Figs. 10, 11 and 12).

Fig. 10 Open-type panicle

Fig. 11 Closed-type panicle

Fig. 12 Blooming in browntop millet

23 Conclusion

Future food and fodder demands will place a high value on small millet crops. It is necessary to create high-yielding genotypes that are highly resistant to pests and diseases. In the coming days, these crops' productivity has to be raised. In this situation, hybridization is essential to develop new varieties, make use of the genetic resources at hand, and create variability. Due to their high degree of self-pollination, these crops are challenging to hybridize. Even so, only a small number of artificial hybridization methods have proven effective. This chapter compiles all the crossing methods that are currently available for the various millet crops. Breeders of small millet will find it helpful to apply various hybridization strategies for various small millet crops.

References

Amit UP, Joshi D, Mishra KK, Lakshmi VK, Anil K (2021) Millets for life: a brief introduction. In: Kumar A, Tripathi KM, Joshi D, Kumar V (eds) Millets and millet technology. Springer Nature, Singapore, pp 1–32

Anonymous (2022) National conference on kharif campaign, 19th April 2022 (Ministry of Agriculture and farmers welfare)

Ayyangar GNR (1934) Recent work on the genetics of millets in India. Madras Agric J 22(1):16–26

Bandyopadhyay T, Muthamilarasan M, Prasad M (2017) Millets for next generation climate-smart agriculture. Front Plant Sci 8:1266. https://doi.org/10.3389/fpls.2017.01266

Basappa GP, Muniyamma M, Chinnappa CC (1987) An investigation of chromosome numbers in the genus Brachiaria (Poaceae: Paniceae) in relation to morphology and taxonomy. Can J Bot 65:2297–2309

Bor NL (1960) The grasses of Burma, Ceylon, India and Paksitan (excluding Bambuseae) (Vol I). Pergamon Press, London, UK, p 767

Chang LP (1958) Studies on flowering and hybridization technique in Setaria. Nungyehhsueh Pao (Jour Agric) Act Agric Sin 9:68–76

Changmei S, Dorothy J (2014) Millet-the frugal grain. Int J Sci Res Rev 3:75–90
Cui WS, Du G, Zhao ZH (1979) The selection and utilization of male sterile line/Shanxi 280 of foxtail millet. Sci Agric Sin 12:43–46. (in Chinese)
Darmency H, Pernes J (1985) Use of wild Setaria viridis (L.) Beauv to improve triazine resistance in cultivated *S. italica* (L.) by hybridization. Weed Res 25:175–179
De Wet JMJ, Rao KEP, Mangesha MH, Brink DE (1983) Domestication of sawa millet (*Echinochloa colona*). Econ Bot 37:283–291
Diao X, Jia G (2017) Origin and domestication of foxtail millet. In: Genetics and genomics of Setaria https://doi.org/10.1007/978-3-319-45105-3_4
Dodake SS, Dhonukshe BL (1998) Variability in floral structure and floral biology of finger millet (*Eleusine coracana* (L.) Gaertn.). Indian J Genet 58:107–112
Doggett H (1989) Small millets: a selective overview. In: Seetharam A, Riley K, Harinaryana G (eds) Small millets in global agriculture, pp 3–18
Dwivedi S, Upadhyaya H, Senthilvel S, Hash C, Fukunaga K, Diao (2012) Millets: genetic and genomic resources. Plant Breed Rev 35:247–375
Emrey TM (2022) Investigation of finger millet floral structure and hand emasculation. J Plant Physiol Pathol 10(6):1–3
Gomeshe SS (2017) Proso millet (*Panicum miliaceum* (L.)): genetic improvement and research needs. In: Patil JV (ed) Millets and sorghum. Biology and genetic improvement. John Wiley & Sons, Ltd., pp 150–179
Gowda BTS (1997) Genetic enhancement and breeding strategies in finger millet (*Eleusine coracana* Gaertn.). In: National seminar on small millets, 23–24 April, 1997, Coimbatore, India, p 16–18
Gowda MVC, Pushpalatha N, Jhadav SS, Satish RG, Boronayaka MB, Sujay V, Pramila CK, Manasa KG, Gowda J, Ravishankar P, Krishnappa M, Narasimhamurthy DN (2014) PS-1 (IC0598201; INGR14015), a finger millet (Eleusine coracana) germplasm with partial sterility, useful in hybridization and easy maintenance. Indian J Plant Genet Resour 27:192
Gupta A (2006) Improvement of millets and pseudo-cereals for rainfed agrculture in hill region. In: Gupta HS, Srivastva AK, Bhatt JC (eds) Sustainable production from agricultural watersheds in north west Himalaya. Vivekananda Parvatiya Krishi Anusandhan Sansthan, Almora, Uttaranchal, India, pp 163–174
Gupta A, Mahajan V, Gupta HS (2010) Genetic resources and varietal improvement of small millets for Indian Himalaya. In: Tewari LM, Pangtey YPS, Tewari G (eds) Biodiversity potentials of the Himalaya. Gyanodaya Prakashan, Nainital, India, pp 305–316
Gupta A, Sood S, Agrawal PK, Bhatt JC (2011) Floral biology and pollination system in small millets. Eur J Plant Sci Biotechnol 6(2):80–86
Gupta SM, Arora S, Mirza N, Pande A, Lata C, Puranik S, Kumar J, Kumar A (2017) Finger millet: a "certain" crop for an "uncertain" future and a solution to food insecurity and hidden hunger under stressful environments. Front Plant Sci 8:643. https://doi.org/10.3389/fpls.2017.00643
Hao XF, Wang JZ, Wang GQ, Wang LY, Wang XY (2009) AFLP analysis of photo sensitive male sterile gene in millet. J Shanxi Agric Sci 37:10. (in Chinese)
Hegde PS, Chandra TS (2005) ESR spectroscopic study reveals higher free radical quenching potential in kodo millet (*Paspalum scrobiculatum*) compared to other millets. Food Chem 92(1):177–182
Hiremath SC, Dandin SB (1975) Cytology of *Paspalum scrobiculatum* Linn. Curr Sci 44:20–21
House LR, Osmanzai M, Gomez MI, Monyo ES, Gupta SC (1995) Sorghum and millets: chemistry and technology (St. Paul, Minnesota, USA: American Association of Cereal Chemists, Inc), pp 27–67
Hulse JH, Laing EM, Pearson OE (1980) Sorghum and the millets. Their composition and nutritional value. Academic Press, New York, p 997
Jasovskij IV (1960) An effective method of hybridization for millet (in Russian.) sel. Semenovod 3:69–70. Plant breeding abstr 31: 4712
Jayaraman N, Suresh S, Nirmala A, Ganeshan NM (1997) Genetic enhancement and breeding strategies in small millets. In: National seminar on small millets, 23–24 April, 1997, Coimbatore, India, pp 19–21

Jiang H, Barbier H, Brutnell T (2013) Methods for performing crosses in *Setaria viridis,* a new model system for the grasses. J Vis Exp 80:e50527. https://doi.org/10.3791/50527

Jijau C (1989) Importance and genetic resources of small millets with emphasis on foxtail millet (*Setaria italica*) in China. In: Seetharam A, Riley KW, Harinarayana G (eds) Small millets in global agriculture, 1st edn. Oxford and IBH Publishing Company, Delhi, India, pp 93–100

Keller W (1952) Emasculation and pollination techniques. International Grass land Congress Proceedings 6:1613–1619

Malm RN, Rachie KO (1971) Setaria millets: a review of the world literature. University of Nebraska, Lincoln, p 133

Manjappa G (2017) Heterotic potential of hybrids based on partial male sterile line (PS1) and development of male sterile lines with seedling markers in finger millet (*Eleusine coracana* L. Gaertn.) Ph.D. Thesis (Unpub.), Univ Agric Sci, Bangalore

Manjappa G, Rangaiah S, Gowda MVC (2019) Assessment of heterotic potential of hybrids using a novel partial male sterile mutant (PS1) in finger millet (*Eleusine coracana* L. Gaertn). Mysore J Agric Sci 49(2):266–269

Manjappa, Gowda MVC, Rangaiah S, Pushpa Latha N, Sachin SJ, Sujay V, Vijaykumar KV (2022) Partial male sterility with virescence seedling marker simplifies hybridization in finger millet. In Proceedings of International conference on harnessing the potential of finger millet for achieving food and nutritional security: Challenges and Prospects, pp 2

Mgonja MA, Lenne JM, Manyasa E, Sreenivasaprasad S (2007) Finger millet blast management in East Africa. Creating Opportunities for Improving Production and Utilization of Finger Millet," In Proceedings of the First International Finger Millet Stakeholder Workshop, Projects R8030 & R8445 UK Department for International Development—Crop Protection Programme, (Patancheru: International Crops Research Institute for the Semi-Arid Tropics) 196

Miyaji Y, Samura T (1954) The influence of atmospheric humidity on flowering and pollination in *Setaria italica*. Bull Fac Agric Kagoshima Univ 1–6

Nagaraja TE, Gowda MVC, Prabhakar, Jayaramegowda, Krishnappa M, Ravishankar P, Krishna TV, Nandini C, Sujata Bhat, Ganapathy KN, Hariprasanna K, Saritha HS, Gazala Parveen S, Vilas A Tonapi (2022) An overview of Research on crop improvement in finger millet (*Eleusine coracana*(L) Gaertn.) at Project coordinating unit (small millets). International conference on "Harnessing the potential of Finger millet for achieving food and nutritional security: Challenges and Prospects, Held at VC Farm Mandya, 78–92, ISBN: 978-93-91355-33-3

Nanda JS, Agrawal PK (2008) Botany of field crops (Vol 1). Kalyani Publisher, India, p 381

Nandini C, Bhat S, Srinathareddy, Jayaramegowda, Prabhakar (2019) Modified crossing (SMUASB) method for artificial hybridization in Proso millet (*Panicum miliaceum* L.) and Little millet (*Panicum sumatrense*). Electron J Plant Breed 10(3):1161–1170

Napper DM (1965) Grasses of Tanganyaika: with keys for identification, Bulletin 18, Ministry of Agriculture, Forest and Wildlife, Tanzania, pp 146

Nelson LA (1984) Technique for crossing proso millet. Crop Sci 21:205–206

Nirmalakumari A, Vetriventhan M (2010) Characterization of foxtail millet germplasm collections for yield contributing traits. Electron J Plant Breed 1(2):140–147

Nugroho RB, Suwarno WB, Khumaida N, Ardie SW (2020) Male-sterile induction method in foxtail millet (Setaria italica). Biodivers J Biol Divers 28(9):21. https://doi.org/10.13057/biodiv/d210951

Popov GI (1946) The importance of diversity in millet (in Russian). Agrobiologiya 2:28–43

Primak NN, Jakovlev AG (1964) Thermal emasculation of millet. Agro-biologija (Agrobiology) 613–614

Raj SM, Mahudeswaran K, Shanmugasundaram A (1964) Observation on the hot water technique of emasculation of ragi flowers (Eleusine coracana Gaertn.). Madras Agric J 51:71–75

Rajput SG, Plyler-harveson T, Santra DK (2014) Development and characterization of SSR markers in proso millet based on switch grass genomics. Am J Plant Sci 5:175–186

Ramakrishna BM, Krishnappa M, Seenappa K, Halaswamy BH, Gowda J, Vasanth KR, Somu G, Gowda BTS, Seetharam A (2002) Evaluation of Kodo millet (Paspalum scrobiculatum) germplasm. All India Co-ordinated Small Millets Improvement Project, University of Agricultural Sciences, Bangalore, 39 pp

Saleh ASM, Zhang Q, Chen J, Shen Q (2013) Millet grains: nutritional quality, processing, and potential health benefits. Compr Rev Food Sci Food Saf 12(3):281–295

Saxena R, Vanga SK, Wang J, Orsat V, Raghavan V (2018) Millets for food security in the context of climate change: a review. Sustain For 10:2228. https://doi.org/10.3390/su10072228

Seetharam A (1998) Small millets research: achievement during 1947–97. Indian J Agric Sci 68(8):431–438

Seetharam A, Riley KW, Harinarayana G (1986) Small millets in global agriculture, Chapter-11. In Proceedings of the First International Small Millets Workshop, Bangalore, India

Seetharam A, Gowda J, Halaswamy JH (2003) Small millets. In: Chowdhury SK, Lal SK (eds) Nucleus and breeder seed production manual. Indian Agricultural Research Institute, New Delhi, India, pp 54–67

Siles MM, Baltensperger DD, Nelson LA (2001) Technique for artificial hybridization of foxtail millet (*Setaria italica* (L.) Beauv.). Crop Sci 41:1408–1412

Siles MM, Russell WK, Baltensperger DD, Nelson LA, Johnson B, Vleck LDV, Jensen SG, Hein G (2004) Heterosis for grain yield and other agronomic traits in foxtail millet. Crop Sci 44:1960–1965

Sood S, Joshi DC, Pattanayak A (2020) Breeding advancements in barnyard millet. In: Gosal SS, Wani HS (eds) Accelerated plant breeding, cereal crops, vol 1. Springer Nature, Switzerland, pp 391–410

Sood S, Joshi DC, Rajashekara H, Tiwari A, Bhinda MS, Kumar A, Kant L, Pattanayak A (2023a) Deciphering the genomic regions governing major agronomic traits and blast resistance using genome wide association mapping in finger millet. Gene 854:147115

Sood S, Mondal T, Ramesh SP, Joshi DC, Kant L, Pattanayak A (2023b) Comparison of dehulling efficiency and grain nutritional parameters of two cultivated barnyard millet species (Echinochloa spp.). Heliyon:e21594

Srivastava DP, Yadav A (1972) Emasculation of flowers of Panicum miliare by hot water treatment. Sci Cuk 38:450

Sundararaj DP, Thulasidas G (1976) Botany of field crops. Macmillan Publisher, India, p 509

Tonapi VA, Bhat V, Sangappa S, Rao D (2022) Strategies for enhancement of production and productivity of millets, Finger millet: natures master grain-A Souvenir, International Conference on Harnessing the potential of finger millet for achieving food and nutritional security: challenges and prospects, pp 1–15

Upadhyaya H, Dwivedi S, Singh S (2014) Forming core collections in barnyard, Kodo and little millets using morpho-agronomic descriptors. Crop Sci 54:1–10

Vetriventhan M, Azevedo VCR, Upadhyaya HD, Nirmala Kumari A, Kanepotaka J, Anitha S, Antony Ceasar S, Muthmilarasan M, Bhat V, Hariprasanna K, Bellundagi A, Cheruku D, Bhagyalaxmi C, Santra D, Vannirajan C, Tonapi VA (2020) Genetic and genomic resources, and breeding for accelerating improvement of small millets: current status and future interventions. Nucleus 63:217–239

Wang TY, Du RH, Hao F (1993) Studies and utilization of highly male sterility of summer millet (*Setaria italica*). Sci Agric Sin 26:88. (in Chinese)

Wang ZM, Devos KM, Liu CJ, Wang RQ, Gale MD (1998) Construction of RFLP-based maps of foxtail millet, Setaria italica (L.) P. Beauv Theor Appl Genet 96:31–36

Yadava HS (1997) Retrospect and prospect of kodo millet in Indian agriculture. In: National Seminar on Small Millets, 23–24 April, Coimbatore, India, pp 7–9

Yang XY, Wan ZW, Perry L, Lu H, Wang Q, Zhao C, Li J, Xie F, Yu J, Cui T, Wang T, Li M, Quansheng G (2012) Early millet use in northern China. Proc Natl Acad Sci 109:3726–3730

Zhao ZH, Cui WS, Du G, Yang SQ (1996) The selection of millet photo- (thermo-) sensitive line 821 and study on the relation of sterility to illumination and temperature. Sci Agric Sin 29:23–31. (in Chinese)

Zhuravel BN (1956) On the biology of millet blooming under conditions of Belorussian SSR. (In Russian.) sb. Nauch. Tr. Akad. Nauk. B. SSR Inst. Sotsialistselkh. 4: 249–264

Diseases and Their Management Strategies in Small Millets

Farooq Khan, Gutha Venkata Ramesh,
T. Tharana Poonacha, Anusha M. Nayak, Poojashree M. S.,
Arthi, Prasanna S. Koti, and K. B. Palanna

Abstract

Small millets, encompassing species like finger millet, foxtail millet, little millet, kodo millet, barnyard millet, and proso millet, are resilient, drought-tolerant grains revered for their nutritional richness and adaptability to diverse agroecologies. Cultivated primarily in semiarid regions of Asia and Africa, these grains offer vital sustenance and economic stability to millions. Despite their nutritional significance, these crops have encountered neglect due to the dominance of major staples. However, recent interest in their health benefits and climate-resilient traits has sparked a resurgence in their cultivation, emphasizing their potential to enhance food security and nutrition worldwide. Despite their commendable traits, small millets contribute to only 2% of the world's cereal production. Their genetic potential faces limitations due to various biotic and abiotic stresses, leading to considerable yield and economic losses. Fungal diseases like blast (*Magnaporthe* spp.), leaf spot/blight (*Helminthosporium* spp.), rusts, and smuts are major constraints in cultivating small millets across key growing regions. Addressing these challenges globally demands a deeper under-

F. Khan · T. T. Poonacha · A. M. Nayak · M. S. Poojashree · Arthi ·
Department of Plant Pathology, University of Agricultural Sciences, GKVK,
Bengaluru, Karnataka, India

G. V. Ramesh
Department of Plant Pathology, Punjab Agricultural University, Ludhiana, India

P. S. Koti
Department of Plant Biotechnology, University of Agricultural Sciences, GKVK, Bengaluru,
Karnataka, India

K. B. Palanna
ICAR-AICRP on Small Millets, Project Coordinating Unit,
University of Agricultural Sciences, Bengaluru, Karnataka, India

D. C. Joshi et al. (eds.), *Minor Millets*,
https://doi.org/10.1007/978-981-96-4265-6_6

standing of diseases, their causal organisms, symptomatology, and epidemiology to devise effective management strategies. This book chapter emphasizes prevalent diseases in small millets, detailing symptoms, survival modes, spread, integrated mitigation practices, and novel approaches. Implementing these strategies offers potential to boost millet production and productivity by effectively managing diseases.

Keywords

Small millets · Cultivation · Production · Diseases · Management

1 Introduction

Nutri-cereals, commonly referred to as millets, have been under cultivation for over 10,000 years, thriving in regions characterized by low rainfall and impoverished soil conditions. These high-energy grains are indispensable in diets worldwide. The term "millet," derived from the French word *Mile* denoting thousand, hints at the abundance within these grains—just a handful contains thousands. Seven principal types of millets are globally cultivated: finger millet, pearl millet, foxtail millet, barnyard millet, proso millet, kodo millet, and little millet (Yenagi et al. 2010). Millets have gained recognition as a rich source of essential micronutrients and protein (Malleshi and Klopfenstein, 1998). In response to modern sedentary lifestyles, the pursuit of healthier diets has found its answer in these small grains. Finger millet's remarkable calcium content (>350 mg/100 g) stands as a highlight, while foxtail, barnyard, and proso millets boast over 10% protein content. Furthermore, little millet and foxtail millet contain higher fat content (>4%). These grains are also abundant in crude fiber, ranging from 6.7% to 13.6%. Barnyard and little millets hold substantial iron quantities (9.3–18.6 mg/100 g), surpassing major cereals like rice, wheat, barley, maize, and sorghum (Dwivedi et al. 2012; Kam et al. 2016).

Though millets are considered as the hardy crops, they are also affected by many biotic and abiotic factors due to climate change. Climate change significantly amplifies the prevalence and severity of plant diseases. Rising temperatures, altered precipitation patterns, and extreme weather events create favorable conditions for pathogen proliferation and spread. Shifts in climate alter ecosystems, favoring new disease vectors and disrupting natural balances. Studies like the IPCC's Fifth Assessment Report highlight these concerns, emphasizing the escalating threat to global agriculture. Increased disease pressure threatens crop yields, food security, and biodiversity. Efforts to mitigate climate change and implement adaptive agricultural practices are crucial to curbing this escalating impact on plant health and global food systems.

Small millets are affected by a wide array of pathogens, viz., fungal, bacterial, viral, nematode, and plant phanerogamic parasites. Millets encounter varied diseases, some persistently severe across diverse climates, inflicting substantial economic losses. Conversely, specific diseases sporadically emerge under certain climatic conditions with lesser impact. Fungal infections outweigh bacterial and viral ailments, encompassing critical millet diseases like grain mold, ergot, smut, anthracnose, downy mildew, blast, rust, charcoal rot, foot rot, banded sheath blight, and sheath rot. These maladies target diverse plant parts—roots, stems, leaves, peduncles, or grains—negatively impacting yield and quality. Their significance spans seed production, certifications, and millet marketing, emphasizing the need for vigilant management strategies (Nagaraja et al. 2007 and Patro et al. 2018).

Therefore, this chapter endeavors to gather the existing information, focusing on prevalent diseases found in small millets. It highlights their common occurrence, symptoms, methods of spread, and survival and introduces updated mitigation practices. These practices aim to control diseases, enhance millet production, and elevate nutritional quality. The major diseases of small millets will be grouped based on the symptoms, plant parts affected, and the common pathogen infecting a diverse array of small millets.

2 Fungal Diseases

2.1 Foliar Diseases

2.1.1 Blast Disease

Blast, caused by *Pyricularia* spp., poses a serious and highly destructive threat in major millet-growing regions worldwide. It serves as a significant production constraint, especially in the cultivation of finger and foxtail millet leading to considerable economic losses due to varying degrees of crop damage under natural conditions. It also infects proso and barnyard millets under specific climatic situations. The initial report of finger millet blast in India originated from the Tanjore delta of Tamil Nadu (McRae, 1920). Foxtail millet blast was initially documented in 1917 by Nishikado in Japan (Nishikado 1917), and its presence in India was noted in 1919, also from Tamil Nadu (McRae 1920). Additionally, occurrences have been documented in Maharashtra, Andhra Pradesh (Sinha and Upadya, 1997), and Uttarakhand (Kumar 2013). Recently blast disease was also reported in little millet also (Rawat et al. 2016). This disease has become widespread across key millet cultivation regions and is extending to new locations and becoming an important constraint in pearl millet, finger millet, and foxtail millet production (Poonacha et al. 2023). Palanna et al. (2023a) reported a large variation both across (44%) and within (56%) Magnaporthe population infecting millets and rice at the molecular level between virulent and avirulent genes. Further, they also divided *Magnaporthe* spp. isolates into four groups using molecular markers regardless of their geographic distribution, host plants, or tissues affected, and it indicates the prevalence of numerous pathotypes and virulence factors at the field level, which may lead to a high degree of pathogenic variation. Das et al. (2023) reported ten pathotypes in the *Pyricularia grisea* population adapted to finger millets. Emerging pathotypes exhibit varying intensities, influenced by the specific cultivar, prevailing weather conditions, and agricultural practices in place (Poonacha et al., 2023).

2.1.1.1 Etiology

Blast occurs as a result of the ascomycetes fungus *P. grisea* (Cooke.) Sacc., previously known as *Pyricularia oryzae* Cavara., which functions as the anamorph of *Magnaporthe grisea* (Hebert) Brar. This filamentous fungus is heterothallic and is known to be pathogenic to nearly 40 plant species spanning 30 genera within the Poaceae family (Ou 1980; Murakami et al. 2000; Inukai et al. 2006).

Pyricularia setariae Y. Nisik. specifically targets foxtail millet. Kulkarni and Patel (1956) classified *P. setariae* into four distinct physiological races, considering pathogenic traits, cultural attributes, physiological aspects, and morphological characteristics of the pathogen. Nevertheless, Gaikwad and D' Souza's (1987) findings indicated that the *P. setariae* strain infecting *Setaria italica* (foxtail millet) differs from the variants affecting rice, pearl millet, and finger millet. Finger millet isolates were shown to be pathogenic on a wide variety of millets, whereas rice isolates were only found to infect rice and wheat (Palanna et al. 2023b).

2.1.1.2 Diagnostic Symptoms

The blast pathogen has the capability to infect crops throughout all stages of growth in both finger and foxtail millet. However, it tends to target young seedlings or germlings more aggressively, leading to the development of dark patches with a burnt appearance in the nursery during severe infections (Parthasarathy et al. 2020). *P. grisea* attacks finger millet from the seedling stage through flowering to heading, resulting in symptoms such as leaf blast, neck blast, and finger blast. On the other hand, in foxtail millet, *P. setariae* targets the leaf lamina, causing symptoms specifically identified as leaf blast (Nagaraja et al. 2007).

Symptoms of foxtail millet leaves mainly developed as from a small water-soaked yellowish dot, which later turned circular to an oval spot with a grayish center surrounded by a brown margin. Spots measured an average 2–5 mm in diameter within 2–3 days. The spots then coalesce and result in drying of leaves. The disease starts with the lower leaves and extends to upper leaves. No symptoms were observed on the neck of foxtail millet (Palaniswami et al. 1970).

Under field conditions in barnyard millet, symptoms manifest primarily on young seedlings. These symptoms present as spindle to circular-shaped spots of different sizes. Initially, these spots exhibit a yellowish margin surrounding a grayish center. As the condition progresses, the centers transition to ash color. In humid conditions, a fungus develops an olive-gray overgrowth specifically at the center of these spots (Nagaraja et al. 2007).

2.1.1.3 Economic Importance

The economic significance of blast in finger and foxtail millet is notably high, marking them as among the most crucial diseases. In contrast, blast affecting proso, barnyard, and little millet is comparatively less prevalent and economically less

Table 1 Yield loss caused by blast disease in different crops

Crop	Yield loss	Place	Reference
Finger millet	42%	Ethiopia	Lule et al. (2014)
	50–100%	TN, India	Sekar and Ramalingam (2019)
	35.78%	Gujarat, India	Prajapati et al. (2013)
	22.57–56.67%	Karnataka, India	Anon (2021)
Foxtail millet	30–40%	India	Nagaraja et al. (2007)
	60%	India	Karthikeyan and Gnanamanickam (2008)
	40%	India	Das et al. (2021)

impactful and confined to specific locations. The yield loss caused by blast in small millets is presented in Table 1.

2.1.1.4 Epidemiology

The blast pathogen poses a threat to the crop across all growth stages, from seedling (vegetative) to grain formation (reproductive). Young seedlings, particularly in nursery and field settings under favorable weather conditions, are highly susceptible to blast (Kumar and Singh 2010). Optimal conditions for rapid disease development and spread involve moderate temperatures (25–30 °C), humidity exceeding 90%, and cloudy days followed by intermittent rainfall that sustains leaf wetness for over 10 h. Continuous rainfall during heading stages can result in finger blast, leading to substantial yield losses in both finger and foxtail millet. Additionally, excessive nitrogen fertilizer application tends to promote heightened blast incidence (Prakash et al. 2007).

2.1.1.5 Disease Cycle

The blast pathogen persists within glumes, straw, and certain grassy weeds. It's inherently seed-borne, dwelling within the pericarp and endosperm (Anitha et al., 2005 and Viswanath and Seetharam 1989). Its life cycle complexity stems from its susceptibility to weather conditions, influencing both survival and the dissemination of inoculum through various means. In the absence of the primary host, it sustains itself on graminaceous weeds, acting as secondary hosts and serving as the primary source of infection. However, airborne spores primarily drive the pathogen's spread, occasionally utilizing seeds as a mode of transmission.

2.1.2 Rusts

Rusts caused by different *Puccinia* and *Uromyces* species in small millets tend to have limited economic impact unless all components of the disease triangle align (Butler 1918). Among the seven small millets, rusts pose an economic threat specifically to foxtail millet caused by *Uromyces setariae-italica*, first documented by Yoshino in Japan. This rust disease consistently manifests in foxtail millet, often in severe forms, yet it generally doesn't cause substantial yield losses. In India, foxtail millet rust is prevalent in key millet-growing states like Karnataka, Maharashtra, Tamil Nadu, Madhya Pradesh, Bihar, and Andhra Pradesh. Recent reports also indicate severe occurrences of rust disease in browntop millet across its predominant cultivation regions.

2.1.2.1 Etiology

Uromyces genus is the prominent rust-causing genus in small millets. It causes rust of foxtail millet (*U. setariae-italica*), finger millet (*Uromyces eragrostidis*), and little and proso millets (*Uromyces linearis*), whereas in kodo millet, rust is caused by *Puccinia substriata.*

The *Uromyces* has both uredial and telial stages. Uredospores germinate producing one or more germ tubes capable of infecting the host. The telia are formed on the leaf blade, leaf sheath, and stem and are larger in size than uredia. The uredospores are pedicellate, oval or globose, echinulate, and yellowish brown having three to four pores. The teliospores are single-celled, pedicellate, oblong, globose or polygonal, and yellowish brown with a smooth thick wall, which is much thicker at the apex than at the base. The uredospores infect the host and produce uredia in 7–10 days. The fungus can perpetuate in uredial stage on collateral hosts. The rust appears within 20–25 days of sowing, and the intensity increases as the plants grow older. The telia appear at the time of maturity of the crop.

The uredospores of *Puccinia* develop in uredia and are subglobose to elliptic, stalked, single celled, finely echinulated, light brown-colored, and measured 31 × 25 μm (28–34 × 22–31 μm) with four equatorial germ pores. Paraphyses have been observed in the uredia. Teliospores are two-celled, oblong, rounded at both ends and with smooth uniform thick wall. Spores measure 40 × 22 μm (28–47 × 19–28 μm). The pedicels are short and colored. Mesospores measuring 25–37 × 15–25 μm are common. Marginal groups of paraphyses are also present in the telia.

2.1.2.2 Diagnostic Symptoms

The pathogen's impact spans across all stages of crop growth, yet severe damage occurs primarily if infection initiates before flowering. The disease begins with small, brown uredosori appearing on both sides of the leaves. As it progresses, mature and ruptured pustules form linear patterns on the upper leaf surface; however, these pustules tend to cover the entire leaf blade rather than maintaining a linear structure. Infection is observable on major foliar plant parts such as leaf sheaths, culms, and stems. It tends to be more severe on upper leaves compared to middle and lower ones. In instances of severe infection, premature drying of leaves and poor grain development are common observations in the field. Each rust variant exhibits distinct spore production morphology: *U. setariae-italica* and *U. linearis* produce light yellow and black teliospores, respectively, while *P. substriata* generates small, brown, oval spots on the upper leaf surfaces. The teliospores in *P. substriata* are dark brown, single-celled, stalked, and oblong to globose in shape, characterized by smooth and thick walls (Kumar and Singh 2010).

2.1.2.3 Economic Importance

Rust disease is common on foxtail millet causing considerable yield loss. In severe form, yield losses may vary from 30% to 40% (Saralamma and Kiranmai 2019). Li et al. (2015) recorded yield loss of 10–30% due to rust in foxtail millet.

2.1.2.4 Disease Cycle

The uredospores persist, spread, and infect the host, generating uredia within 7–10 days. Collateral hosts, primarily various graminaceous weed species, likely contribute significantly to the disease cycle by aiding in its perpetuation. Rust infection typically emerges within 20–25 days of sowing, contingent on the prevalence of inoculum and favorable weather conditions. The intensity of infection escalates as the plants mature, with dark telial pustules developing at crop maturity (Nagaraja et al. 2007).

2.1.3 Leaf Spots/Blights

Leaf blights, also referred to as leaf spots, brown spots, or seedling blights, represent common diseases affecting small millets following blast occurrences. These blights caused by various pathogens—*Helminthosporium* spp., *Alternaria alternata*, and *Cercospora eleusinis*, each targeting specific millet hosts. Among these, *Helminthosporium* spp.-induced leaf blights are gaining significance due to their escalating severity, wider distribution, and altered virulence as they adapt to diverse agroecological regions. This blight poses a serious threat across all small millet crops, with finger millet notably experiencing heightened damage initially observed by Butler in India back in 1918. Recent studies by Ramesh et al. in 2021a and 2021b highlight *Bipolaris setariae*-induced leaf blight as the most critical and destructive disease affecting browntop millet (Table 2).

Table 2 List of small millet leaf spot and leaf blight diseases and their causal organisms

Crop	Disease	Causal organism	Reference
Finger and little millet	Brown spot	*Drechslera nodulosum* (Sacc.) subram and Jain (PS: *Cochliobolus nodulosus*)	Viji et al. (2000)
Foxtail and browntop millet	Leaf spot	*Bipolaris setariae* (PS: *Cochliobolus setariae*)	Ramesh et al. (2021a)
Proso millet	Leaf spot	*Bipolaris panici-miliacei*	Lee-Du Hyung (1997)
Barnyard millet	Leaf spot or blight	*Exserohilum crus-galli*	Nagaraja et al. (2007)
Little and Kodo millet	Alternaria leaf blight	*Alternaria alternata* (Fr.) Keissl.	Praveen et al. (2021)
Finger millet	Cercospora leaf spot	*Cercospora eleusinis*	Munjal et al. (1961)
Finger millet	Curvularia leaf spot	*Curvularia lunata* (Walker) Boedijin	Shaw (1921)

Fig. 1 Symptoms of small millet diseases. Finger millet (**a**–**h**): (**a** and **b**) leaf blast; (**c**) finger blast; (**d**) neck blast; (**e**). Brown spot; (**f**) banded blight; (**g**) foot rot and (**h**) Cercospora leaf spot. Foxtail millet (**i**–**l**): (**i**) blast; (**j**) brown spot; (**k**) rust; and (**l**) banded blight. Barnyard millet (**m**–**p**): (**m**) leaf blast; (**n**) banded blight; (**o**) grain smut; and (**p**) head smut. Browntop millet (**q**–**r**): (**q**) rust and (**r**) leaf blight; kodo millet (**s**) head smut; and little millet (**t**) leaf blight

2.1.3.1 Etiology

The genus *Helminthosporium* falls under the phylum *Ascomycota*, class *Dothideomycetes*, subclass *Pleosporomycetidae*, order *Pleosporales*, and family *Pleosporaceae*. Its septate hyphae conidiophores are brown to dark brown, characterized by parallel walls, erect structure, and halting elongation upon the formation of the terminal conidium. The conidia are multicellular (with six or more cells), relatively large (measuring 9–40 μm), solitary, club-shaped and range in color from pale to dark brown. They are positioned along the sides of the conidiophores, with their wider end oriented toward the conidiophore. *Helminthosporium* distinguishes itself from *Bipolaris*, *Drechslera*, and *Exserohilum* due to its characteristic formation of parallel-walled, upright conidiophores (Alcorn 1988) (Fig. 1).

2.1.3.2 Diagnostic Symptoms

Seeds infected with the pathogen may fail to germinate due to preemergence seed rot. If they do germinate, postemergence rot is frequently observed. In instances where seeds or seedlings manage to survive, or when healthy seedlings face inoculum from external sources, a distinct symptom emerges on the leaf lamina—manifesting as brown to dark brown spots. The pathogen tends to infect leaves more readily from the upper surface or between the leaf and leaf sheath. Symptoms are noticeable on the leaf sheath as well (see Figure), particularly in mature plants, where a fuzzy fungal growth becomes apparent at the lesion's center, particularly under elevated humidity conditions (Nagaraja et al. 2007).

2.1.3.3 Epidemiology

Regions characterized by warm humidity, with temperatures typically ranging from 15 to 30 °C throughout the cropping season, provide favorable conditions for symptom development. In humid environments, pathogen growth becomes noticeable on older spots. Moderately high temperatures coupled with elevated relative humidity (RH) create an ideal environment for leaf blight disease (Nagaraja et al. 2007).

The region where the leaf sheath meets the leaf blade is commonly affected, leading to a dark brownish discoloration. Neck infections display initial dark tan lesions that could expand both upward and downward. In severe cases, this can lead to neck breakage, causing the affected portion to remain attached to the plant. The pathogen's impact might extend to the ear head, fingers, and grains. Affected grains might fail to mature fully and shrink, resulting in substantial crop losses (Kumar and Singh 2010). Symptom manifestation can differ across different crops.

2.1.3.4 Disease Cycle

The pathogen survival spans over 18 months in soil, while spores on grains maintain viability for up to a year (Vidhyasekaran 1971 and Nagaraja et al. 2007). Additionally, the fungus remains viable on plant debris and stubble. Airborne conidia act as a secondary source of inoculum. The optimal temperature for initiating infection ranges around 30–32 °C, but disease occurrence can span from 10 to 37 °C (McRae 1922). Elevated humidity and sporadic rainfall during ear emergence and preceding grain formation stages significantly increase ear infections, leading to substantial yield losses. In severe infections, young seedlings or germlings might succumb within 3 days postinfection, while mature plants may perish within 15 days after infection.

2.2 Earhead Diseases (Smuts)

Various smut pathogens, including *Ustilago* spp., *Melanopsichium* spp., *Sporisorium* spp., and others, are recognized for their ability to infect all small millets, leading to diverse symptoms like grain smut, head smut, and kernel smut. Grain smut is commonly found in finger, foxtail, and barnyard millets, while head smut prevails in kodo, barnyard, and proso millets. Generally, smuts affecting small millets are of

Table 3 List of small millet smut diseases and their causal organisms

Crop	Disease	Causal organism	Reference
Finger millet	Grain smut	*Melanopsichium eleusinis* (Kulk.) Mundk. and Thirum.(Syn: *Ustilago eleusines* Kulk.)	Thirumalachar and Mundkur (1947)
Foxtail and Proso millet	Grain smut	*Ustilago crameri*	Pall et al. (1980)
Barnyard millet	Grain smut	*Ustilago panici-frumentacei* Bref.	Vasudeva (1954)
Little millet	Grain smut	*Macalpinomyces sharmae* Vanky (Syn: *Tolyposporium sharmae*)	Jain et al. (2006)
Kodo millet	Head smut	*Sorosporium paspali-thunbergii* (Henn.) S. Ito (Syn: *Sorosporium paspali* Mc Alp.)	Viswanath and Seetharam (1989)
Proso millet	Head smut	*Sporisorium destruens* (Schltdl.) Vánky (Syn: *Sphacelotheca destruens*)	Sinha and Upadhyay (1997)
Barnyard millet	Head smut	*Ustilago crus-galli* Tracy and Earle	Pall et al. (1980)
Foxtail millet	Head smut	*Ustilago crameri*	Kumar (2011)
Barnyard millet	Kernel smut	*Ustilago paradoxa*	Viswanath and Seetharam (1989)

limited concern due to their sporadic occurrence and minimal economic impact (Table 3).

2.2.1 Grain Smut

Among smut diseases affecting small millets, grain smuts prevail as the most commonly occurring disease. They are widespread in finger, little, barnyard, foxtail and proso millet caused by various genera. The first report of ragi grain smut dates back to 1922 in Kolhapur (Kulkarni 1922). Subsequently, its presence was documented by several scientists including Coleman (1920), McRae (1924), Narasimhan (1934), Mundkur (1939), and Venkatarayan (1947) in the Mysore state province.

2.2.1.1 Etiology

Ragi grain smut, caused by *Melanopsichium eleusinis*, generates globose or subglobose spores that are auburn in color, measuring between 7 and 11 μ with an average size of 9.5 μ. These spores possess a rough, minutely pitted epispore (Mundkur 1939). They initiate germination by forming a septate promycelium that produces both lateral and terminal sporidia. Initially, the promycelium emerges as a small papilla, gradually extending into a stout germ tube. The optimal temperature for spore germination is noted to be 25 °C.

2.2.1.2 Diagnostic Symptoms

Symptoms of grain smut become apparent during the grain formation stage, emerging within a few days after flowering. Initially, affected ovaries transform into velvety greenish smut sori, initially not larger than the normal grain size. Over time, the spore balls (sori) grow, exerting pressure that pushes apart the glumes. These sori are enveloped by a delicate membrane, which later ruptures, exposing the sorus. As the sorus matures, its greenish outer tunica transitions to a pinkish-green hue and eventually to a dingy black shade (Thirumalachar and Mundkur 1947). When wind blows, loosened spores are easily dispersed, leaving the glumes empty. In some instances, infected grains of certain crops develop late in the season, maintaining a greenish color and gradually enlarging; upon pressing, these grains release spores (Sharma and Khare 1987).

2.2.1.3 Economic Importance

The impact of *U. crameri* results in losses ranging from 8% to 50% of the grain yield (Nagaraja et al. 2007), with grain infection reaching up to 75% (Sundararaman, 1921). Jain and Tripathi (2002) revealed a reduction in grain yield by 9.80–53.55% and a decrease in panicle length by 6.40–38.90% in little millet.

2.2.1.4 Epidemiology

Cool temperatures paired with high relative humidity create favorable conditions for the rapid development and dissemination of the disease. The fungus tends to target the majority of grains within an ear by forming sori in flowers and the basal parts of the palea, though the infection often spares the terminal portion of the spike. These sori appear pale greyish and typically measure between 2 and 4 mm in diameter. Upon maturation, the sori rupture, releasing a dark powdery mass of spores. These spores are dark brown, ranging from angular to round in shape, with smooth walls and measuring approximately 7–10 μm in diameter (Das 2017).

2.2.1.5 Disease Cycle

Grain smuts primarily spread through external seed contact and airborne transmission. Flower infection happens via secondary inoculum, specifically airborne spores. There's been relatively a limited systemic research on grain and head smuts, although instances of soilborne infection have been observed. The fungus remains as dormant dikaryotic mycelium within the seed tissues, acting as the primary source of inoculum for the following season (Wang 1943).

2.2.2 Head Smut

Head smut in small millets generally poses minor economic concerns unless it proliferates unchecked under conditions that fulfill the disease triangle. It is more prevalent in countries such as Europe, Australia, and Eastern Asia. In India, it is widespread among kodo, barnyard, and proso millets, consistently reported in regions like Karnataka, Madhya Pradesh (Pall et al. 1980), Tamil Nadu, and Uttarakhand (Kumar et al. 2007).

2.2.2.1 Etiology

Sporisorium paspali's teliospores are formed in loose, ball-like masses. These spore balls gradually split into individual spores with minimal pressure. The individual spores are globose, varying from angular to roughly pear-shaped, and display colors ranging from dark to yellowish brown, possessing thick, smooth walls. Germination of these spores occurs through the production of septate, single or branched hyphae that are narrowed at the septum. These hyphae bear both lateral and terminal sporidia.

2.2.2.2 Diagnostic Symptoms

A defining symptom of head smut is the complete transformation of the entire panicle into a single sorus, resulting in the deformation of the inflorescence into a smut ball. The pathogen responsible for head smut exclusively infects the ovaries of the plant. Additionally, head smut leads to distinctive symptoms such as gall-like swellings on the stem, nodes of young shoots, and in the axils of older leaves. Occasionally, twisted and deformed clusters of leafy shoots with aborted ears may also emerge. The affected ovaries appear as hairy, round, and grey sacs initially without significant size increase. These gall-like swellings are enveloped by a rough, hairy membrane composed of host tissue (Mundkur 1943). Upon rupture, these membranes expose the spore mass, which is dispersed by the wind.

2.2.2.3 Economic Importance

Sporisorium paspali has become established in certain areas, although its distribution and severity fluctuate based on favorable conditions and specific cultivars. Viswanath (1992) documented a grain yield reduction of 30–40% attributed to *S. paspali*. Jain and Yadava (1997) noted yield losses ranging from 13.15% to 32.98% which varied in correspondence with the incidence of smut.

2.2.2.4 Epidemiology

Head smut is transmitted through external seed contact, often appearing sporadically late in the crop season, close to the crop's maturity. An optimum temperature range of 20–25 °C facilitates the initiation and colonization of the infection.

2.2.2.5 Disease Cycle

The disease primarily spreads through external seed contact, initiating seedling infection through germ tube penetration from seed-borne teliospores, penetrating the cell wall. Once inside the seedling tissue, the hyphae extend both between and within cells, resulting in systemic infection. Eventually, this infection reaches the meristematic tissues and ultimately affects the ear by the time the crop matures (Kumar and Singh, 2010).

2.2.3 Kernel Smut

Kernel smut, an infrequent disease, sporadically impacts barnyard millet in unprotected cultivation. Its initial observation traces back to Italy. In India, Viswanath and Seetharam (1989) recorded instances of kernel smut occurring in Bihar, Maharashtra, and Tamil Nadu.

2.2.3.1 Etiology

Kernel smut is caused by *Ustilago paradoxa* Syd., P. Syd. and E.J. Butler. Chlamydospores produced by *U. paradoxa* are smooth, olive brown in color, and round which measures up to 7–11 μm in diameter.

2.2.3.2 Diagnostic Symptoms

Fungi initiate infection during flowering, transforming floral components into fungal structures that resemble swollen greenish bodies. The infection typically occurs sporadically, affecting only a limited number of grains within an ear, with up to 25 grains/ear forming smut sori measuring 1.5–4 × 1–2 mm (Nagaraja et al. 2016).

2.2.4 Udbatta Disease

Previously considered widespread and of minor significance in specific paddy cultivation areas, udbatta disease, identified by Butler and Bisby in 1931, is caused by *Ephelis oryzae* Syd. and reaches its perfect stage in *Balansia oryzae-sativae* Hashioka. This disease has been observed across various host plants such as grasses like *Cynodon dactylon*, *Pennisetum* sp., and *Eragrostis tenuifolia* (Govindu and Thirumalachar 1961 and Ranganathaiah 1972), as well as in little millet (Mohanti et al. 1969).

During the rainy season of 2008, sporadic occurrences of the disease were observed in experimental plots of kodo millet at the ICAR-All India Coordinated Research Project on Small Millets, GKVK farm in Bengaluru, Karnataka. Among the kodo millet varieties, RBK 155 recorded the highest disease incidence at 19.56%.

2.2.4.1 Etiology

The disease was caused by *Ephelis oryzae* Syd. having perfect stage *Balansia oryzae sativae* Hashioka. The mycelium of the fungus is septate, hyaline to dirty grey, and later becomes dark-colored. The conidia of the fungus are needle-shaped, hyaline, and septate measuring 13–35 × 1–2 μm (Table 4).

2.2.4.2 Diagnostic Symptoms

The affected plants stand out due to their deformed flower clusters, which display greyish-white fungal growths. Within these infected clusters, the individual sections of the flower stalks are fused together and attached to the main stem by sticky masses of spores that solidify into a hardened covering. Additionally, dark masses called sclerotia develop on mature clusters. A healthy plant typically showcases an open flower cluster about 30- to 40-cm-long. In contrast, the infected plant

Table 4 List of small millet udbatta disease and their causal organisms

Crop	Disease	Causal organism	Reference
Kodo millet	Udbatta	*Ephelis japonica* P. Henn.	Butler and Bisby (1931)
Foxtail millet	Udbatta	*Ephelis oryzae* Syd.	Mohanty and Mohanty (1957)
Little millet	Udbatta	*Ephelis oryzae* Syd.	Nagaraja et al. (2007)
Proso millet	Udbatta	*Ephelis oryzae* Syd.	Nagaraja et al. (2007)

experiences a transformation where the flower sections stick together, forming a cylindrical shape, reducing the length of the cluster to 15–23 cm (Nagaraja et al. 2007).

2.2.5 Downy Mildew

Downy mildew also known as crazy top or green ear disease represents a significant threat to finger and foxtail millet across major growing regions worldwide. In India, green ear infection in ragi was initially documented by Venkatarayan (1947) in the former Mysore state, expanding subsequently to Tamil Nadu and Uttarakhand (Kumar et al. 2007). Meanwhile, foxtail millet's downy mildew has been reported in various Indian regions like Karnataka, Tamil Nadu, Maharashtra, Bihar, Andhra Pradesh, and Kashmir (Rangaswami and Mahadevan 1999). This disease can lead to substantial crop failure due to underdeveloped ears, resulting in yield losses of up to 50% (Pall et al. 1980).

2.2.5.1 Etiology

Downy mildew is caused by *Sclerophthora macrospora* (Sacc.) Thirum. [Syn. *Sclerospora macrospora* Sacc.] in finger millet, whereas *Sclerospora graminicola* (Sacc.) J. Schröt. infects foxtail millet. It has a wide host range which includes finger millet, foxtail millet, pearl millet, maize, *Eleusine indica* (Ulstrup 1955), wheat, oat, *Eragrostis pectinacea* and *Digitaria marginata*.

2.2.5.2 Diagnostic Symptoms

Early signs of downy mildew in affected plants include seedling leaf chlorosis. Plants show distinct symptoms of stunted growth with shortened internodes and increased tillering. Infected leaves display pale yellow, translucent lesions. Under humid conditions, a whitish bloom of sporangiophores and sporangia becomes prominent on the leaf surface. Mildly infected plants may form ears that appear malformed, resembling green leafy structures, known as the "green ear" symptom. While the pathogen often affects the entire ear, sometimes only sections are involved, with the rest producing normal grains. Upon maturity, chlorotic patches appear on the upper leaf surface, accompanied by downy fungus growth on the lower surface (Kumar and Singh 2010; Nagaraja et al. 2016).

Typically, downy mildew in finger millet doesn't exhibit the typical symptom of a white cottony growth on the underside of leaves. As a result, the disease becomes identifiable only after ear formation (during the reproductive stage). Eventually, parts or the entire ear, including the palea, lemma, and glumes, transform into leafy structures. Spikelet growth starts at the base and spreads upward, leading to a bush-like appearance in the mature ear, showcasing the distinct "green ear" symptom (Thirumalachar and Narasimhan 1949).

2.2.5.3 Epidemiology

Optimal conditions for disease progression involve moderately high temperatures between 25 and 30 °C and elevated relative humidity. Night temperatures around

22–25 °C promote sporangia production and aid zoospore release from these structures.

2.2.5.4 Disease Cycle

The downy mildew fungus, an obligate parasite, resides internally and externally within seeds, exhibiting a wide host range. Safeeulla (1955) detailed the pathogen's life cycle in small millets, primarily originating from soil or seed-borne oospores. Disease severity is chiefly influenced by temperature, soil moisture, and sowing time, requiring an optimal soil temperature of 20–21 °C (with a range of 12–13 °C minimum and 30 °C maximum) for infection initiation. Early sown crops are more susceptible compared to late or timely sowing. High relative humidity and soil moisture accelerate disease development (Nagaraja et al. 2007). Recent observations noted a high incidence of downy mildew in two prominent finger millet varieties, GPU 28 (resistant to blast) and PR 202 (susceptible to blast), in Tumkur district, Karnataka.

2.3 Soilborne Diseases

Majorly in this group two diseases are important, (a) banded sheath blight and (b) foot rot.

2.3.1 Banded Sheath Blight

The first documented case of banded sheath blight (BSB) was identified on finger millet in Vellayani, Kerala, India (Das and Girija 1989). Subsequently, severe instances of the disease were also documented in experimental fields at Birsa Agricultural University in Ranchi (Dubey 1995). Following these initial reports, the disease has become a common occurrence in regions characterized by hot and humid climates where small millets are grown.

2.3.1.1 Etiology

Banded sheath blight stems from *Rhizoctonia solani* (basidial state: *Thanatephorus cucumeris*). This fungus generates initially dull, cream-colored mycelial colonies that eventually shift to a light brown color. Identifying *R. solani* involves noting its distinct branching pattern at right angles and specific constrictions at the branching points. It infects all small millet crops, viz., finger, foxtail, barnyard, proso, kodo and little millet.

2.3.1.2 Diagnostic Symptoms

The symptoms of the disease manifest initially on the lower leaves and leaf sheath in oval to irregularly shaped spots. These spots start as light grey to dark brown patches, eventually developing a white center surrounded by a narrow reddish-brown border. They are widespread across the leaf surface. In favorable conditions, these spots may grow and merge, covering the entire leaf sheath and surface and creating distinct brown bands across the plant leaf. This leads to the withering and drying of the entire leaf. Additionally, symptoms can also be noticed on peduncles, fingers, and glumes. In humid settings, one may observe mycelial growth and sclerotia on these lesions (Das et al. 2016).

2.3.1.3 Economic Importance

In most cases, BSB doesn't lead to significant economic losses. However, when all conditions in the disease triangle align, it has the potential to cause substantial economic damage. If the infection spreads to the peduncle, it can cause neck rot, affecting grain production by hindering proper grain filling and resulting in the formation of undersized, shriveled grains. This process can significantly impact overall yield and quality.

2.3.1.4 Epidemiology

The pathogen persists through various weather conditions by forming sclerotia in the soil, serving as the primary source of infection for the following season. The disease thrives in conditions characterized by moderate temperatures (around 26 ± 20 °C) and high humidity levels (>80%) (Dubey 1995). Late maturing varieties show lower disease severity because some plants manage to escape the disease in autumn when cooler temperatures prevail (Patro et al. 2018).

2.3.1.5 Disease Cycle

The *R. solani* persists in a dormant state as mycelium, which then germinates into infective mycelium when weather conditions are favorable, such as high humidity and moderate temperatures. The initial symptoms appear as small, irregular, or oval brown spots on the lower leaf surface, which later grow and merge to create bands. Many sclerotia become visible on the affected leaves, leading to their eventual death. The fungus also spreads between fields via irrigation water, contaminated soil, and leftover plant debris.

2.3.2 Foot Rot

The foot rot/wilt disease was initially documented by Coleman in 1920 in Mysore state. Subsequently, it surfaced in various regions across India where finger millet is grown. In the case of ragi, field losses could potentially climb to 50% when weather conditions are favorable for the disease (Basta and Tamang 1983). Primarily observed in Karnataka, Tamil Nadu, Gujarat, and Odisha, this ailment was initially deemed minor due to its sporadic occurrence. However, recent research indicates a rising prevalence of the disease, especially under irrigated conditions.

2.3.2.1 Etiology

Sclerotium rolfsii Sacc., known in its perfect stage as *Athelia rolfsii* (Curzi) C.C. Tu and Kimbr., or *S. rolfsii* (*Pellicularia rolfsii* in its perfect state), thrives in its imperfect stage across both field and laboratory settings. The perfect stage, observed less frequently, manifests under favorable conditions, occasionally yielding fertile sporophores. *S. rolfsii* has a broad host range (Aycock 1966), but among the small millets, finger millet is the only host reported so far.

2.3.2.2 Diagnostic Symptoms

Typically, plants face attacks during their inflorescence or seed formation stage. Infection takes place around the soil line, specifically in the plant's collar region. Initially, the affected area appears water-soaked, later darkening to brown or dark brown, leading to eventual shrinkage that triggers wilting across the entire plant. The fungus displays a cottony white growth on various plant parts, eventually forming roundish structures resembling velvety mustard seeds known as sclerotial bodies.

2.3.2.3 Epidemiology

During the winter, the fungus persists as sclerotia within the soil, serving as the main source of infection. Sandy loam soils with minimal moisture tend to increase the occurrence of the disease. Optimal disease development occurs in warm temperatures and high relative humidity, conditions commonly found during the monsoon season.

2.3.2.4 Disease Cycle

The fungus persists in the soil as sclerotia, serving as the primary source of infection. When the rainy season begins, these sclerotia germinate, producing basidia. Each basidium generates four haploid basidiospores. The fungus targets the plant's collar region, causing infection that eventually leads to the plant's demise. As the cropping season ends and nutrients become scarce while unfavorable weather sets in, the fungus transforms into tightly packed masses of mycelia known as sclerotia. These visible bodies, indicative of aging and imminent death, are easily spotted on the plant. These sclerotia accumulate within the soil alongside plant debris and are transported across fields via rainwater, aiding the fungus in spreading from one area to another.

3 Bacterial Diseases

In addition to fungal diseases, various bacterial diseases sporadically affect small millet cultivation. Bacterial pathogens pose a lesser threat compared to fungi unless specific environmental conditions, susceptible cultivars, and significant inoculum levels coincide during the cropping season. These bacterial diseases have been observed across different small millet varieties in certain growing areas, but they have not emerged as predominant factors under natural conditions. The information regarding bacterial disease of small millets and their causal organism is given in Table 5.

4 Viral Diseases

Viral diseases hold minimal significance in small millets. While fungal and bacterial infections pose sporadic challenges, viral ailments have notably lower impact. These diseases rarely emerge as major threats in small millet cultivation.

Table 5 List of small millet bacterial diseases and their causal organisms

Crop	Disease	Causal organism	Reference
Finger millet	Bacterial leaf spot	*Xanthomonas eleusinae*	Rangaswami et al. (1961)
	Bacterial leaf blight	*Xanthomonas axonopodis pv. Coracanae*	Desai et al. (1965)
	Leaf stripe	*Pseudomonas eleusineae*	Billimoria and Hegde (1971)
Kodo millet	Bacterial leaf streak	*Xanthomonas* species	Nema et al. (1978)
Proso millet	Bacterial leaf stripe	*Pseudomonas syringae* Van Hall pv. *panici*	Nagaraja et al. (2007)
	Bacterial leaf streak	*Pseudomonas avenae*	Nagaraja et al. (2007)
Barnyard millet	Bacterial leaf streak	*Pseudomonas avenae*	Nagaraja et al. (2007)
Foxtail millet	Bacterial leaf streak	*Pseudomonas avenae*	Nagaraja et al. (2007)
	Bacterial leaf spot	*Pseudomonas albo-precepitans* Rosen	Rosen (1922)

Table 6 List of small millet viral diseases

Crop	Disease	Reference
Finger millet	Ragi mottle streak disease: *Ragi mottle streak virus*	Mariappan et al. (1973)
	Ragi severe mosaic: *Sugarcane mosaic virus*	Subbayya and Raychaudhuri (1970)
	Ragi streak disease: *Eleusine* strain of *maize streak virus*	Anonymous (1975)
Barnyard millet	*Wheat streak virus*, *sugarcane mosaic virus* and *Eleusine virus 2*	Sill and Agusiobo (1955)
Proso millet	*Wheat streak virus*, *rice dwarf or stunt virus* and *maize leaf streak virus*	Sill and Agusiobo (1955)

Consequently, their impact on small millet production and cultivation remains limited, with farmers encountering fewer issues related to viral diseases in these crops. The information regarding bacterial disease of small millets and their causal organism is given in Table 6.

5 Diseases Caused by Nematodes

There is limited available information on the yield loss resulting from nematode diseases. However, numerous reports have documented the presence of nematodes in small millets, as detailed in Table 7.

Table 7 List of nematode species reported in small millets

Crop	Nematode reported	Reference
Finger millet	*Heterodera gambiensis*	Setty (1975)
	Rotylenchulus reniformis Linform and Oliveira	Chandrasekaran (1964)
	Helicotylenchus, *Trichodorus*, and *Pratylenchus* species	Nagaraja et al. 2007
	Criconemoides ornatus	Mohanty and Das (1976)
Kodo millet	*Meloidogyne incognita* (Kofoid and white) Chitwood	Alam et al. 1973
Little millet	*Heterodera delvi*	Krishnaprasad et al., 1980)
	Helicotylenchus abunaamai	Padhi and Das (1982)
Proso millet	*Aphelenchoides besseyi*	Gokte et al. (1992)

6 Plant Phanerogamic Parasite

The witch weed (*Striga* spp.) exhibits a broad ecological range and has been discovered in association with kodo millet plants. In India, its presence has been documented in Andhra Pradesh, Karnataka, and Madhya Pradesh. Field surveys conducted during 2001 and 2003 reported varying average incidences of *Striga* species, ranging from 1.6% to 2.0%. In particular, the districts of Rewa, Satna, and Sidhi in Madhya Pradesh recorded incidence frequencies between 66.7% and 100.0% during these years (Nagaraja et al. 2007).

6.1 Etiology

Striga asiatica (L) Kuntze and *Striga densiflora* Benth. were found associated with the roots of kodo millet, which is a partial root parasite.

6.2 Diagnostic Symptoms

Striga species infestations become evident in the field once the *Striga* plants emerge from the soil. These parasitic plants connect their underground sections to the roots of the host plant through haustoria, drawing water and nutrients from the host. Affected plants display stunted growth with limited aerial development, resulting in sparse panicles. In severe cases, if infestation happens early, the plants might wither and dry up before reaching the flowering stage (Yadava and Jain 2006).

6.3 Economic Importance

The extent of grain yield losses caused by *Striga* species infestations is influenced by both the quantity of attacking *Striga* plants and the level of resistance in the host crop. According to Jain and Tripathi (2005), infestations of *Striga densiflora* in kodo millet led to reported losses ranging from 42.4% to 65.8% in grain yield per plant.

7 Complex Diseases

Diseases involving multiple pathogens pose complex challenges, often complicating diagnosis and management. Interactions between multiple pathogenic agents can intensify disease severity, impacting crop health. Understanding these interactions is crucial for effective control strategies in agriculture and plant health management. Small millets have few complex diseases (Table 8).

7.1 Management Strategies of Major Diseases in Small Millets

Integrated disease management blends diverse strategies which includes cultural, biological, and chemical measures to control diseases as well as use of resistant cultivars. It harmonizes practices like crop rotation, resistant varieties, biological agents, and precise chemical use. This holistic approach mitigates risks, enhances crop health, and sustains ecosystems while minimizing reliance on any single control method.

(i) **Cultural Methods Used in Disease Control.**

Several cultural method such as, strategic sowing, optimum plant populations and spacing, timely weeding, proper fertilizer application, crop rotation, thorough summer plowing, residue removal, bund cleaning postharvest, diseased plant removal, and careful irrigation aid in minimizing disease inoculum and incidence. These cultural practices play a pivotal role in managing millet diseases:

(a) **Crop rotation**: Primarily manages seed and soilborne pathogens like downy mildew, ergot, smut, banded blight, and sheath rot.
(b) **Deep summer plowing:** Reveals dormant pathogen spores to manage diseases such as downy mildew, smut, and several fungal and bacterial leaf infections.

Table 8 List of complex disease of small millets along with their causal organism

Crop	Disease	Causal organism	Reference
Finger millet	Grain mold	*Fusarium moniliforme*, *Alternaria alternata*, *Curvularia lunata*, *Phoma sorghina*, *Aspergillus* spp.	Das et al. (2016)
Kodo millet	Kodo millet poisoning	*Phomopsis paspali*, *Aspergillus flavus*, *A. tamari*	Pendse (1974), Nagaraja et al. (2007)

(c) **Adjustment of date of sowing**: Early sowing reduces blast and rust severity.
(d) **Optimization of plant population:** High plant population encourages disease development, hence maintaining an optimal plant count aids in managing diseases like downy mildew, blast, rust, and grain mold in millets.
(e) **Use of disease-free seeds:** The most environmentally friendly approach to controlling diseases involves using disease-free seeds, which effectively manage downy mildew, banded sheath blight, foot rot, ergot, and blast in millets.
(f) **Sanitation:** This method aids in reducing primary source of inoculum and surviving structures of the pathogen. Downy mildew, banded sheath blight, foot rot, ergot, and blast can all be managed using this approach.
(g) **Eradication of alternate and collateral hosts:** Timely removal effectively mitigates diseases such as ergot, downy mildew, blast, rust, leaf spots, as well as bacterial and viral infections.
(h) **Fertilizer management**: Nutrients play a crucial role in the crop-pathogen relationship. Adjusting nitrogenous fertilizer levels reduces the occurrence of diseases like blast and downy mildew.
(i) **Vector management**: Several viral, bacterial, and fungal diseases might enter via their vectors visiting hosts. Therefore, achieving optimal pathogen eradication involves incorporating measures to address these vectors for the best results.

(ii) **Host Plant Resistance**

Genetic resources or germplasm harbor unique genes and alleles that offer resistance to both biotic and abiotic factors while enhancing various yield-related traits. Exploring these resources is crucial for developing disease-resistant varieties, especially for low-value crops like small millets. Notably, numerous disease-resistant varieties have been identified and released for various millet-growing regions in India, detailed in Table 9 (www.aicrpsm.res.in).

(iii) **Biological Control**

Recently, biological control of plant diseases has gained significance over synthetic chemical pesticides due to numerous benefits for ecosystems and human health. These methods ensure crop protection against various biotic stresses while aiding in plant growth promotion and safeguarding against both biotic and abiotic pressures in a sustainable manner. Biocontrol agents like *Trichoderma* and *Pseudomonas* serve as effective remedies for soilborne diseases in millets (such as foot rot in finger millet and banded sheath blight in small millets), particularly in cases where sufficient host resistance is absent and can be applied through seed treatment and soil application.

(iv) **Chemical Control**

(a) **Downy mildew**: Treating seeds using Ridomyl-MZ at 6 g/Kg seed and applying one or two sprays as necessary at 3 g/L effectively reduce disease incidence.
(b) **Blast**: Applying carbendazim at 1 gm/kg of seed followed by necessary fungicide sprays such as carbendazim (0.2%), iprobenfos (IBP) (0.1%), premixture fungicide (carbendazim+mancozeb) (0.1%), ediphenfos (0.1%), propiconazole (0.1%), or tricyclazole (0.1%) effectively manages diseases. Initiate the first

Table 9 Disease-resistant varieties identified and released for the different sorghum growing areas of India (2000–2018)

Crop	Disease	Resistant or tolerant cultivars
Finger millet	Blast	GPU 26, GPU 28, GPU 45,GPU 48, KMR 204, Chilika (OEB 10), VL 315, PRM 1, Bharathi (VR 762), Srichaitanya, GN-5, VL Mandua-348, KOPN 235, OEB 526, OEB 532, PPR 2700 (Vakula), VL 352, GNN-6,KMR 340, Dapoli-2 (SCN-6) and CO 15
Foxtail millet	Downy mildew	Meera (SR 16), SiA 3085, RAU (Rajendra Kauni 1–2)
	Rust	TNAU 196, RAU (Rajendra Kauni 1–2)
	Blast	RAU (Rajendra Kauni 1–2),SiA 3085
	Brown spot, smut and leaf blight	RAU (Rajendra Kauni 1–2)
Barnyard millet	Grain smut	VL Madira
	Blast	Tarini (OLM 203), GNV-3
	Grain smut	Tarini (OLM 203), OLM 217, and GNV-3
	Rust	OLM 217
	Brown spot	Kolab (OLM 36)
	Sheath blight	Kolab (OLM 36), GNV-3
	Head smut	Jawahar Kutki 4 (JK 4)
Proso millet	Brown spot	GPUP 8
	Leaf blight	PRC 1
Kodo millet	Head smut	Jawaharkodo 155 (RBK 155), Jawaharkodo 48 (JK 48), JK 106, JK 65, JK 98, Jawaharkodo 137, JK 13

spray upon symptom detection, with additional sprays at flowering stage at 10- to 15-day intervals to control neck and finger infections in finger millet.

(c) **Grain mold**: Apply one of the listed fungicides if there are intermittent rains during earhead emergence, a week later, and during the milky stage (mancozeb+captan at 0.2%, thiram+carbendazim at 0.2%, or propiconazole at 0.1%). Apply two to three sprays to minimize grain mold.

(d) **Smuts:** Seed dressing with sulfur @ 4 g/kg seed.

(e) **Banded sheath blight:** Treating seeds with mancozeb at 2–3 g/kg and employing necessary fungicide sprays such as propiconazole at 1 ml/liter, hexaconazole at 2 ml/liter, or validamycin at 2 ml/liter yields highly effective results.

(f) **Rust:** Applying foliar sprays of mancozeb (0.2%), hexaconazole (0.1%), difenoconazole (0.1%), and propiconazole (0.1%) effectively controls the disease. Administering two sprays at 15-day intervals right after symptom onset is recommended for improved disease management.

(g) **Leaf spot/leaf blight**: Treating seeds with carbendazim at 2 g/kg or mancozeb at 0.2% followed by necessary sprays as needed.

8 Modern Strategies for Enhanced Control of Small Millet Diseases

The evolving climate amid small millet cultivation has led to a surge in new diseases and heightened aggressiveness of existing pathogens. While current management strategies offer distinct advantages, they also present limitations. Addressing small millet diseases requires not just conventional but also innovative, cost-effective, and efficient mitigation strategies, integrating them into integrated disease management (IDM). Considering the rising costs due to agrochemicals, exploring eco-friendly, pioneering methods becomes crucial. These include nanomolecule formulations, endophyte biopriming, utilizing rhizosphere and phyllosphere-derived bioagents, adopting biofumigation, plant immunization, genome-assisted breeding, genome editing, and multi-omics approaches. These novel green technologies serve as viable alternatives, complementing existing methods rather than replacing them.

Harnessing germplasm The primary aim of plant disease management is to establish enduring resistance, reducing reliance on chemicals. Scrutinize novel germplasm and potential wild relatives to identify major gene sources for resistance. Integrating these genes into elite cultivars using molecular marker-based breeding techniques facilitates the rapid development of resistant varieties.

Embracing microbiome principles The recent evolution of utilizing microbiomes in plant disease management has transformed our understanding of how these microbiomes interact with plants and their environment. This shift may steer research and practices away from conventional approaches. While the potential benefits of employing microbiomes in disease management are evident, further research is necessary for effective exploitation (Jeger et al. 2021).

Utilizing genomic tools (CRISPR/Cas9) for enhanced resistance/control The CRISPR/Cas9 system has significantly impacted plant pathology and breeding by enabling precise genome editing through RNA-guided DNA endonuclease activity. This technology has bolstered disease resistance in various commercial crops like rice, wheat, tomato, and grape. Beyond crop genome editing, targeting fungal and fungi-like pathogens using CRISPR/Cas9 offers eco-friendly insights into plant disease management. Effective disease control involves leveraging multiple plant disease resistance mechanisms via CRISPR/Cas9. Exploring fungal and oomycete genomes using this system offers potent approaches (Paul et al. 2021), serving a dual purpose of deepening pathogen understanding and formulating management strategies. CRISPR/Cas9 has efficiently and precisely facilitated targeted mutagenesis in plants, augmenting resistance against fungal diseases.

Holistic multi-omics approach This method enables a comprehensive grasp of plant tissue interactions with microbial communities, shedding light on their abilities and effects. Through multi-omics, diverse plant defense pathways and the pathogen's responses, including the production of various toxic metabolites, are studied. Numerous pathways within plants, such as the salicylic acid (SA) signaling pathway, hold paramount importance, alongside critical plant hormones like abscisic acid and jasmonic acid. These pathways regulate acquired resistance in plants and mediate interactions among members of plant microbial communities (Crandall et al., 2020).

Assessing the current disease landscape in small millets, it's evident that fungal diseases present a substantial challenge to both current and future production. This chapter aims to briefly outline the critical aspects of some of the most impactful diseases in small millets, jeopardizing their production potential. While recognizing the presence of additional minor diseases like bacterial and viral infections, this chapter focuses primarily on fungal diseases due to their significant real-world threat and their considerable impact on yield. It serves as a reference for field pathologists, offering a concise yet comprehensive understanding of the disease complexity and encouraging a holistic perspective in their evaluation, despite space limitations.

References

Alam MM, Khan MA, Saxena SK (1973) Some new host records of root knot nematode, *Meloidogyne incognita* (Kofoid & White, 1919) Chitwood, 1949. Indian J. Nematol 3:159–160

Alcorn JL (1988) The taxonomy of "Helminthosporium" species. Annual Review of Phytopathol 26(1):37–56

Anitha K, Chakrabarty SK, Rao RDVJ, Prasada Babu B, Sarath AB, Varaprasad KS, Khetarpal RK (2005) Quarantine processing of exotic cereals and millets germplasm during 1986–2003. IntJPlant Prot 33:105–110

Anonymous (1975) Annual report of the virologist, AICRP small millets 1974–75. UAS, Bangalore

Anonymous (2021) Annual Progress Report of Annual Progress Report of Indian Council of Agricultural research (ICAR)-All India Co-Ordinated Research Project (AICRP) on Small Millets. https://www.millets.res.in/.

Aycock R (1966) Stem rot and other diseases caused by *Sclerotium rolfsii*. North Carolina Agricultural Experiment Station Bull. No. 174

Basta BK, Tamang DB (1983) Preliminary report on the study of millet diseases in Nepal. In: Maize and finger millet. 10th Summer Workshop 23–28 Jan, 1983, held at Rampur, Chitwan, Mysore

Billimoria KN, Hegde RK (1971) A new bacterial disease of ragi, Eleusine coracana (Linn) Gaertn in Mysore state. Curr. Sci. 40:611–612

Butler EJ (1918) Fungi and diseases in plants. Thaker Spinck and Co., Calcutta, p 547

Butler EJ, Bisby GR (1931) Fungi of India. Sci. Monogra. XVIII, Calcutta

Chandrasekaran NM (1964) Studies on the reniform nematode *Rotylenchulus reniformis* with special reference to its pathogenicity on castor (*Ricinus communis*) and ragi (*Eleusine coracana*). M.Sc. (Agri.) thesis, Univ. Of Madras

Coleman LC (1920) The cultivation of ragi in Mysore. Bull Dep Agric Mysore Gen Ser:11–12

Crandall SG, Gold KM, Jiménez-Gasco MDM et al (2020) A multi-omics approach to solving problems in plant disease ecology. PLoS One 15(9):e0237975. https://doi.org/10.1371/journal.pone.0237975

Das IK (2017) Millet diseases: current status and their management. Millets and sorghum: biology and genetic improvement, pp 291–322

Das L, Girija VK (1989) Sheath blight of ragi. Curr Sci 58:681–682

Das IK, Nagaraja A, Tonapi VA (2016) Diseases of millets- a ready reckoner. Indian Institute of Millets Research, Rajendranagar, Hyderabad 500030, Telangana pp 67

Das IK, Krishnan S, Tonapi VA (2021) Current status and management of foxtail millet [*Setaria italica* (L.) Beauv.] blast disease. In: Blast disease of cereal crops. Springer, Cham, pp 201–209

Das IK, Palanna KB, Patro TSSK, Kumar S, Ganapathy KN, Elangovan M, Rawat L, Raveendra HR, Patil H, Kannababu N, Tonapi VA (2023) Pathogenic variations of Pyricularia grisea population causing blast of finger millet (Eleusine coracana) in India. Indian Phytopathology. https://doi.org/10.1007/s42360-023-00598-w

Desai SG, Thirumalachar MJ, Patel MK (1965) Bacterial blight disease of *Eleusine coracana* Gaertn. Indian Phytopath 28:384–386

Dubey SC (1995) Banded blight of finger millet caused by *Thanatephorus cucumeris*. Indian J Mycol Pl Pathol 25:315–316

Dwivedi SL, Upadhyaya HD, Senthilvel S, Hash CT, Fukunaga K, Diao X, Prasad M (2012) Millets: genetic and genomic resources. Plant Breed Rev 35:247–375

Gaikwad AP, D' Souza TF (1987) A comparative study on *Pyricularia* spp. J Maharashtra Agril Univ 12:134–135

Gokte N, Mathur VK, Rajam AL (1992) *Panicum miliaceum*- a new host record for *Aphelenchoides besseyi*. Indian Phytopathol 45

Govindu HC, Thirumalachar MJ (1961) Studies on some species of *Ephelis* and *Balansia* occurring in India. Mycopathologia 14:189–197

Inukai T, Vales MI, Hori K, Sato K, Hayes PM (2006) RM1 confers blast resistance in barley and is located within the complex of resistance genes containing Mla, a powdery mildew resistance gene. Mol Plant-Microbe Interact 19:1034–1041

Jain AK, Tripathi SK (2002) Pathogenic variability in the isolates of *Sorosporium paspali-thunbergii* causing head smut of kodo millet. In: Rodrigues BF, Gour HN, Bhatt DJ, Kamat N (eds) Advances in fungal diversity and host pathogen interactions. Goa, Dept. of Botany, Goa University, pp 85–89

Jain AK, Tripathi SK (2005) Management of grain smut (*Macalpinomyces sharmae*) in little millet. Indian Phytopath. 60(4):467–471

Jain AK, Yadava HS (1997) Recent approaches in disease management of Small Millets. Proc. Nat. Semi. on Small Millets—Current Research Trends and Future Priorities as Food, Feed and in processing for Value Addition, held at TNAU, Coimbatore (T.N.) from 23–24 April, 1997 pp 31–33

Jain AK, Tripathi SK, Singh RP (2006) *Macalpinomyces sharmae*—A new threat for the cultivation of little millet in Madhya Pradesh. Proc. Nat. Symp. on Emerging Plant Diseases, their diagnosis and management. N.B.U., Siliguri (W.B.) pp 31–32

Jeger M, Beresford R, Bock C et al (2021) Global challenges facing plant pathology: multidisciplinary approaches to meet the food security and environmental challenges in the mid-twenty-first century. CABI Agric Biosci 2:20. https://doi.org/10.1186/s43170-021-00042-x

Kam J, Puranik S, Yadav R, Manwaring HR, Pierre S, Srivastava RK, Yadav RS (2016) Dietary interventions for type 2 diabetes: how millet comes to help. Front Plant Sci 7:1454

Karthikeyan V, Gnanamanickam SS (2008) Biological control of Setaria blast (Magnaporthe grisea) with bacterial strains. Crop Prot 27(2):263–267

Krishnaprasad KS, Krishnappa K, Setty KGH, Reddy BMR, Reddy HR (1980) Susceptibility of some cereals to ragi cyst nematode, *Heterodera delvi*. Curr. Res. 9(6):114–115

Kulkarni GS (1922) The smut of nachani or ragi (*Eleusine coracana*). Ann Appl Biol 9:184–186

Kulkarni S, Patel MK (1956) Study of the effect of nutrition and temperature on the size of spores in *Pyricularia setariae* Nishikado. Indian Phytopath 9:31–38

Kumar B (2011) First record of smut disease of foxtail Millet caused by *Ustilago crameri*. Korn J Mycol Pl Pathol 41(3):459–461

Kumar B (2013) Diseases of small millets in Uttarakhand and their management. In: Singh KP, Prajapati CR, Gupta AK (eds) Innovative approaches in plant disease management-crop diseases and their management. LAP Lambert Academic Publishing, Germany, pp 257–287

Kumar B (2016) Status of small millets diseases in Uttarakhand. International Journal of Plant Protection 9(1):256–263

Kumar B, Singh KP (2010) Important small millets diseases in India and their management. Plant Pathology Section, College of Forestry and Hill agriculture, Hill Campus, Ranichauri, Tehri Garhwal, Uttarakhand

Kumar VBS, Amruta Bhat S, Nagaraju (2007) Virulence analysis of *Pyricularia grisea* (Cke.) Sac. on different finger millet genotypes and cultivars to determine race differential lines. Environ Ecol 25(1):190–192

Lee-Du Hyung (1997) Morphological characters and seed transmission of *Bipolaris panici-miliacei* causing leaf spot of common millet. Korean J Pl Pathol 13(1):18–21

Li ZY, Wang N, Dong L, Bai H, Quan JZ, Liu L, Dong ZP (2015) Differential gene expression in Foxtail Millet during incompatible interaction with *Uromyces setariae italicae*. PLOS ONE 10(4):e0123825. https://doi.org/10.1371/journal.pone.0123825

Lule D, de Villiers S, Fetene M, Bogale T, Alemu T, Geremew G et al (2014) Pathogenicity and yield loss assessment caused by *Magnaporthe oryzae* isolates in cultivated and wild relatives of finger millet (*Eleusine coracana*). Indian. Journal of Agricultural Research. 48(4):258–268

Malleshi NG, Klopfenstein CF (1998) Nutrient composition, amino acid and vitamin contents of malted sorghum, pearl millet, finger millet and their rootlets. Int J Food Sci Nutr. 49:415e22

Mariappan V, Natarajan C, Kandaswamy TK (1973) Ragi streak disease in Tamil Nadu. Madras Agric J 60:451–453

McRae W (1920) Detailed administration report of the Government Mycologist for the Year 1919–20. Madras Agric. Dept, India

McRae W (1924) Economic Botany, Part III—Mycology Annul Rept Bot Sci Advice, India (1922–23) pp 31–35

Mohanti NN, Padhi NN, Mohanti AK (1969) A new record of *Ephelis* on *Panicum miliare*. Indian Phytopath 22:138–139

Mohanty KC, Das SN (1976) Free amino acids in the roots of finger millet plants infected with ring nematodes. Indian Phytopath 29:434–436

Mohanty UN, Mohanty NN (1957) *Ephelis* on some new hosts. Sci Cult 22:465–468

Mundkur BB (1939) A contribution towards a knowledge of Indian Ustilaginales. Trans Br Mycol Soc 23:105

Mundkur BB (1943) Studies in Indian cereal smuts. VI. The smuts on sawan, *Echinochloa frumentacea*. Indian J Agric Sci 13:631–633

Munjal RL, Lall G, Chona BL (1961) Some *Cercospora* species from India VI. Indian Phytopath 14:179–190

Murakami J, Tosa Y, Kataoka T, Tomita R, Kawasaki J, Chuma I, Sesumi Y, Kusaba M, Nakayashiki H, Mayama S (2000) Analysis of host species specificity of *Magnaporthe grisea* toward wheat using a genetic cross between isolates from wheat and foxtail millet. Phytopathol 90:1060

Nagaraja A, Kumar J, Jain AK, Narasimhudu Y, Raghuchander T, Kumar B, Hanumanthe Gowda B (2007) Compendium of small millets diseases. GKVK Campus Bangalore, Project coordination cell, All India Coordinated Small Millets Improvement Project, UAS, p 80

Nagaraja A, Kumar B, Jain AK, Patro TS, Nageswar Rao TG (2016) Diseases of small millets. Diseases of field crops and their management. Indian Phytopathological Society, New Delhi, pp 295–371

Narasimhan MJ (1934) Report of the work done in Mycology section. Dept Agri, Mysore, pp 1932–1933

Nema AG, Kulkarni SN, Pall BS (1978) Bacterial leaf streak of kodo (*Paspalum scrobiculatum* L.). Sci Cult 45(9):365–366

Nishikado Y (1917) Studies on the rice blast fungus, (I). Berichte des Ohara Instituts für Landwirtschaftliche Forschungen. 1(2):171–218

Ou SH (1980) Pathogen variability and host resistance in rice blast disease control. Annu Rev Phytopathol 18:167–187

Padhi NN, Das SN (1982) Host range of the spiral nematode, *Helicotylenchus abunaamai*. Indian J Nematol 12(1):53–59

Palaniswami A, Govindaswamy CV, Vidhyasekaran P (1970) Studies on blast disease of Tenai. Madras Agricultural Journal 57:686–693

Palanna KB, Vinaykumar HD, Prasanna SK, Rajashekara H, Devanna BN, Anilkumar C, Jeevan B, Raveendra HR, Khan F, Bhavana CHS, Upadhyay V, Patro TSSK, Rawat L, Rajesh M, Saravanan PT, Netam P, Rajesha G, Das IK, Patil HE, Jain AK, Saralamma S, Nayaka SC, Prakash G, Nagaraja TE (2023a) Exploring the diversity of virulence genes in the Magnaporthe population infecting millets and rice in India. Front. Plant Sci. 14:1131315. https://doi.org/10.3389/fpls.2023.1131315

Palanna KB, Vinaykumar HD, Koti PS, Jeevan B, Rajashekara H, Raveendra HR et al (2023b) Geographic distribution, host preference and phylogenetic relationships among Pyricularia species inciting millet and rice blast disease in India. Plant Pathology 00:1–14. https://doi.org/10.1111/ppa.13830

Pall BS, Jain AC, Singh SP (1980) Diseases of lesser millets. JNKVV, Jabalpur, pp 53–54

Parthasarathy S, Thiribhuvanamala G, Prabakar K (2020) Diseases of field crops and their management. CRC Press

Patro TSSK, Meena A, Palanna KB, Jain AK et al. (2018) Text book on diseases of small millets. Agriculture Research Station, ANGRAU, Vizianagarm, 1-72pp

Paul NC, Park SW, Liu H et al (2021) Plant and fungal genome editing to enhance plant disease resistance using the CRISPR/Cas9 system. Front Plant Sci 12:700925. https://doi.org/10.3389/fpls.2021.700925

Pendse GS (1974) A note of new chemical compound isolated from a fungus hitherto unknown. Experiential 30:107–108

Poonacha TT, Bhavana CS, Farooqkhan, Ramesh GV, Gavayi N, Koti PS, Palanna KB, Rajashekara H, Rajesh G, Das IK (2023) Blast disease of millets: Present status and future perspectives. Millets-Rediscover Ancient Grains

Prajapati VP, Sabalpara AN, Pawar DM (2013) Assessment of yield loss due to finger millet blast caused by *Pyricularia grisea* (Cooke) Sacc. Trends in Biosci 6(3):876–788

Prakash P, Ravishankar CR, Prasad N (2007) Blast incidence as affected by nitrogen levels in finger millet (*Eleusine coracana*) genotypes. J Mycol Pl Pathol 37:590

Praveen B, Nagaraja A, Prasanna Kumar MK et al (2021) First report of *Alternaria alternata* causing leaf blight on little millet (*Panicum sumatrense*) in India. Pl Dis 105(4):1202

MC Rae W (1922), Report of the Imperial Mycologist. *Agric. Res. Inst. Pusa Scientific Reports* 1921–22, pp.44–50.

Ramesh GV, Palanna KB, Vinaykumar HD, Koti PS, Mahesha HS, Nagaraja TE, Tonapi VA, Jeevan B (2021a) Occurrence and characterization of *Bipolaris setariae* associated with leaf blight of browntop millet (*Brachiaria ramosa*) in India. J Phytopathol 169(10):613–622

Ramesh VG, Palanna KB, Praveen B, Vinaykumar HD, Koti PS, Sonavane P, Tonapi VA (2021b) First confirmed report of leaf blight on browntop millet caused by *Bipolaris setariae* in Southern Peninsular India. Plant Dis. https://doi.org/10.1094/PDIS-11-20-2445-PDN(NAAS-9.53)

Ranganathaiah KG (1972) Investigations on seed borne nature and cultural aspects of *Ephelis oryzae* Syd., incitant of udbatta disease of Rice (*Oryzae sativa* L.), Ph.D thesis submitted to the University of Mysore, Mysore, 138p

Rangaswami G, Mahadevan A (1999) Diseases of cereals. In: Diseases of crop plants in India, 4th edn. Prentic Hall of India, Pvt. Ltd., New Delhi, pp 160–264

Rangaswami G, Prasad NN, Eswaran KSS (1961) Two new bacterial diseases of sorghum. Andhra Agric J 8(6):269–272

Rawat L, Kukreti A, Bisht TS (2016) First record of leaf blast disease in browntop millet in Mid Hills of Uttarakhand.

Rosen HR (1922) A bacterial disease of foxtail (*Chaetochloa lutescens*). Annals of the Missouri Botanical Garden 9(4):333–402

Safeeulla KM (1955) Comparative morphological and cytological studies in some species of the genera *Albugo*, *Sclerophthora* and *Sclerospora*. Ph.D. Thesis, Univ. of Mysore, India. pp 179

Saralamma S, Kiranmai MJ, CVCM R (2019) Screening of genotypes of *Setaria italica* (L.) against rust disease under field conditions during rabi season. Journal of Research ANGRAU 47(4):84–87

Sekar J, Ramalingam P (2019) Potential of Finger Millet Indigenous. Harnessing Useful Rhizosphere Microorganisms for Pathogen and Pest Biocontrol, 2:277–103

Setty KGH (1975) Studies on ragi cyst nematode *Heterodera* Sp. Report of UNDP/ICAR Research Project, University of Agril. Sciences, Hebbal, Bangalore

Sharma PN, Khare MN (1987) Two new smut diseases in little millet (*Panicum sumatrense*) from India. Acta Botanica Indica 15:143–144

Shaw FJF (1921) Report of the Imperial Mycologist Scient. Reports Agric. Res. Inst. Pusa 1920–21:34–40

Sill WH, Agusiobo PC (1955) Host range studies of the wheat streak-mosaic virus. Plant Dis Reptr 39:633–642

Sinha AP, Upadhyay JP (1997) *Millet ke rog*. Directorate of Publication. G.B. Pant University of Agriculture and Technology, Pantnagar, Uttarakhand, p 180

Subbayya J, Raychaudhuri SP (1970) A note on a mosaic disease of ragi (*Eleusine coracana*) in Mysore, India. Indian Phytopath 23:144–148

Sundararaman S (1921) *Ustilago crameri* Koern on *Seatria italica*. Beauv. Bull. Agric. Res. Inst., Pusa 97(11)

Thirumalachar MJ, Mundkur BB (1947) Morphology and the mode of transmission of the ragi blight. Phytopathol 37:481–486

Thirumalachar MJ, Narasimhan MJ (1949) Downy mildew of *Eleusine coracana* and *Iseilema laxum* in Mysore. Indian Phytopathol 2:47–51

Ulstrup AJ (1955) Crazy top of some wild grasses and the occurrence of the sporangial stage of the pathogen. Plant Dis Rep 39:839–841

Vasudeva RS (1954) The fungi of India. Sci Monogra ICAR pp 12

Venkatarayan SV (1947) Diseases of ragi (*Eleusine coracana*). Mysore Agric J 24:50–57

Vidhyasekaran P (1971) Saprophytic survival of *Helminthosporium nodulosum* and *H. tetramera* in soil. Indian Phytopath 24:347–353

Viji G, Gnanamanickam SS, Levy M (2000) DNA polymorphisms of isolates of *Magnaporthe grisea* from India that are pathogenic to finger millet and rice. Mycol Res 104:161–167

Viswanath S (1992) Management of biotic factors (diseases). In 6th Annual Small Millets Workshop at BAU, Ranchi-Kanke (Bihar) from 30 April-2nd May 1992

Viswanath S, Seetharam A (1989) Diseases of small millets and their management in India. In: Small Millets in Global Agriculture (Eds.) Seetharam A, Riley KW and Harinarayana G, Oxford & IBH Publishing Co. Pvt. Ltd., New Delhi, India. pp 237–253.

Wang CS (1943) Studies on the cytology of *Ustilago crameri*. ibid 33:1122–1133

Yadava HS, Jain AK (2006) Recent advances in kodo millet. ICAR, New Delhi: 84p

Yenagi NB, Handigol JA, Ravi SB, Mal B, Padulosi S (2010) Nutritional and technological advancements in the promotion of ethnic and novel foods using the genetic diversity of minor millets in India. Indian J Plant Genet Res 23(1):82–86

Insect Pests of Millet Crops and Their Integrated Management

Amit Umesh Paschapur, Ipsita Samal, Prabhu C. Ganiger, P. N. Guru, and Ashish Kumar Singh

Abstract

Insect pests pose a significant threat to millet crops, undermining their resilience and productivity despite their natural hardiness and adaptability to harsh conditions. Millets, including species like little millet, kodo millet, barnyard millet, and pearl millet, are essential staples in semi-arid and arid regions, particularly in Africa and Asia. These crops are crucial for food security, providing essential nutrients and supporting the livelihoods of millions, especially in rural areas where agriculture is the primary economic activity. However, millet crops face significant challenges from insect pests throughout their growth cycle. Pests such as stem borers, shoot flies, aphids, and armyworms have evolved mechanisms to attack millet plants, causing extensive damage and reducing yields. The impact of these pests extends beyond agricultural losses, affecting the economic and social well-beings of smallholder farmers. Reliance on chemical pesticides exacerbates environmental and health risks and promotes pesticide resistance. Therefore, a holistic approach integrating various pest management strategies within the framework of integrated pest management (IPM) is essential. This chapter deals with the major and minor insect pests of millet, their biology, ecology, and behaviour and explores comprehensive IPM strategies, emphasizing sustainable practices. The aim is to equip farmers, researchers, and policymakers

A. U. Paschapur (✉) · A. K. Singh
ICAR-Vivekananda Parvatiya Krishi Anusandhan Sansthan, Almora, Uttarakhand, India

I. Samal
ICAR-National Research Centre on Litchi, Muzaffarpur, Bihar, India

P. C. Ganiger (✉)
PC Unit, ICAR-AICRP on Small Millets, UAS, Bangalore, India

P. N. Guru
ICAR-Central Institute of Post-Harvest Engineering and Technology, Ludhiana, Punjab, India

D. C. Joshi et al. (eds.), *Minor Millets*,
https://doi.org/10.1007/978-981-96-4265-6_7

with the knowledge and tools to enhance millet production sustainably and promote food security.

Keywords

Millet insect pests · Integrated pest management · Ecological engineering · Crop resilience · Pest population dynamics · Pest ecology

1 Introduction

Insect pests pose a significant threat to millet crops, undermining their resilience and productivity despite their natural hardiness and adaptability to harsh conditions (Venkateswarlu and Shanker 2011; Dutta et al. 2024). Millets, encompassing a diverse group of small-seeded grasses such as little millet (*Panicum sumatrense*), kodo millet (*Paspalum scrobiculatum*), barnyard millet or Japanese millet (*Echinochloa esculenta*), browntop millet (*Brachiaria ramosa*), foxtail millet (*Setaria italica*), proso millet (*Panicum miliaceum*), finger millet (*Eleusine coracana*), pearl millet (*Pennisetum glaucum*) have long been cultivated as staple foods in regions characterized by semi-arid and arid climates, notably in Africa and Asia (Paschapur et al. 2021; Rakshit et al. 2023). These crops play a vital role in ensuring food security, providing essential nutrients, and supporting the livelihoods of millions of people, particularly in rural areas where agriculture is a primary economic activity (Tiwari et al. 2022).

Despite their historical and nutritional significance, millet crops face formidable challenges from insect pests throughout their growth cycle (Sharma et al. 2010). These pests, ranging from stem borers and shoot flies to aphids and armyworms, have evolved various mechanisms to attack millet plants, causing damage to stems, leaves, and grains. Stem borers, for instance, tunnel into the stems of millet plants, weakening their structural integrity and impeding nutrient transport, ultimately leading to stunted growth and decreased grain yield (Degri et al. 2014). Shoot flies target young seedlings, laying eggs on the central shoot, which results in the characteristic "dead-heart" symptom, where the central whorl dies, compromising plant vigour and density (Sathish et al. 2017). Aphids, on the other hand, feed on the sap of millet plants, causing yellowing, curling, and stunting of leaves, while also excreting honeydew that promotes the growth of sooty mould and transmitting viral diseases (Serba and Michaud 2019). Armyworms, with their voracious appetite, can defoliate millet plants rapidly, leading to substantial yield losses if left unchecked (Sharma and Sullivan 2000).

The impact of insect pests on millet crops is not only agronomically significant but also economically and socially consequential. Smallholder farmers, who often rely on millet cultivation for their livelihoods, are particularly vulnerable to the ravages of pest infestations. Crop losses resulting from pest damage can have severe repercussions, affecting food availability, income generation, and overall household well-being (Kahane et al. 2013). Moreover, the reliance on chemical pesticides as a primary means of pest control can exacerbate environmental degradation, human health risks, and pesticide resistance amongst pest populations, highlighting the urgency of adopting sustainable pest management practices (Gill and Garg 2014; Rani et al. 2021).

Addressing the challenges posed by insect pests in millet cultivation requires a holistic approach that integrates various pest management strategies within the framework of integrated pest management (IPM). IPM emphasizes the use of multiple complementary tactics, including cultural, biological, mechanical, and chemical controls, to minimize pest damage while maximizing environmental and economic sustainability (Karlsson Green et al., 2020). By employing IPM practices, farmers can mitigate the adverse effects of insect pests on millet crops while minimizing reliance on chemical inputs, conserving natural resources, and safeguarding ecosystem health (Ehi-Eromosele et al. 2013).

1.1 Significance of the Study

It is documented that the total number of publications on "insect pests of millets and their management" increased from 2000 as described in (Fig. 1a), showing their prominence, though they contribute more to agricultural, veterinary, and food sciences, followed by biological sciences and crop and pasture production as discussed in (Fig. 1b). Further, it was also observed that the studies on "insect pests of millets and their management" contributed significantly to SDG 2: Zero Hunger followed by SDG 15: Life on Land and SDG 13: Climate Action (Fig. 1c). Furthermore, majority of publications since 2000 have been found to be in form of Articles (18,810) followed by chapters (18,095) and edited books (7011) (Fig. 1d). Moreover, Fig. 2a (co-occurrence map) revealed that terms like "larvae", "adult", "exposure", "inhibitor", "plant growth", and "climate change" were found to have significantly higher occurrence (Fig. 2a). Furthermore, the co-authorship analysis of countries (minimum threshold of five documents per author and five citations per document) revealed that, amongst total countries, "the United States", "China", "the United Kingdom", and "India" had significant contributions (Fig. 2b) towards "insect pests of millets and their management". Considering this, in this chapter, we will deal into an in-depth exploration of the major and minor insect pests that afflict millet crops, examining their biology, ecology, and behaviour. We will also discuss comprehensive IPM strategies for effectively managing these pests, drawing upon the latest research findings, case studies, and practical applications. By providing a thorough understanding of insect pests and their management, this chapter aims to empower farmers, extension agents, researchers, and policymakers with the knowledge and tools needed to sustainably enhance millet production and promote food security in regions where millets are essential crops. Through collaborative efforts and a commitment to innovation, we can mitigate the challenges posed by insect pests and realize their full potential.

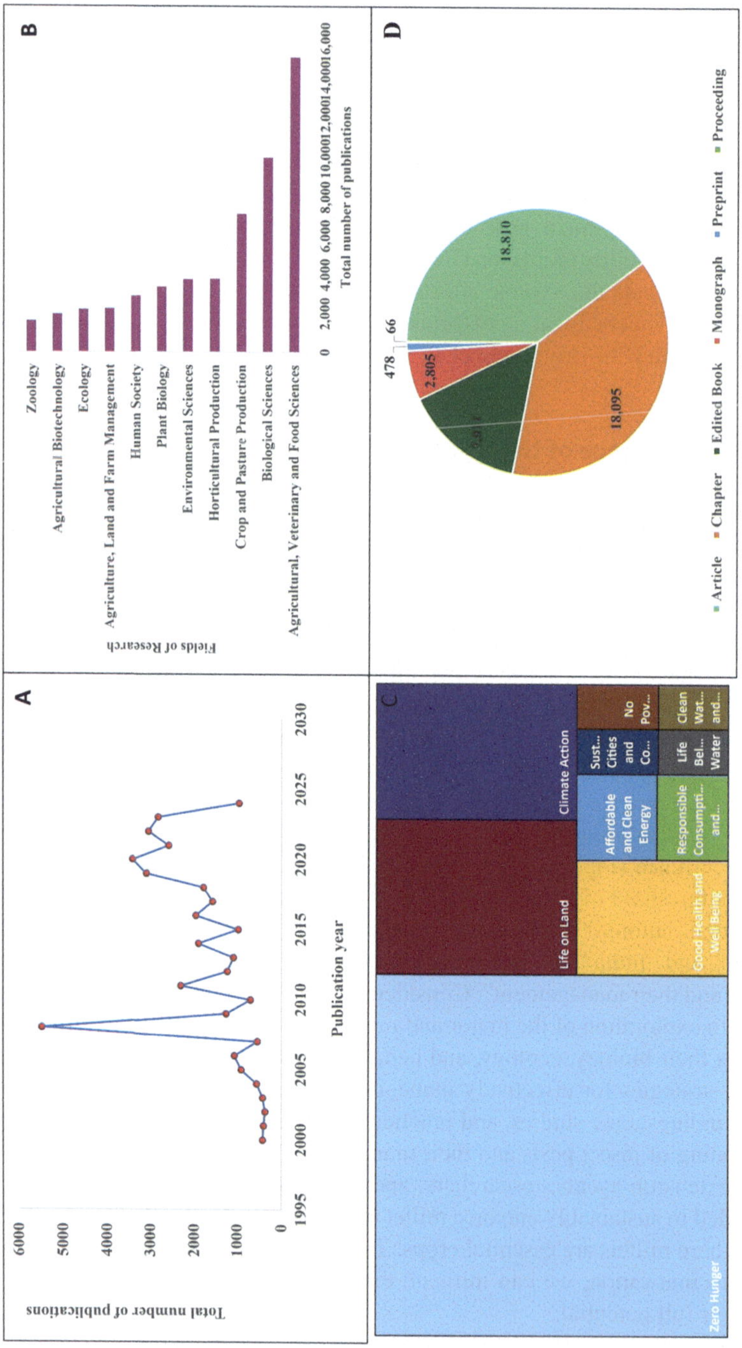

Fig. 1 (a–d) Total number of publications on "Insect pests of millets and their management"

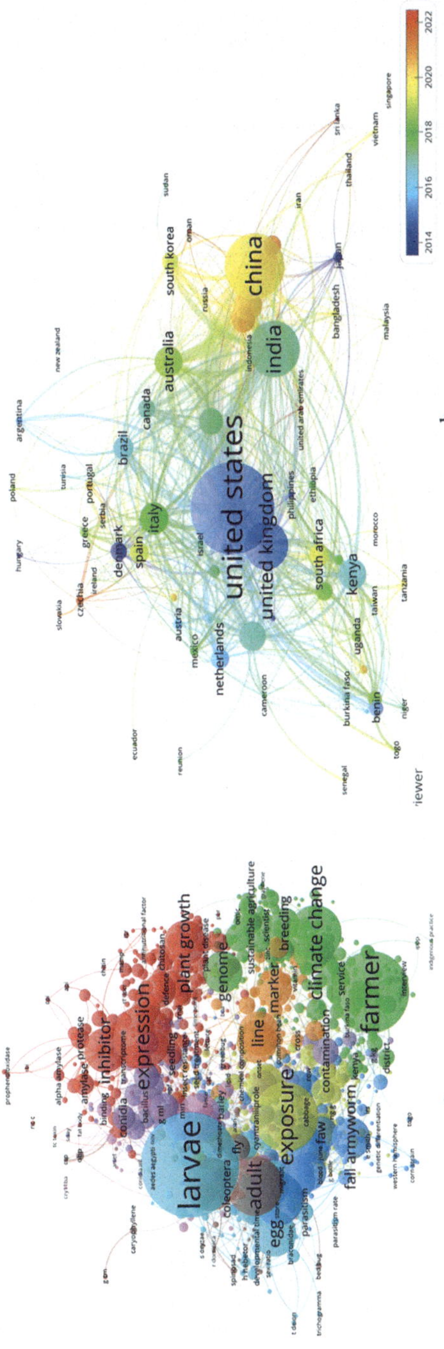

a

b

Fig. 2 (a–b) Key words used and countries publishing works on insect pests of millets and their management

2 Major Insect Pests of Millet Crops

Stem borer complex (*Chilo* spp., *Saluria* spp., *Sesamia inferens*, *Scirpophaga* spp.)

2.1 Pink Stem Borer (*Sesamia inferens*)

Description The pink stem borer (Fig. 3) is a significant pest of millet, particularly pearl millet. The larvae are creamy white with a pinkish tinge and a brown head capsule. Adults are moths with light brown wings.

Damage Larvae bore into the stem, causing internal tunnelling that weakens the plant and disrupts nutrient flow. This damage leads to stunted growth, reduced grain yield, and potential plant death. Infestation signs include dead hearts, yellowing of leaves, and broken stems.

2.1.1 Management

Monitoring Regular field scouting and pheromone traps to monitor adult moth populations. For instance, pheromone traps should be installed at the rate of ten traps per hectare, checked weekly to count and remove captured moths. Action thresholds can be set, such as capturing five to ten moths per trap per week, for timely interventions.

Biological Control Release of parasitoids such as *Trichogramma* spp., which target borer eggs. These parasitoids can be released at a rate of 50,000–100,000 per hectare during the early stages of the borer population build-up.

Cultural Practices Planting resistant millet varieties, practicing crop rotation, and destroying crop residues post-harvest to break the pest life cycle. For instance, removing and burning crop residues immediately after harvest can prevent overwintering of larvae.

Chemical Control Targeted application of insecticides like chlorantraniliprole, indoxacarb, flubendiamide, etc. Insecticides should be applied as soil treatment or

Fig. 3 Egg, larva, dead heart, and adult pink stem borer affecting minor millets

Fig. 4 Dead heart or central shoot of shoot fly attack in small millets

foliar sprays, following economic threshold levels and integrated with other IPM tactics to avoid resistance.

2.2 Millet Shoot Fly (*Atherigona soccata, Atherigona approximata, Atherigona* spp.)

Description The millet shoot fly is a major pest during the early stages of crop growth. Adults are small, greyish flies that lay cigar-shaped eggs on lower side of the leaves on young millet plants (Paschapur et al. 2022).

Damage Larvae damage the central shoot, leading to the characteristic "dead-heart" symptom (Fig. 4) where the central whorl dies. Infestation can lead to significant yield losses if not managed early. At later stages, the plants form excessive tillering and yield is drastically reduced.

2.2.1 Population Dynamics of Shoot Flies (Dead Heart) on Millets

The number of dead-heart/20 plants/m^2 area was recorded at 14, 21, 28, 45, and 60 days of the crop growth by sampling dead heart plants in different millets, viz., kodo millet, little millet, foxtail millet, barnyard millet, proso millet, finger millet, and sorghum at ZARS, UAS, GKVK Bengaluru starting from September week I to November week II. Shoot fly activity was more during September week I to September week III, i.e., 21–28 days after sowing (Fig. 2). Total number of eggs used for infestation × LPI (%) millets under Bengaluru conditions revealed that shoot fly population was more abundant on little millet (78%) followed by proso millet (46%), and kodo and foxtail millets had 33% and 28% dead hearts, respectively (Fig. 5).

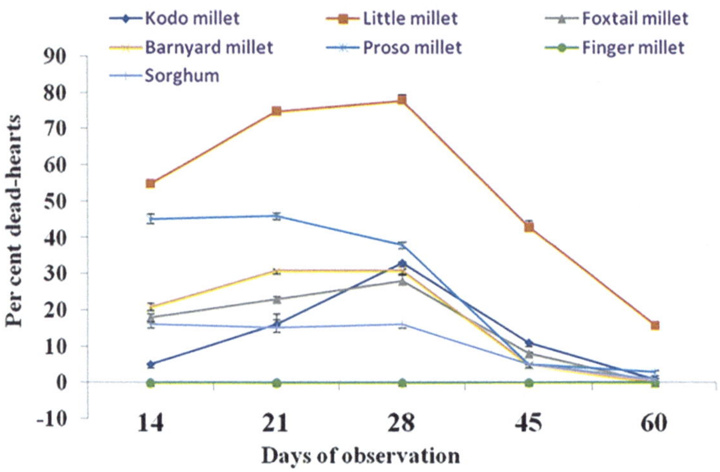

Fig. 5 Shoot fly infestation dynamics in different minor millet species during the growing season

Management Key management strategies include the following:

Cultural Practices Timely sowing to avoid peak fly populations. Early planting can help avoid the peak population of flies. Increase the seed rate. Removal of affected plants and destruction can help in reducing the pest population.

Seed Treatment Use of insecticide-treated seeds to protect young plants. For example, treating seeds with imidacloprid at 5 g per kg of seed can provide protection against early infestation.

Biological Control Utilization of parasitoids like *Spalangia* spp. and predators such as ground beetles. Augmentative releases of parasitoids can be done when fly populations are observed.

Chemical Control Application of insecticides such as cypermethrin or deltamethrin when adult flies are detected in significant numbers. Sprays should be timed when adults are active, usually in the morning or late afternoon, and should target the base of the plants and lower side of the leaves, where flies are likely to lay eggs. Whirl application of chlorantraniliprole granules for maggot control is also recommended.

2.3 Aphid (*Schizaphis graminum*)

Description Aphids are small, green, soft-bodied insects that suck the plant sap of millets (Fig. 6).

Fig. 6 Leaf or shoot aphid affecting minor millets

Damage Aphids cause direct damage by sucking sap, leading to yellowing, stunting, and leaf curling. They also excrete honeydew, promoting sooty mould growth, and can transmit viral diseases.

Management Effective control measures include the following:

Biological Control Encouraging natural predators such as lady beetles, lacewings, and parasitic wasps like *Aphidius* spp., Lady beetles can be released at rates of 1000–10,000 per hectare, depending on aphid densities.

Botanical Insecticides Using neem-based products and insecticidal soaps. Neem oil (at 3–5%) can be sprayed on infested plants to deter feeding and disrupt aphid reproduction.

Cultural Practices Planting resistant varieties and maintaining field hygiene to reduce aphid habitats. Avoiding excessive nitrogen fertilization can also make plants less attractive to aphids.

Chemical Control Application of selective insecticides such as diafenthiuron or pymetrozine, ensuring minimal impact on beneficial insects. Foliar applications should be based on regular monitoring and applied only when aphid populations exceed economic threshold levels.

2.4 Fall Armyworm (*Spodoptera frugiperda*)

Description The fall armyworm is an emerging and highly destructive pest with a larval stage that can devastate millet crops (Fig. 7). Larvae are brownish with characteristic stripes on the abdomen and a Y-shaped mark on the head (Suby et al. 2020).

Fig. 7 Larva and adult of fall armyworm affecting minor millets

Damage Larvae feed voraciously on leaves, creating large ragged holes, and can damage reproductive parts of the plant, severely impacting yields.

Management Key strategies for managing fall armyworm include the following:

Monitoring Using pheromone traps to detect adult moths and establish thresholds for intervention. Traps should be placed at the rate of 8–10 per hectare and checked weekly.

Biological Control Employing natural enemies like parasitoids of eggs, amongst the species, *Trichogramma pretiosum* Riley and *Trichogramma atopovirilia* Oatman and Platner were potential egg parasitoids of FAW and have been used for augmentative biological control and entomopathogenic nematodes. *Bacillus thuringiensis* (Bt) and nucleopolyhedrovirus (NPV) are effective biopesticides. *Bt* can be applied at 1 kg per hectare, while NPV should be applied at 250 LE (larval equivalents) per hectare.

Cultural Practices Implementing crop rotation and planting trap crops to divert armyworms. Intercropping with cowpeas or other non-host plants can reduce armyworm damage.

Chemical Control Targeted use of insecticides like spinosad or emamectin benzoate or flubendiamide or chlorantraniliprole or indoxacarb, considering resistance management practices. Applications should be based on larval monitoring and timed to target early instar larvae for maximum effectiveness.

Fig. 8 Grasshoppers affecting minor millets

2.5 Grasshoppers (Orthoptera: Acrididae)

Description Grasshoppers are significant pests of millet crops (Fig. 8), with a broad range of species including the *Chrotogonus* spp., *Acrida* spp., *Melanoplus* spp., and *Sphingonotus* spp. They have a robust body, large hind legs for jumping, and wings for flying. Grasshoppers can rapidly increase in population and migrate over long distances, posing a serious threat to millet fields (Paschapur et al. 2022).

Damage Grasshoppers feed voraciously on millet leaves, stems, and grains, causing extensive defoliation and reducing photosynthetic capacity. Severe infestations can lead to stunted growth, reduced yield, and, in extreme cases, complete crop loss. Their feeding creates irregular, ragged holes in the leaves and can also damage the reproductive parts of the plant, significantly impacting millet productivity.

2.5.1 Management Practices

Monitoring Using visual inspections and sweep nets to regularly monitor grasshopper populations in millet fields is essential for early detection. Predictive models that incorporate weather data can help anticipate outbreaks.

Biological Control Promoting natural enemies such as birds, spiders, and parasitoid wasps can help reduce grasshopper populations. Pathogens like *Beauveria bassiana* (fungus) and *Nosema locustae* (bacterium) can also be effective biological control agents. Application of *Beauveria bassiana* should be at a rate of 1–2 kg per hectare to ensure adequate coverage.

Cultural Practices Rotating millet with non-host crops can disrupt the grasshopper life cycle. Deep ploughing can destroy grasshopper egg pods in the soil. Removing weeds and crop residues reduce alternative food sources. Adjusting the planting date of millets to avoid peak grasshopper activity can reduce the risk of infestation.

Chemical Control Using insecticides should be a targeted approach, especially during severe outbreaks. Recommended insecticides include the following: dust formulations (10 kg/acre) work better against hoppers, viz., malathion, methyl parathion, and fenvalerate, etc., whereas bait formulations, containing carbaryl or bran baits laced with insecticide can effectively target grasshoppers. Sprays, such as indoxacarb, deltamethrin, or emamectin benzoate can be applied at early nymph stages for maximum effectiveness. Spinosad can be applied on the field bunds at a rate of 0.5–1.0 kg per hectare. Applications should be based on nymph monitoring and timed to target early instar grasshoppers, ensuring resistance management practices are in place to prevent the development of resistance in pest populations.

2.6 Root Aphid (*Hysteroneura setariae*, *Tetraneura nigriabdominalis*)

Description The root aphid (Fig. 9) is a small, soft-bodied insect that feeds on the roots of millet plants. These aphids are typically yellow to brown in colour and form dense colonies on the roots. Their feeding can cause significant damage to the plant's root system, leading to stunted growth and reduced yield.

Damage Root aphids feed by sucking sap from the roots of millet plants, which can result in weakened plants, stunted growth, and reduced nutrient and water uptake. Infested plants often exhibit yellowing and wilting, and severe infestations can lead to plant death. Additionally, root aphids can transmit plant pathogens, exacerbating the damage.

2.6.1 Management

Monitoring Regular inspection of millet fields for signs of aphid infestation is crucial. Sampling roots and examining for the presence of aphids can help in early detection. Using sticky traps placed near the soil surface can also aid in monitoring aphid activity.

Biological Control Promoting natural enemies like predatory beetles (e.g., lacewings and ladybugs) and parasitic wasps can help control root aphid populations. Applying entomopathogenic fungi such as *Beauveria bassiana* can be effective. These fungi can be applied at a rate of 1–2 kg per hectare to ensure adequate coverage and infection of the aphids.

Cultural Practices Rotating millet with non-host crops can break the life cycle of root aphids. Removing crop residues and weeds that may harbour aphids can reduce initial infestation sources. Maintaining optimal soil moisture levels can help reduce aphid colonization, as aphids prefer dry conditions.

Chemical Control Using insecticides should be considered a last resort and applied with caution to avoid disrupting beneficial insects. Soil-applied systemic

Fig. 9 Root aphids affecting minor millets

insecticides, such as imidacloprid or thiamethoxam, can be effective against root aphids.

2.7 White Grubs or Root Grubs

Description White grubs are the larvae of scarab beetles (Coleoptera: Scarabaeidae) and are significant soil-dwelling pests of millet crops. These larvae are typically C-shaped, creamy white with brown heads and can vary in size depending on their development stage. Common species affecting millets include *Phyllophaga* spp., *Holotrichia* spp., *Anomala* spp., and *Schizonycha* spp. (Shyam Prasad and Bhagwat 2015).

Damage White grubs feed on the roots and underground parts of millet plants, leading to significant root damage. Symptoms of infestation include the following:

Wilting and Yellowing Infested plants often show wilting and yellowing of leaves due to impaired water and nutrient uptake.

Stunted Growth Severe root pruning can result in stunted plant growth and reduced tillering.

Plant Death High grub populations can kill young seedlings and reduce plant stands, leading to patchy fields.

2.7.1 Management

Monitoring Regular soil sampling and examination for white grubs can help in early detection. This involves digging around the root zone of plants and inspecting for grubs. Observing plant symptoms like wilting and yellowing can also indicate potential white grub presence.

Biological Control Encouraging natural predators like birds, nematodes (e.g., *Heterorhabditis* spp.), and fungi (e.g., *Metarhizium anisopliae*) can help control grub populations. Entomopathogenic nematodes can be applied at a rate of one billion infective juveniles per hectare for effective control.

Cultural Practices Rotating millet with non-host crops can reduce white grub populations by disrupting their life cycle. Deep ploughing before planting can expose and kill grubs, reducing their numbers in the soil. Applying organic amendments such as neem cake can have a repellent effect and reduce grub infestations.

Chemical Control Soil-applied insecticides like cyantraniliprole 0.4G can be used for severe infestations. These should be applied as per the manufacturer's recommendations, ensuring minimal impact on nontarget organisms.

3 Minor Insect Pests of Millet Crops

3.1 Grain Midge (*Stenodiplosis* (Contarinia) *sorghicola* (Coquillett))

The grain midge is a small fly that lays eggs on developing millet grains. The larvae feed on the developing kernels, causing damage that reduces grain size and weight, leading to yield losses (Gahukar and Reddy 2019).

3.2 Leafhoppers (*Cicadulina* sp.)

Leafhoppers are small, sap-sucking insects that can feed on millet leaves. Their feeding activity can cause stippling, yellowing, and curling of leaves, leading to reduced photosynthesis and overall plant vigour. They are known to cause hopper burn symptoms (Kalaisekar and Padmaja 2017).

3.3 Cutworms (*Agrotis* spp.)

Cutworms are nocturnal caterpillars that feed on millet seedlings by cutting them off at the base. Their feeding activity can lead to stand reduction and patchy emergence, resulting in uneven crop growth and yield variability (Kalaisekar and Padmaja 2017; Gahukar and Reddy 2019).

3.4 Millet Bugs (*Eurystylus* spp.)

Millet bugs are small, sap-feeding insects that pierce millet seeds and suck out their contents. Their feeding activity can reduce seed quality and germination rates, leading to yield losses and reduced grain quality (Sharma and Davies 1988).

3.5 Sorghum Headworm (*Helicoverpa armigera*)

Although primarily a pest of sorghum, the sorghum headworm can also infest millet heads, feeding on developing grains. Their feeding activity can result in grain damage, reduced yield, and increased susceptibility to fungal infections (Makwana et al. 2017).

3.6 Thrips (Thysanoptera)

Thrips are tiny, slender insects that feed on millet leaves by rasping the surface and sucking out cell contents. Their feeding activity can cause silvering, stippling, and distortion of leaves, leading to reduced photosynthesis and plant growth (Kamakshi et al., 2021).

3.7 Mealybugs (Pseudococcidae)

Mealybugs are small, soft-bodied insects that feed on millet plants by piercing and sucking sap from stems and leaves. Their feeding activity can weaken plants and lead to stunted growth, particularly in young seedlings (Khairani et al. 2021).

3.8 Beetles (Chrysomelidae)

Various beetle species, such as flea beetles (*Chaetocnema* spp.) and grain beetles (*Oryzaephilus* spp.), can infest millet crops and feed on leaves, stems, or grains. While their damage may not be severe, high populations can contribute to yield losses and reduce grain quality (Kaplin and Antonov 2006).

3.9 Earwigs (Dermaptera)

Earwigs are nocturnal insects that feed on a variety of plant materials, including leaves, flowers, and seeds. While they are not considered major pests of millets, high populations can cause damage to millet heads and developing grains, particularly during periods of high humidity (Rawat et al. 2021).

3.10 Millet Stem Maggot (*Chlorops oryzae*)

The millet stem maggot is a small fly whose larvae bore into millet stems and feed internally. Infestations can lead to stem breakage, lodging, and reduced grain yield. While not widespread, millet stem maggots can be problematic in some regions, especially during periods of high rainfall (Vasanthi et al. 2014).

3.11 Wireworms (Coleoptera: Elateridae)

Wireworms are the larvae of click beetles that feed on plant roots and underground plant parts. While they are more commonly associated with cereal crops like wheat and maize, wireworms can occasionally infest millet fields, causing damage to seedlings and reducing plant vigour (Sharma et al. 2020).

3.12 Millet Sawfly (*Acantholyda erythrocephala*)

Millet sawflies are small, sawfly larvae that feed on millet leaves and stems. While they are not typically considered major pests, localized outbreaks can occur, leading to defoliation and reduced photosynthesis.

3.13 Mites (Acari)

Various species of mites, including spider mites and eriophyid mites, can infest millet plants and feed on plant tissues. While they are generally more problematic in dry and hot conditions, mite infestations can cause leaf stippling, bronzing, and reduced plant growth (Sajwan et al. 2022).

3.14 Seedcorn Maggot (*Delia platura*)

The seedcorn maggot is a small fly whose larvae feed on germinating seeds and seedlings. While more commonly associated with legumes and vegetables, seedcorn maggots can occasionally infest millet fields, leading to stand reductions and poor crop establishment (Getu et al. 2006).

3.15 Frit Fly (*Oscinella frit*)

Frit flies are small, greyish flies whose larvae feed on millet leaves and stems. While they are not considered major pests, high populations can cause leaf mining damage and reduce plant vigour, particularly in young seedlings (Nartshuk 2014).

3.16 Leafminers (Diptera)

Various species of leaf-mining flies can infest millet crops and feed on leaf tissues, causing characteristic serpentine mines. While leafminer damage is generally not severe, heavy infestations can reduce photosynthesis and plant growth.

3.17 Cetonid beetles

Cetonids are minor pests of millets, and their damage usually ranges between 5% and 15%. They mainly attack the crop during flowering stage and feed on floral parts. Due to their damage, the seed set is drastically affected. Their damage is noticed only during the periods of low rainfall and high-temperature seasons. Due to changing rainfall and temperature regimes over the year, the pest population density is crossing Economic threshold levels (ETL) levels (>10%) and necessitating the use of insecticides for their management in Indian Himalayas (Paschapur et al. 2022).

These minor insect pests, along with the major ones, highlight the diverse range of insect threats that millet crops face worldwide.

4 Integrated Pest Management (IPM) Strategies

4.1 Cultural Control

4.1.1 Crop Rotation

Rotating millet with non-host crops disrupts the life cycles of pests, reducing their populations and minimizing damage (Bragg et al. 2016). For example, rotating millet with legumes or oilseeds can help in breaking the life cycle of specific pests such as stem borers and shoot flies.

4.1.2 Intercropping

Intercropping millet with pest-repellent plants or other cereals can reduce pest incidence and spread by creating a more diverse habitat less favourable to specific pests. For instance, intercropping millet with sorghum or cowpea can reduce shoot fly and aphid populations (Poddar et al. 2019).

4.1.3 Sowing Time and Seed Treatment

Adjusting sowing times to avoid peak pest populations and treating seeds with insecticides or biopesticides can minimize early infestations and protect young plants. For example, sowing millet early in the season can help avoid peak periods of shoot fly activity, while seed treatments with insecticides like imidacloprid or thiamethoxam provide early protection against soil and seedling pests (Karuppuchamy and Venugopal 2016).

4.2 Biological Control

4.2.1 Natural Predators and Parasitoids

Encouraging or releasing natural enemies such as lady beetles, lacewings, and parasitoids like *Trichogramma* spp. helps in controlling pest populations naturally and reducing reliance on chemical pesticides. For instance, releasing *Trichogramma* wasps at 50,000–100,000 per hectare can help control stem borer populations by parasitizing their eggs (Kumar et al. 2013).

4.2.2 Biopesticides

Using biopesticides such as neem oil, *Bt*, and entomopathogenic fungi provides an eco-friendly alternative to chemical pesticides, preserving beneficial insects and reducing environmental impact. Neem oil can be applied at 3–5% concentration to deter feeding and disrupt reproduction of pests like aphids and armyworms. *Bt* formulations can be applied at 1 kg per hectare to target lepidopteran larvae, while entomopathogenic fungi like *Beauveria bassiana* and *Metarhizium anisopliae* can be used at 1 kg per hectare to control soil-dwelling pests like white grubs (Thakur et al. 2020).

4.3 Mechanical and Physical Control

4.3.1 Handpicking and Destroying Infested Plants

Manual removal of infested plants and larvae can reduce pest populations, especially in smallholder farms where such labour-intensive methods are feasible. For example, handpicking and destroying larvae of stem borers or armyworms can significantly reduce their populations (Grzywacz et al. 2014).

4.3.2 Traps and Barriers

Using pheromone traps to monitor and capture adult pests and installing physical barriers such as netting or row covers can prevent pest entry and infestation, protecting millet crops. For instance, pheromone traps for fall armyworm can help in early detection and mass trapping, while row covers can protect young seedlings from shoot flies and aphids.

4.4 Chemical Control

4.4.1 Insecticides

Applying selective insecticides when pest populations exceed economic threshold levels can protect millet crops. It is essential to follow recommended dosages and application timings to minimize environmental impact and resistance development. Integrated use of insecticides as part of an IPM strategy ensures their effectiveness and sustainability. For example, using insecticides like spinosad or emamectin benzoate for fall armyworm management should be based on monitoring and applied during early larval stages for maximum effectiveness.

4.5 Host Plant Resistance

Breeding and using pest-resistant millet varieties are a sustainable and long-term strategy. Ongoing research focuses on enhancing resistance to major pests through conventional breeding and biotechnological approaches, including genetic engineering and marker-assisted selection (Sanjana Reddy et al. 2017). For example, developing and planting millet varieties resistant to stem borers and aphids can significantly reduce pest damage and reliance on chemical control methods.

Effective management of insect pests in millet crops requires a comprehensive approach integrating cultural, biological, mechanical, and chemical methods. IPM strategies not only help in reducing pest populations and damage but also promote environmental sustainability and economic viability for farmers. Continuous research, extension services, and farmer education are crucial for the successful implementation and adaptation of IPM practices in diverse agro-ecological regions. The use of pest-resistant varieties, biopesticides, and natural predators is pivotal in maintaining ecological balance and ensuring long-term pest management solutions.

References

Bragg DE, Silvia IR, John G, Uma S, Abrol DP (2016) Integrated pest management in tropical cereal crops. In: Integrated Pest Management in the Tropics. New India Publishing Agency, New Delhi, pp 249–273

Degri MM, Mailafiya DM, Mshelia JS (2014) Effect of intercropping pattern on stem borer infestation in pearl millet (Pennisetum glaucum L.) grown in the Nigerian Sudan Savannah. Adv Entomol 28(2014):81

Dutta G, Paramanik B, Bhabai B, Layek J, Choudhury A, Dutta S, Bhattacharjee S, Rahman FH (2024) Climate change impacts and adaptation strategies for agronomic crops. In: Climate change impacts on soil-plant-atmosphere continuum. Springer, Singapore, pp 383–404

Ehi-Eromosele CO, Nwinyi OC, Ajani OO (2013) Integrated pest management. In: Weed and pest control-conventional and new challenges. IntechOpen

Gahukar RT, Reddy GV (2019) Management of economically important insect pests of millet. J Integ Pest Manag 10(1):28

Getu E, Tadesse A, Negeri M, Tefera T, Tsaheye H, Dejene A (2006) Review of entomological research on maize, sorghum and millet maize, sorghum and millet. In: Increasing crop production through improved plant protection, vol I, p 167

Gill HK, Garg H (2014) Pesticide: environmental impacts and management strategies. Pesticides-Toxic Aspects 8(187):10–5772

Grzywacz D, Stevenson PC, Mushobozi WL, Belmain S, Wilson K (2014) The use of indigenous ecological resources for pest control in Africa. Food Secur 6:71–86

Kahane R, Hodgkin T, Jaenicke H, Hoogendoorn C, Hermann M, Keatinge JD, d'Arros Hughes J, Padulosi S, Looney N (2013) Agrobiodiversity for food security, health and income. Agron Sustain Dev 33:671–693

Kalaisekar A, Padmaja PG (2017) Insect pests of millets and their host plant relations. In: Millets and sorghum: biology and genetic improvement, pp 267–290

Kamakshi N, Pullaibai P, Devi VS, Sharma AS, Padmalatha Y (2021) Seasonal incidence and estimation of yield losses due to insect pest in small millets at scarce rain fall zone of Andhra Pradesh. J Entomol Zool Stud 9(2):464–467

Kaplin VG, Antonov PV (2006) The biology, ecology, and harmfulness of the flea-beetle Chaetocnemaaridula Gill (Coleoptera, Chrysomelidae) in the forest-steppe of the middle Volga Area. Entomol Rev 86:176–187

Karlsson Green K, Stenberg JA, Lankinen Å (2020) Making sense of Integrated Pest Management (IPM) in the light of evolution. Evol Appl 13(8):1791–1805

Karuppuchamy P, Venugopal S (2016) Integrated pest management. In: Ecofriendly pest management for food security. Academic Press, pp 651–684

Khairani HS, NurulaliaA L, ArdieB SW (2021) Pests and diseases of foxtail millet (Setaria italicaL. Beauv.) cultivated in the greenhouse. J Trop Crop Sci 8(3):146

Kumar P, Sekhar JC, Kaur J (2013) Trichogrammatids: integration with other methods of pest control. In: Biological control of insect pests using egg parasitoids, pp 191–208

Makwana AB, Parmar GM, Detroja AC, Rathwa MG (2017) Population dynamics of Helicoverpa armigera in pearl millet

Nartshuk EP (2014) Grass-fly larvae (Diptera, Chloropidae): diversity, habitats, and feeding specializations. Entomol Rev 94:514–525

Paschapur AU, Joshi D, Mishra KK, Kant L, Kumar V, Kumar A (2021) Millets for life: a brief introduction. Millet Millet Technol:1–32

Paschapur A, Subbanna AR, Gupta J, Parihar M, Mishra KK (2022) Insect pest scenario in Uttarakhand Himalayas, India, under changing climatic conditions. Int J Biometeorol 66(7):1445–1460

Poddar N, Yele Y, Kumari A (2019) Cropping system approach in IPM. In: Prasad D, Lal G, Ahmad I (eds) Adaptive crop protection management strategies. Write and Print Publications, New Delhi, pp 358–371

Rakshit S, Prabhakar, Kumar P (2023) Maize and millets. In: Trajectory of 75 years of Indian agriculture after independence. Singapore, Springer, pp 163–187

Rani L, Thapa K, Kanojia N, Sharma N, Singh S, Grewal AS, Srivastav AL, Kaushal J (2021) An extensive review on the consequences of chemical pesticides on human health and environment. J Clean Prod 283:124657

Rawat L, Gaikwad MB, Kukreti A (2021) Insect fauna associated with small millets in mid hills of Uttarakhand. J Entomol Zool Stud 9(1):1494–1502

Sajwan R, Rawat L, Joshi M, Kumar P, Mamgain A, Kharola G (2022) Importance, taxonomy, bionomics and damaging symptoms of various insect pests associated with millets. Vigyan Varta 3(10):50–63

Sanjana Reddy P, Bhagwat VR, Prasad GS, Tonapi VA (2017) Breeding for insect resistance in sorghum and millets. In: Breeding insect resistant crops for sustainable agriculture. Springer, Singapore, pp 231–264

Sathish R, Manjunatha M, Rajashekarappa K (2017) Incidence of shoot fly, Atherigona pulla (Wiedrmann) on proso millet at different dates o sowing. J Entomol Zool Stud 5:2000–2004

Serba DD, Michaud JP (2019) Sugarcane aphid resistance in pearl millet. Kans Agric Exp Stat Res Rep 5(6):23

Sharma HC, Davies J C (1988) Insect and other animal pests of millets

Sharma HC, Sullivan DJ (2000) Screening for plant resistance to oriental armyworm, Mythimna separata (Lepidoptera: Noctuidae) in pearl millet, Pennisetum glaucum. J Agric Urban Entomol 17(3):125–134

Sharma HC, Srivastava CP, Durairaj C, Gowda CL (2010) Pest management in grain legumes and climate change. In: Climate change and management of cool season grain legume crops, pp 115–139

Sharma A, Jaronski S, Reddy GV (2020) Impact of granular carriers to improve the efficacy of entomopathogenic fungi against wireworms in spring wheat. J Pestic Sci 93(1):275–290

Shyam Prasad G, Bhagwat VR (2015) Insect pests of millets and their management

Suby SB, Soujanya PL, Yadava P, Patil J, Subaharan K, Prasad GS, Babu KS, Jat SL, Yathish KR, Vadassery J, Kalia VK (2020) Invasion of fall armyworm (Spodoptera frugiperda) in India. Curr Sci 119(1):44–51

Thakur N, Kaur S, Tomar P, Thakur S, Yadav AN (2020) Microbial biopesticides: current status and advancement for sustainable agriculture and environment. In: New and future developments in microbial biotechnology and bioengineering. Elsevier, pp 243–282

Tiwari H, Naresh RK, Kumar L, Kataria SK, Tewari S, Saini A, Yadav RK, Asati R (2022) Millets for food and nutritional security for small and marginal farmers of North West India in the context of climate change: a review. Int J Plant Soil Sci 34(23):1694–1705

Vasanthi N, Saleena LM, Raj SA (2014) Silicon in crop production and crop protection-a review. Agric Rev 35(1):14–23

Venkateswarlu B, Shanker AK (2011) Dryland agriculture: bringing resilience to crop production under changing climate. In: Crop stress and its management: perspectives and strategies. Springer, Dordrecht, pp 19–44

Breeding Advances in Minor Millets: Achievements and Limitations

N. Anuradha, Sk Abdul Salam, T. S. S. K. Patro,
P. Kranthi Priya, M. Bala Bharathi, M. Divya, D. Ratna Babu,
L. Madhavi Latha, U. Triveni, and Y. Sandhya Rani

Abstract

Minor millets represent a group of promising yet underutilized food crops in semiarid and tropical locations globally. They are classified as famine crops due to their capacity to yield satisfactorily in the most severe environmental conditions globally. Notwithstanding the promise of minor millets, genetic enhancement initiatives significantly trail those of major crops. The primary breeding objectives for enhancing competitiveness of minor millets in modern agroecosystems involve the systematic characterization of active breeding materials, parental lines, and germplasm, focusing on grain and fodder yield, yield-attributing traits, grain quality, and consumer-friendly characteristics. Nonetheless, the primary obstacles hindering the advancement of breeding in minor millets include inadequately comprehended genetic mechanisms, laborious hybridization processes, and narrow genetic base. Recombination breeding remains inadequately applied for gene introgression due to the lack of effective emasculation and hybridization methods. The genomic resources for facilitating genetic improvement in these crops are insufficiently documented. This chapter encapsulates the advances achieved in cultivar development of small millets and the constraints hindering the acceleration of genetic enhancement.

N. Anuradha (✉) · S. A. Salam · T. S. S. K. Patro · P. K. Priya · M. Divya · U. Triveni
Y. S. Rani
Agricultural Research Station, Acharya NG Ranga Agricultural University, Vizianagaram, India

M. B. Bharathi · D. R. Babu
Agricultural College, Acharya NG Ranga Agricultural University, Bapatla, India

L. M. Latha
Agricultural Research Station, Acharya NG, Ranga Agricultural University, Ananthapuram, Andhra Pradesh, India

D. C. Joshi et al. (eds.), *Minor Millets*,
https://doi.org/10.1007/978-981-96-4265-6_8

Keywords

Breeding advancements · Genetic base · Hybridization techniques · Minor millets · Recombination breeding

1 Introduction

Millets, often referred to as the "miracle crops," are the "next-generation smart crops" because of their ability to address climate change and ensure food and nutritional security. Millets are C_4 plants with greater C sequestration that can grow in harsh environments with limited water and nutrient resources, but they accumulate higher amounts of nutrients in their grains. They have several health benefits. Magnesium (Mg) in millets helps prevent diabetes and reduces the effects of migraines and heart attacks. Their high dietary fiber content also aids in repressing obesity. The occurrence of cancer is mitigated due to the presence of phenols. Additionally, gluten-sensitive individuals can safely consume millets without the fear of developing celiac disease. Sorghum and pearl millets are considered major millets among the various types, with all other millets classified as minor millets based on the volume of cultivation. Major and minor millets were the earlier versions; later, minor millets were renamed as small millets and are now collectively referred to as nutri-cereals, which includes all millets. In Ayurveda, millets were addressed as "Trina dhanya," while the contemporary term for millets in India is "Siri-anna" due to their richness in nutrition. This chapter focuses on seven small millets; their scientific names, origins, and chromosome numbers, and so on are given in Table 1 (Upadhyaya and Vetriventhan 2018; Pramitha et al. 2023; Srikanth 2023).

2 Nutritional Importance of Small Millets

The evolving dietary patterns that embrace multigrains have prompted a significant interest in food and nutritional security. There is a growing emphasis on the nutritional value of small millets, and processed foods designed to fulfill daily nutritional needs are now being provided to growing children (Muthamilarasan and Prasad 2021). In adverse conditions, small millets, also recognized as "wonder cereals," not only thrive but also boast a wide range of health benefits. Micronutrients, essential amino acids, and vitamin B complex are abundant in small millets, nutrients that are typically lacking in our everyday staple diets. The gluten-free and higher fiber attributes of millets are well-received by patients dealing with diabetes and celiac disease as part of their diet.

Among millets, finger millet boasts the highest concentration of calcium (300–350 mg/100 g), along with significant levels of phosphorous, iron, and zinc content. Additionally, its greater caloric value is associated with its abundant carbohydrate content. It possesses the richest mineral content. Both its protein (6–8%)

Table 1 Small millets origin, scientific names, and Ayurveda names

S. no.	Name of the crop	Scientific name	Origin	Chromosome no. (2*n*)	Name as mentioned in Ayurveda	Picture
1	Finger millet	*Eleusine coracana* (L.) Gaertn.	Uganda and Ethiopia	4× = 36	Nartaki	
2	Foxtail millet	*Setaria italica* (L.) P. Beauv.	China	2× = 18	Kangu	
3	Little millet	*Panicum sumatrense* Roth. Ex Roem. & Schuit	Eastern Ghats of India	4× = 36		
4	Proso millet	*Panicum miliaceum* (L.)	North-east China	4× = 36	Cheenaka	
5	Barnyard millet	*Echinocloa esculenta* (A.) *Echinocloa frumentaceae*	Japan and India	6× = 54	Shyamaka	
6	Kodo millet	*Paspalum scrobiculatum* (L.)	India	2*n* = 4× = 40	Koradusha	
7	Browntop millet	*Brachiari aramosa*	India	2× = 18; 4× = 16, 72		

and fat (1.5–2%) contents are comparatively lower. Sulphur-rich amino acids contribute to the uniqueness of finger millet proteins. Acknowledged for their exceptional malting characteristics, these grains are widely utilized in the preparation of weaning foods, given their significant antioxidant activity. Foxtail millet stands out as a nutritious choice with protein levels nearly double that of rice. Laden with minerals such as copper and iron, this grain not only provides essential nutrients but is also renowned for its delightful sweet, nutty flavor. Moreover, it is considered one of the most digestible grains, making it a favorable choice for those seeking a nonallergic option. Kodo millet possesses a wealth of B vitamins, notably niacin, pyridoxin, and folic acid, along with essential minerals such as iron, potassium, magnesium, and zinc. Its elevated lecithin content contributes to the strengthening of the nervous system. Notably, Kodo millet excels in phosphorous content and remarkable radical scavenging activity, attributed to its rich phenol content. Regular consumption of Kodo millet is associated with a decreased risk of cardiovascular dysfunction (Rao et al. 2017; Deepak and Thakur 2018). The nutritional profile of barnyard millet is characterized by a low glycemic index and elevated levels of phosphorous and magnesium (Pramitha et al. 2021). Furthermore, the grains also contain essential functional components such as gamma amino butyric acid (GABA) and beta-glucan, which function as antioxidants and aid in the reduction of blood lipid levels. Little millet is well-regarded for its high levels of magnesium, phosphorous, and protein content. Its unique characteristic is its richness of PUFA and flavonoids (Indirani and Devasena 2021). Proso millet exhibits a robust protein profile, second only to foxtail millet. It further distinguishes itself with higher levels of magnesium and niacin. Additionally, it surpasses other major cereals in essential amino acids, including lysine, leucine, isoleucine, and methionine (Pramitha et al. 2021). It lowers cholesterol levels, mitigating the risk of heart diseases. Browntop millet is acknowledged for the abundant fiber content in its grains. Being gluten-free, the grains of browntop millets contain an array of phytochemicals such as flavonoids, tannins, quinones, and resin. Consumption of these grains aids in the prevention of various noncommunicable ailments.

3 Germplasm Resources

Variability is the prerequisite for any crop breeding program. Germplasm, with its extensive diversity, meets this basic necessity (Joshi et al. 2021) and is rightfully considered the real wealth of the country. It is estimated that almost 65% of the released varieties have been developed through the process of germplasm selection (Nagaraja et al. 2023). A multitude of gene banks across the globe are dedicated to the preservation of diverse germplasm collections, to ensure their availability for future utilization. The global network of 11 Consultative Group on International Agricultural Research (CGIAR) gene banks is responsible for preserving 0.74 million accessions, including cereals, grain legumes, root and tuber crops, forages, and tree species. Nearly 0.15 million small millet accessions (finger millet, foxtail millet, kodo millet, proso millet, barnyard millet, little millet, and browntop millet) are currently safeguarded across

Table 2 Database of small millets global germplasm

Sl. no.	Database	Contact details	Developed by
1.	PGR Database	PGR Databases (ernet.in)	NBPGR, New Delhi
2.	GENESYS	www.genesyspgr.org	Genesis, ICRISAT
3.	FAO-WIEWS (FAO—World Information and Early Warning System)	http://apps3.fao.org/wiews/wiews.jsp?i_l=EN	FAO
4.	SINGER (System-wide Information Network for Genetic Resources)	http://singer.cgiar.org/	CGIAR and associated partners
5.	EURISCO (European Plant Genetic Resources Search Catalogue)	http://eurisco.ecpgr.org/	Bioversity International, Rome, Italy (from 2003 to 2014) Institute of Plant Genetics and Crop Plant Research (IPK), Gatersleben, Germany (from 2014)
6.	GBIF (Global Biodiversity Information Facility)	https://www.gbif.org/species/search	GBIF network of participating countries and organizations

Asia, Africa, Europe, and Oceania (Vetriventhan et al. 2020). In India, the collection and preservation of small millet germplasm is undertaken by organizations such as the International Crop Research Institute for the Semi-Arid Tropics (ICRISAT); National Bureau of Plant Genetic Resources, New Delhi; Indian Institute of Millet Research (ICAR-IIMR), Hyderabad; Project Coordinating Unit at the University of Agricultural Sciences, Bengaluru. Globally, the data on small millets germplasm is stored in various data bases as given in Table 2.

4 Varietal Improvement Through Traditional Breeding

Varietal improvement involves the development of new varieties of crops though the combination of desirable traits. These new crop varieties are better adapted to specific local conditions and demonstrate improvement of yield, other characteristics such as resistance to diseases and drought, optimized water usage, nutrient content, and more. The primary emphasis in the development of small millet varieties is focused on optimizing grain yield and addressing biotic stress tolerance factors. This includes resistance to blast disease in finger millet, sheath blight in various small millets, smut in kodo, proso, and barnyard millet, as well as shoot fly in proso and kodo millet.

The small millet varieties released for cultivation in the last 25 years are listed in Table 3. Finger millet received most attention and 68 varieties were released during last 25 years in India. In the same timeframe, kodo millet witnessed the release of 22 varieties, while 21 in little millet and foxtail millet, whereas only 14 and 10 in proso millet and barnyard millet, respectively. Browntop research initiated only a decade ago, and had very little variability. Hence, only a single variety was introduced during this period.

Table 3 Varieties of small millets released for cultivation during the last 25 years

S. no.	Name of the variety	Year of release	Recommended domain
(i) Finger millet			
1	Champavathi (VR708)	1998	Andhra Pradesh, Uttar Pradesh, Tamil Nadu, Karnataka, and Odisha
2	MR1	1998	Karnataka
3	PR230 (Maruthi)	1998	Andhra Pradesh
4	GPU28	1998	Karnataka, Tamil Nadu, Chhattisgarh, Odisha, and Jharkhand
5	BM9-1	1999	Karnataka, Andhra Pradesh, Odisha, Madhya Pradesh, and Maharashtra
6	OUAT2	1999	All ragi growing areas of the country
7	L5	1999	Karnataka
8	GPU26	2000	Karnataka
9	GPU45	2001	Madhya Pradesh, Gujarat, Jharkhand, Karnataka, and Maharashtra
10	Chilika (OEB10)	2001	Odisha, Madhya Pradesh, Gujarat, Andhra Pradesh, and Tamil Nadu
11	VL315	2004	Uttaranchal
12	TNAU946 (CO 14)	2004	Tamil Nadu
13	GPU48	2005	Karnataka
14	GN4	2006	Gujarat
15	VL332	2006	All ragi growing areas of the country
16	PRM1	2006	Hills of Uttarakhand
17	VL324	2006	Lower and mid-hills of Uttarakhand
18	VR762 (Bharathi)	2006	Andhra Pradesh
19	Paiyur2	2008	Tamil Nadu
20	ML365	2008	Zone 5 of Karnataka
21	GN5	2009	Gujarat
22	Srichaithanya (VR847)	2009	Andhra Pradesh
23	GPU66	2009	Karnataka
24	KMR301 (Gowri)	2009	Southern dry zone of Karnataka
25	GPU67	2009	Chhattisgarh, Jharkhand, Karnataka, Maharashtra, Tamil Nadu, and Uttarakhand
26	PRM2	2010	Hills of Uttarakhand
27	VL347	2010	All ragi growing areas of the country
28	KOPN235 (PhuleNachni)	2011	Submountain and Western Ghat zones of Maharashtra
29	OEB526 (Arjuna)	2011	Odisha, Bihar, Chhattisgarh, Karnataka, and Tamil Nadu
30	OEB532	2012	Odisha, Chhattisgarh, Karnataka, and Maharashtra, and Tamil Nadu
31	KMR204	2012	Karnataka
32	VR936 (Hima)	2012	Andhra Pradesh

(continued)

Table 3 (continued)

S. no.	Name of the variety	Year of release	Recommended domain
33	Vakula (PPR2700)	2012	Andhra Pradesh
34	IndiraRagi1	2012	Chhattisgarh
35	VL352	2012	All ragi growing areas of the country
36	GNN 6	2013	Gujarat
37	GN 5	2013	Gujarat
38	VL Mandua 348	2013	Uttarakhand
39	Dapolii (SCN 6)	2013	Maharashtra
40	KMR 240	2016	Karnataka
41	GNN 6	2016	Gujarat
42	Vakula (PPR 2700)	2016	Karnataka
43	DHFM-78-3	2017	Karnataka
44	GPU-66	2017	Karnataka
45	VL Mandua 379 (VL 379)	2017	Uttarakhand, Bihar, Jharkhand, and Madhya Pradesh
46	Chhattisgarh Ragi-2 (BR-36)	2017	Chhattisgarh
47	VL Mandua 376 (VL 376)	2018	Uttarakhand, Bihar, Madhya Pradesh, Jharkhand, Gujarat, Andhra Pradesh, Karnataka, Tamil Nadu, Odisha, and Maharashtra
48	GN-8 (WN-585)	2019	Gujarat
49	Tirumala (PPR 1012)	2019	Andhra Pradesh
50	Vegavathi (VR 929)	2019	Andhra Pradesh, Karnataka, Tamil Nadu, Odisha, Jharkhand, Gujarat, Maharashtra, and Puducherry
51	Gowthami (PR 10-45)	2019	Andhra Pradesh
52	KMR- 630	2019	Karnataka
53	VR 988	2019	Andhra Pradesh
54	KALUA (OEB 532)	2019	Odisha
55	CFMV 2 (FMV 1118) (GN9/Gira)	2020	Andhra Pradesh, Maharashtra, Gujarat, Chhattisgarh, and Odisha.
56	CFMV 1 (Indravathi) (FMV1116) (VR 1101) (Finger Millet)	2020	Andhra Pradesh, Karnataka, Odisha, Puducherry, and Tamil Nadu
57	VL Mandua 378	2021	Uttarakhand.
58	VL Mandua 382	2021	Uttarakhand.
59	CFMV-3 (Ekvijay)	2021	Gujarat, Telangana, Tamil Nadu, Maharashtra, and Andhra Pradesh.
60	Chhattisgarh Ragi-3 (FMV-1102) (BR-14-3)	2021	Chhattisgarh, Bihar, Uttarakhand, Jharkhand, Assam, and Madhya Pradesh
61	ATL 1 (TNEc 1285)	2021	Tamil Nadu
62	Dapoli-3 (DPLN-2)	2021	Maharashtra
63	Birsa Marua 3	2021	Jharkhand

(continued)

Table 3 (continued)

S. no.	Name of the variety	Year of release	Recommended domain
64	VL Mandua 400 (CFMV 5)	2022	Karnataka, Andhra Pradesh, Madhya Pradesh, Bihar, Gujarat, Chhattisgarh, and Jharkhand
65	CFMV 4 (FMV 1166)	2022	Andhra Pradesh, Maharashtra, and Tamil Nadu
66	Gosthani (VR 1099)	2022	Andhra Pradesh
67	FMV 1191 (VL Mandua 408)	2023	Madhya Pradesh and Uttarakhand
68	FMV 1198 (OEB 610)	2023	Maharashtra, Chhattisgarh, and Tamil Nadu
(ii) Little millet			
1	Paiyur2	2000	
2	Kolab (OLM36)	2001	MP, Odisha, Chhattisgarh, Bihar, Karnataka, and Gujarat
3	Tarini (OLM203)	2001	Odisha, AP, Bihar, and Tamil Nadu
4	Sabara (OLM20)	2003	All little millet growing areas of the country
5	Co4	2005	Tamil Nadu
6	JK36	2009	Madhya Pradesh
7	Soura (OLM208)	2009	Odisha, Gujarat, and Chhattisgarh
8	OLM217	2009	Odisha, Gujarat, and Chhattisgarh
9	GV-2	2015	Gujarat
10	Phule Ekadashi (KOPLM 83)	2015	Maharashtra
11	Jawahar Kutki 4 (JK4)	2015	Madhya Pradesh
12	Co 15	2016	Tamil Nadu
13	Chhattisgarh Kutki-1 (BL-6)	2018	Chhattisgarh, Karnataka, Odisha, Tamil Nadu, and Madhya Pradesh
14	DHML-36-3	2018	Karnataka
15	GNV-3	2018	Gujarat
16	ATL 1 (TNPsu 177)	2019	Tamil Nadu
17	CLMV 1 (JAICAR Sama-1/ LMV 518)	2020	Telangana, Andhra Pradesh, Puducherry, Tamil Nadu, and Maharashtra
18	Kalinga Saun 217 (OLM217)	2021	Odisha
19	GV-4 (Ambika)	2021	Gujarat
20	Hagari Same-1 (HS-1) (IIMR LM-8437-17)	2021	Karnataka
21	GPUL 6	2021	Karnataka
(iii) Proso millet			
1	GPUP8	2001	Karnataka and Tamil Nadu
2	GPUP21	2003	Karnataka and Tamil Nadu
3	Pratap Cheena	2006	All proso millet growing areas of the country
4	Co5	2007	Tamil Nadu
5	TNAU145	2007	All proso millet growing areas of the country
6	PRC1	2008	Uttarakhand hills
7	TNAU151	2008	Tamil Nadu

(continued)

Table 3 (continued)

S. no.	Name of the variety	Year of release	Recommended domain
8	TNAU164	2009	Andhra Pradesh, Bihar, Karnataka, Maharashtra, Tamil Nadu, and Uttaranchal
9	TNAU202	2011	Tamil Nadu, Karnataka, Chhattisgarh, Madhya Pradesh, Gujarat, Andhra Pradesh, and Bihar
10	DHPM-2769	2018	Karnataka
11	PMV 442 (GPUP 25)	2021	Karnataka, Bihar, Tamil Nadu, Puducherry, Telangana, and Andhra Pradesh
12	Hagari Barugu-1 (HB-1)	2021	Karnataka
13	GPUP 28	2021	Karnataka
14	PMV466 (DHPM-60-4)	2023	Karnataka and Tamil Nadu
(iv) Foxtail millet			
1	TNAU186	1998	Andhra Pradesh, Tamil Nadu, and Karnataka
2	PS4	1998	All foxtail millet growing areas of the country
3	Meera (SR16)	2000	Rajasthan
4	Srilakshmi	2002	All foxtail millet growing areas of the country
5	Prathap Kangani (SR1)	2003	Rajasthan
6	Co7 (TNAU196)	2005	Tamil Nadu
7	HMT100-1	2008	Karnataka
8	SiA3085	2011	All foxtail millet growing areas of the country
9	Suryanandi	2012	All foxtail millet growing areas of the country
10	SiA3156	2012	Andhra Pradesh, Bihar, Gujarat, Karnataka, Madhya Pradesh, Tamil Nadu, and Uttarakhand
11	Rajendra Kauni (RAU 2)	2016	Bihar
12	Surya Nandi (SiA3088)	2017	Karnataka
13	DHFt-109-3	2017	Karnataka
14	Garuda (SiA3222)	2019	Andhra Pradesh
15	Renadu (SiA3223)	2019	
16	Hagari Navane-46 (HN-46)	2019	Karnataka
17	ATL 1 (TNSi 331)	2019	Tamil Nadu
18	Mahanandi (SiA3159)	2022	Andhra Pradesh
19	GPUF 3	2022	Karnataka
20	FXV645 (IIMR FxM-7)	2023	Andhra Pradesh and Karnataka (major foxtail millet producing states)
(v) Kodo millet			
1	KK1	1999	Uttar Pradesh
2	JK155 (RBK155)	2000	Karnataka and Madhya Pradesh
3	JK48 (DPS48)	2001	Madhya Pradesh
4	KK2	2002	Uttar Pradesh
5	JK439	2004	Chhattisgarh and Madhya Pradesh
6	JK13	2007	Madhya Pradesh, Chhattisgarh, Uttar Pradesh, Andhra Pradesh, Karnataka, and Tamil Nadu
7	JK106	2009	Madhya Pradesh
8	JK65	2009	Madhya Pradesh, Gujarat, Uttar Pradesh, and Tamil Nadu

(continued)

Table 3 (continued)

S. no.	Name of the variety	Year of release	Recommended domain
9	JK36	2009	Madhya Pradesh
10	JK98	2010	Madhya Pradesh
11	DPS9-1	2011	Madhya Pradesh
12	Indira Kodo1	2012	Chhattisgarh
13	TNAU86	2012	Andhra Pradesh, Chhattisgarh, Gujarat, Karnataka, Madhya Pradesh, Tamil Nadu, and Uttar Pradesh
14	RK390-25	2012	Andhra Pradesh, Chhattisgarh, Gujarat, Karnataka, and Madhya Pradesh
15	Gujarat Anand Kodra-3 (GAK3)	2020	Gujarat
16	CKMV 1 (ATL 2)	2021	Tamil Nadu, Madhya Pradesh, Jharkhand, Chhattisgarh, Gujarat, Karnataka, Andhra Pradesh, and Telangana
17	Dahod Kodo (CKMV 2)	2021	Dahod Kodo (CKMV 2)
18	Chhattisgarh Kodo-03 (BK-36)	2021	Chhattisgarh, Jharkhand, Andhra Pradesh, Karnataka, Gujarat, Tamil Nadu, and Madhya Pradesh
19	Gujarat Kodo millet 4 (Dahod Kodra)	2021	Gujarat
20	ATL 1 (TNPsc 176)	2021	Tamil Nadu
21	Dahod Kodo 1 (CKMV 3)	2022	Andhra Pradesh, Chhattisgarh, Jharkhand, Madhya Pradesh, and Tamil Nadu
22	KMV568 (TNPsc 308)	2022	Gujrat, Chhattisgarh, and Andhra Pradesh
(vi) Barnyard millet			
1	VLMadira172	2000	Uttar Pradesh, Gujarat, and Karnataka
2	Sushrutha (RAU 11)	2000	Zone 8 of Karnataka
3	VLMadira181	2001	Bihar, Karnataka, Madhya Pradesh, and Tamil Nadu
4	PRJ1	2003	Uttarakhand
5	Pratap Sawan1/ER64	2008	Rajasthan
6	VLMadira207	2008	All barnyard millet growing states except Tamil Nadu and Gujarat
7	CO2	2009	Tamil Nadu
8	MDU-1	2017	Tamil Nadu
9	DHBM-93-2	2017	Karnataka
10	BMV611 (IIMR BM-3-1920)	2023	Telangana, Tamil Nadu, Madhya Pradesh, Karnataka, and Andhra Pradesh
(vii) Browntop millet			
1	Hagari Browntop-2 (HBr-2)	2022	Karnataka

Source: Indian Institute of Millets Research (IIMR), http://seednet.gov.in/seedgo/index.htm#Year%20-%202013

Small millet breeding is mainly oriented toward the development of disease and pest-resistant varieties, as the damage caused by these factors is quite significant. In India, numerous resistant/tolerant varieties have been released and are currently being cultivated in the farmers' fields (Fig. 1a–e).

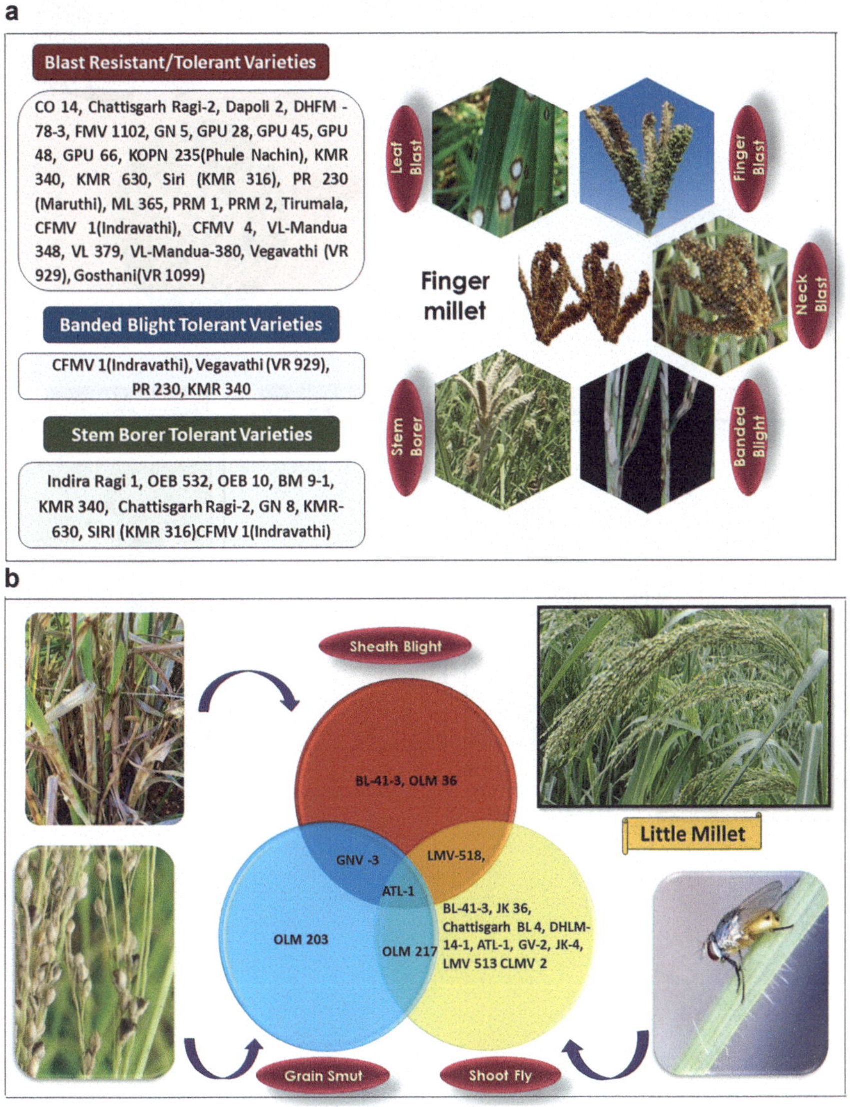

Fig. 1 (**a**) Resistant/tolerant varieties developed in finger millet. (**b**) Resistant/tolerant varieties developed in little millet. (**c**) Resistant/tolerant varieties developed in foxtail millet. (**d**) Resistant/tolerant varieties developed in kodo millet. (**e**) Resistant/tolerant varieties developed in barnyard millet and proso millet

Fig. 1 (continued)

Fig. 1 (continued)

5 Biofortification in Small Millets

Micronutrient malnutrition has alarmingly increased over the past two decades, contributing to several deaths and poor health among people worldwide. A financially viable strategy for addressing micronutrient malnutrition is to biofortify staple crops. HarvestPlus collaborated with various organizations to introduce biofortification into different staple crops through traditional breeding methods. Despite millets being nutritionally superior to other cereals due to their richness in essential amino acids, minerals, vitamins, and proteins, there is still untapped potential to further enhance their nutritional profile. The HarvestPlus group is actively engaged in the enhancement of millet biofortification and contributed to the introduction of the first biofortified millet variety (pearl millet) in India. This variety is suitable for conventional cultivation and serves as a valuable tool to combat iron deficiency. In India, four varieties of finger millet and one little millet have been released for cultivation within the category of small millets (Table 4). All varieties being bred through traditional breeding. The little millet variety, CLMV1, is rich in iron and zinc, while the finger millet varieties, CFMV 1 and 2, are rich in calcium, iron, and zinc. These biofortified varieties were released by the Prime Minister of India on the occasion of the 75th anniversary of the Food and Agriculture Organization (FAO) and the United Nations, during which he dedicated a total of 17 biofortified crop varieties to the country (Anonymous 2022). Later on, few more biofortified varieties were released and it accounts to 11 varieties to date.

Finger millet exhibits a close genetic resemblance to cereals; and the utilization of comparative genomics has proven effective in the identification of quantitative trait loci and genes associated with protein quality. The results of a transcriptomics

Table 4 Biofortified small millet varieties in India

S. no.	Variety	Duration	Grain yield (q/ha)	Rich in (mg/100 g)	Picture
Finger millet					
1.	Vegavathi (VR 929)	115–120	38–40	Iron (13.2)	
2.	Indravathi (CFMV 1)	105–110	32–34	Calcium (428), iron (5.8), and zinc (4.4)	
3.	Gira (CFMV 2)	115–120	36–38	Calcium (454), iron (3.9), and zinc (2.5)	
4.	Gosthani (VR 1099)	115–120	35–38	Iron (5.8) and zinc (3.6)	
5.	PR-10-45 (Gowthami)	122–125	35–37	Zinc (2.7) and calcium (342)	
6.	VL Mandua 400 (CFMV 5)	100–102	34–35	Grain contains higher calcium (399.5) and protein content (8.5%)	

(continued)

Table 4 (continued)

S. no.	Variety	Duration	Grain yield (q/ha)	Rich in (mg/100 g)	Picture
Little millet					
7.	CLMV 1 (LMV 518)	98–102	14–17	Iron (5.9) and zinc (3.5)	
8.	Hagari Same-1 (HS-1) (IIMR LM-8437-17)	90–95	15–16	Iron (3.1), zinc (1.8), and crude fiber (9%)	
Proso millet					
9.	Hagari Barugu-1(HB-1)	65–70	12–14	Iron (5.5), zinc (3.5), and crude fiber (11%)	
Foxtail millet					
10.	RAU (Rajendra Kauni 1-2)	80–83	23–25	Iron (15.5) and zinc (5.0)	
Browntop millet					
11.	Hagari Browntop-2 (HBr-2)	95–100	18–20	Iron (7.0), zinc (4.2), and crude fiber (13%)	

study revealed the presence of numerous calcium sensor genes in finger millet, actively participating in the mechanisms of calcium uptake, translocation, and accumulation. Furthermore, the incorporation of zinc transporters through transgenic expression led to the enhancement of high grain zinc levels in the grains. Effective biofortification of millets is limited by the inhibitory effect of antinutrients such as phytic acid, polyphenols, and tannins. These antinutrients can be reduced by the utilization of RNA interference and genome editing tools such as clustered regularly interspaced short palindromic repeats (CRISPR), transcription activator-like effector nucleases (TALENs), and zinc finger nucleases (ZFNs) (Vinoth and Ravindhran 2017).

6 Molecular Markers Utilized in the Small Millets

Field crop research has been greatly enriched by the evolution of molecular markers, which have been pivotal in the identification and tagging of numerous agronomically significant genes and QTLs. This has greatly contributed to the development of genetic maps, gene identification, and applications in plant breeding, as well as enhancing our comprehension of genetic diversity. Consequently, there is a significant need for the development and application of genome-wide markers to uncover the genetic diversity within a crop gene pool. This can then be efficiently applied in conservation and plant breeding practices. Restriction fragment length polymorphism (RFLP) and SSRs are the most popular and reliable molecular markers for construction of genetic maps. Conversely, SNPs and array-based molecular markers are identified as next-generation high-throughput markers. In the domain of molecular markers, SSRs were found to be predominantly employed in small millets and were also applied in comparative genomics for lineage analysis (Pramitha et al. 2023). Nonetheless, the absence of a complete genome sequence in several small millets acts as a hindrance to the identification of genetic and genomic resources associated with resistance to abiotic or biotic factors. Regardless of environmental conditions and the developmental stage of plants, the robustness of DNA marker technology remains unaffected, allowing for its application in the assessment of several novel evolutionary and taxonomy concerns. Identification of genetic variation at the DNA level is achieved through DNA markers, which include genes and specific DNA segments. Utilizing molecular marker techniques for diversity assessment streamlines the evaluation of germplasm, resulting in efficiencies that translate into time, labor, and land savings. Moreover, the utilization of codominant DNA markers for genetic characterization of germplasm is a pivotal stage in the development of stable and productive cultivars featuring diverse desirable traits. Genetic relationship studies in small millets involved the use of a variety of molecular markers.

6.1 Finger Millet

The application of genome-wide DNA markers to analyze the genetic diversity of finger millet is a valuable tool for breeders. This assists in development of cultivars featuring a range of desirable traits, such as increased yield and enhanced stress tolerance. A widely employed next-generation sequencing (NGS) method, genotyping-by-sequencing (GBS) is engineered for the simultaneous exploration of novel markers and the genotyping of target germplasm. Single nucleotide polymorphism (SNP) markers are the predominant outcome of the GBS technique, making them the most widely adopted DNA markers. In their study, Panwar et al. (2011) employed EST-based resistance gene analogue (RGA) markers to establish molecular markers for the enhancement of fungal blast disease resistance in finger millet through marker-assisted selection (MAS) and breeding strategies.

6.2 Foxtail Millet

Given its extensive use in genomic analysis and the abundance of genome-wide sequence resources, foxtail millet has established itself as a model crop for the exploration to broaden millet biology. Foxtail millet genetic research has advanced through the development and integration of DNA markers such as SSRs, SNPs, EST-SSRs, ILPs, and microRNA-based molecular markers. Among various millets, including proso millet, barnyard millet, little millet, and kodo millet, the foxtail millet DNA markers displayed substantial (>85%) cross-genera transferability (Kumari et al. (2013). The demonstrated higher cross-transferability percentages suggest that these markers have potential applicability in germplasm characterization, marker-trait association, and marker-assisted breeding in millets with limited or no existing genomic resources. Moreover, an exclusive web-based repository known as the Foxtail millet Marker Database (FmMDb) was developed (Suresh et al. 2013) to provide breeders and researchers with access to the available marker resources. In addition to the molecular markers, gene families such as NAC, WD40, AP2/ERF, C2H2 zinc finger, MYB, DCL, AGO, RDR, WRKY, LecRLK, ADP-ribosylation factors, ATG, heat shock proteins factors, CDPK, and LIM genes were also identified. In genome-wide research, Pandey et al. (2013) found 28 342 microsatellite repeat-motifs covering 405.3 Mb of the foxtail millet genome which may be useful for crop improvement in near future.

6.3 Little Millet

Limited research efforts have been directed toward little millet in comparison to other minor millets, leading to a shortage of genomic and molecular information available for this crop. Consequently, there is a shortage of molecular markers designed specifically for the crop. Therefore, random molecular markers such as RAPD or ISSR are selected as the primary markers of choice for the investigation of genetic diversity in the crop. Among these two alternatives, the ISSR marker is the preferred tool for analyzing the genetic diversity of little millet due to its abundance, high reproducibility, and significant polymorphism. Moreover, in the absence of SSR markers for little millet, ISSR markers are the favored option for conducting preliminary studies on genetic diversity (Shankar and Anjani 2022).

6.4 Proso Millet

The genome of proso millet is comparatively compact, spanning 923 megabases. The molecular analysis of DNA polymorphism in proso millet germplasm is primarily conducted using molecular markers such as RAPD, AFLP, ISSR, and SSR. Limited genomic resources for proso millet led to the development of most SSR markers from the genomic resources of closely related grass species (Zargar et al. 2023). A total of 25 SSR markers were generated from the proso millet BAC library (Cho et al. 2010). Presently, genetic and breeding studies in millets have access to approximately 100 SSR markers (Zargar et al. 2023). Numerous SNPs were detected in both wild and mutant types (Francis et al. 2023).

6.5 Barnyard Millet

The comprehensive analysis of genetic diversity, linkage mapping, and QTL identification in molecular breeding studies demands a substantial quantity of molecular markers. The Indian barnyard millet germplasm exhibited high level of molecular variability (Hilu 1994). The absence of genome sequence data in barnyard millet underscores the importance of leveraging comparative genomics. The cross-transferability between foxtail millet and barnyard millet surpassed 90%, and categorized them into the same group as foxtail millet varieties and species (Yadav et al. 2014). The knowledge obtained from previous foxtail millet research will contribute to the molecular techniques in the enhancement of barnyard crop genetics. Five primer pairs of microsatellite loci were created by Danquah et al. (2002) to investigate genetic diversity. Their research demonstrated the efficacy of these microsatellites in distinguishing the three species (*crus-galli* (L.) Link., *colona* (L.) Beauv., and *crus-pavonis* (Kunth.)). Moreover, the microsatellites were found to be valuable tools for species classification within the complex genus *Echinochloa*.

7 Molecular Mapping/Mapping of Traits

The foundation of gene mapping, cloning, and marker-assisted breeding programs is established through the construction of a genetic map using molecular markers. Molecular markers can significantly streamline the process of identifying causative genomic regions (QTLs) associated with complex traits through the correlation of genotype (loci) and phenotype (trait) (Gupta et al. 2005). The process of QTL mapping and trait mapping in small millets is currently in its early phases (Pramitha et al. 2023). In the context of small millet mapping, F_2–F_6 populations and RILs were the key genetic materials utilized. The potential of mapping populations such as NILs and double haploids is yet to be investigated (Vetriventhan et al. 2020; Zhang et al. 2021). Numerous potential markers indicative of disease resistance have been discovered through genomic studies in small millets.

7.1 Finger Millet

The inadequate pool of genetic resources has posed a challenge to in-depth research on finger millet. Studies have considered the secondary and tertiary gene pools as potential sources to overcome the genetic limitations caused by the inbreeding nature of the crop. The presence of recently established draft genome and linkage maps provides a pathway forward to undertake large-scale genomics-assisted breeding to enhance crop improvement. Despite being a rich source of micronutrients such as amino acids, dietary fiber, and phenolic acids in finger millet, genetic improvement is still in its initial stages. The identification of a gene that strongly influences gamete fertility on chromosome 8 was first achieved through the construction of an RFLP-based map by Wang et al. (1998). The construction of the first genetic map for A and B genomes of finger millet genome was accomplished by Dida et al. (2007), who utilized a blend of RFLP, AFLP, and SSR markers. Recently, Pendergast et al. (2022) developed a high-density linkage map using SNPs. QTLs associated with finger blast, neck blast, and leaf blast were identified by Babu (2014a) and Ramakrishnan et al. (2016). Association mapping was employed to identify markers-associated blast resistance in finger millet (Ramakrishnan et al. 2016).

Moreover, studies on finger millet have also explored QTLs related to P efficiency (Ramakrishnan et al. 2017), agronomic traits (Sharma et al. 2018), and biochemical traits including APX, CAT, GR, SOD, and POD activities (David et al. 2021). GWAS was employed to identify marker-trait associations for diverse micronutrients. The markers S1_5895347, S1_55418346, and S1_47207745 were found to be associated with iron, potassium, and sodium, respectively (Puranik et al. 2020). The study conducted by Tiwari et al. (2020) highlighted a specific SNP in aspartyl protease as a potential candidate affecting protein content variation, and the SNP at ATP synthase was established as the principal candidate gene that influences both maturity and yield variations. Variations ranging from 33 to 90% were observed in the amino acid sequences of 19 Ca^{2+} transporter genes, identified in the transcriptome data of developing finger millet seed, as compared to rice Ca^{2+} transporter genes (Kumar et al. 2015). Regulating the activities of CaX exchanger and calmodulin (CAM) proteins for increased grain calcium accumulation involves the participation of two genes, *EcCBP* and *EcCIPK7* (Sharma et al. 2022). In finger millet, QTLs linked to abiotic stresses, key agronomic traits, and nutrient density were identified (Ceasar et al. 2018) (Fig. 2).

7.2 Foxtail Millet

Foxtail millet is characterized by genetic maps constructed using molecular markers. The initial linkage map for foxtail millet was developed by integrating SSR and RFLP marker data (Jia et al. 2009). A high-density haplotype map for foxtail millet was created using ~one million SNPs (Jia et al. 2013). The identification of SNPs associated with ten yield-contributing micronutrients and yield-contributing agronomic traits in foxtail millet was achieved through GWAS (Jia et al. 2013; Jaiswal et al. 2019a). In foxtail millet, QTLs associated with traits such as tiller number, branching, spikelet count, bristles, panicle length and weight, plant height, grain weight/plant, and pericarp color have been mapped (Aggarwal et al. 2022). The initial study on GWAS in this crop revealed the presence of a shattering-associated gene known as "*shattering*1 gene (Si37789m)" (Jia et al. 2013). Through the application of double digest restriction-associated DNA (ddRAD) sequencing approach, a set of 27 candidate genes were identified to be located within the 25-Kb limit region. These genes encode genomic regions that are associated with grain yield

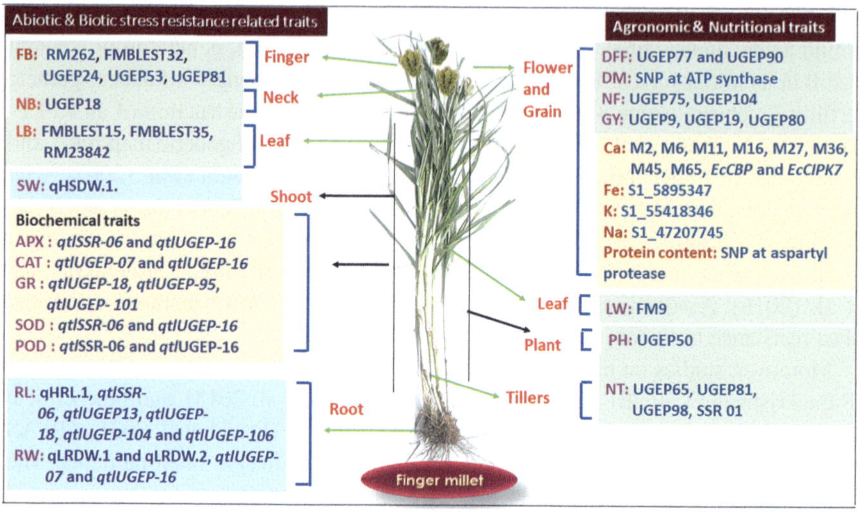

Fig. 2 QTLs or genes identified for various traits in finger millet. **Agronomic traits:** *DFF* days to flowering (50%), *DM* days to maturity, *NF* number of fingers, *GY* grain yield, *LW* flag leaf blade length, *PH* plant height, *NT* number of tillers. **Nutritional traits:** *Ca, Fe, K, Na* calcium, iron, potassium, and sodium content in grain, respectively. **Biotic stress resistance traits:** *FB* finger blast, *NB* neck blast, *LB* leaf blast. **Abiotic stress resistance traits:** *RDW* root dry weight, *RL* root length, *APX* ascorbate peroxidase, *GR* glutathione reductase, *POD* peroxidase, *SOD* superoxide dismutase, *CAT* catalase (CAT)

(such as translation initiation factor 3, RNA recognition motif, RBR family ring finger, pyridine nucleotide-disulfide oxidoreductase domain-containing protein 2, IBR domain, LEA protein and glycosyltransferases), flag leaf width (AP2 domain, amino-oxononanoate synthase, ubiquitous protein, peptidases/proteases, pollen allergen1, and arabinonate dehydratase/hydrolase), and thousand–grain weight (including sentrin/sumo-specific protease, DNA polymerase delta subunit 4 and ATP-binding cassette transporter) (Jaiswal et al. 2019b). Dehydration stress tolerance in foxtail millet seedlings of diverse germplasms is correlated with SNP markers within the SiDREB2 gene (Lata et al. 2011). Marker-trait associations for some micronutrients such as boron, magnesium, zinc, and iron were identified in addition to the quantitative traits (Jaiswal et al. 2019a).

7.3 Proso Millet

The development of the first linkage maps in proso millet incorporating SNP markers marked a significant milestone through Rajput et al.'s (2016) groundbreaking research. This achievement was accomplished using genotyping-by-sequencing (GBS), a next-generation sequencing (NGS) method. Genome-wide association studies facilitated the identification of marker-trait associations, particularly for seed-related and agronomic traits.

8 Genome Sequence of Small Millets

The challenges in the genome sequencing of millets can be ascribed to factors such as high ploidy levels and a significant abundance of repetitive DNA. The anticipated resolution of these bottlenecks is expected to come in due course, with the advent of third-generation sequencing systems integrated with advanced high-throughput data analysis platforms. The genomes of foxtail millet, finger millet, proso millet, and Japanese barnyard millet have all been sequenced among small millets. Additionally, the complete chloroplast genome sequences of foxtail millet, proso millet, little millet, and barnyard millet are accessible. Foxtail millet, having its genome completely sequenced before all other millets, stands out as a model for C_4 crop species due to its small diploid genome, short duration, and self-pollinating characteristics (Vetriventhan et al. 2020). Bennetzen et al. (2012) and Zhang et al. (2012) independently completed the first sequencing of foxtail millet, using two different cultivars. Ni et al. (2017) subsequently released an updated sequence of this specific millet. Sequenced by BGI in China, the genome sequence of "Zhang gu" cultivar of foxtail millet is now accessible in the Foxtail Millet Database (http://foxtailmillet.genomics.org.cn/page/species/index.jsp). Hittlamani et al. (2017) and Hatakeyama et al. (2018) contributed to the successive publication of the whole genome sequence of finger millet. In China, Guo et al. (2017) and Zou et al. (2019) independently completed and published the genome sequences of barnyard millet and proso millet, respectively. The genome of foxtail millet is the smallest (423–510 Mb) among millets, whereas finger millet has the largest genome size of 1.5 Gb, followed by barnyard millet (*E. crus-galli*) at 1.27 Gb.

9 Omics Approaches in Small Millets

Despite significant progress in millet research through genetics and, to a certain extent, genomics, other omics approaches are relatively novel in these crops, and there is limited documentation on the application of alternative omics tools in millet studies. The functional genomic approach of transcriptome-based gene expression profiling allows for the characterization of candidate genes that regulate a range of biological processes. DNA markers derived from transcriptomic sequences demonstrate greater efficiency, cost-effectiveness, flexibility, and informativeness compared to markers sourced from genomic sequence data, enabling gene-based interpretation and the identification of functional variation (Brhane et al. 2022). Another widely embraced omics approach is proteomics, currently applied in millet research, contributing to a better understanding of climate resilience and traits linked to nutrition. Since there is no direct correlation between transcript abundance and its expression with protein/metabolome levels, proteomics and metabolomics play a crucial role in studying posttranslational modifications. The integration of extensive data from diverse omics approaches would simplify the identification of potential candidate genes, allowing for their manipulation through innovative gene-editing tools to enhance crop development.

9.1 Finger Millet

The identification of a salinity-responsive transcriptome was achieved in finger millet (Rahman et al. 2014). Sequencing the transcriptome of the "ML-365" variety was conducted under both well-water and low moisture stress conditions, revealing the identification of 12,893 unigenes that exhibited specific expression patterns in response to low moisture stress (Hittalmani et al. 2017). Transcriptome analysis of leaf samples under drought conditions has led to the identification of two candidate genes, CIPK31 and TAF6, suggesting their potential role in drought tolerance (Parvathi et al. 2019). Through the application of proteomics, an elevation in the levels of calcium binding proteins (CaBPs) synthesis during the later stages of grain filling stages was observed, indicating the sequestration of calcium within the endoplasmic reticulum.

9.2 Proso Millet

There is a scarcity of literature available on transcriptome research in proso millet. The initial report on RNA-seq data proved to be an invaluable sequence resource for gene discovery and the development of markers in proso millet (Yue et al. 2016). The study of leaf transcriptomics provided insights into the molecular mechanisms of drought tolerance (Zhang et al. 2019). The seed protein analysis of four Korean cultivars employing 2-D electrophoresis and mass finger printing might be the earliest report in proso millet regarding proteomics (Roy et al. 2017).

9.3 Foxtail Millet

The transcriptome of total RNA from the root, stem, leaf, and spike of the "Zhang gu" strain marked the first-ever sequencing among small millets, specifically in foxtail millet (Zhang et al. 2012). The transcriptome of foxtail millet was sequenced during *Sclerospora graminicola* infection by Li et al. (2020), leading to the identification of numerous candidate genes for subsequent functional characterization. Remarkably, the research has identified a gene encoding an amidase, *Seita*.2G313400, which demonstrated a substantial 30-fold upregulation in response to the infection.

9.4 Barnyard Millet

Barnyard millet is currently undergoing transcriptome analysis. Transcriptome analysis played a key role in the identification of the gene regulatory mechanisms associated with drought tolerance, Fe and Zn accumulation (Jayakodi et al. 2019), as well as submergence tolerance (Nah et al. 2015).

10 Genome Editing

Diverse crops have embraced the clustered regularly interspaced short palindromic repeat (CRISPR)/CRISPR-associated protein 9 (Cas9) nuclease system as an effective gene-editing tool, demonstrating its broad applicability. The application of this gene-editing tool remains unexplored in small millet crops, with foxtail millet being an exception. The CRISPR system was employed to create a foxtail millet mutant called "xiaomi," characterized by a shortened generation time. A foxtail millet haploid inducer line was successfully developed through CRISPR-Cas9 mediated manipulation of the *SiMTL* gene. The platform provided by doubled haploid lines allows for the exploitation of complex traits, such as climate resilience, in C4 crops. The majority of research has been conducted on the wild progenitor of foxtail millet, *S. viridis*, resulting in the current lack of a robust CRISPR/Cas9 protocol in *S. italica*. The new-generation breeding strategy for small millets involves gene-editing-assisted de novo domestication, targeting polygenic traits such as climate resilience and nutritional aspects.

11 Biotic Stress Tolerance

Small millets, despite their climate resilience and nutritional richness, are confronted with several constraints that can impact their production. The primary factors impeding small millet production on a global level are pests and diseases. Millets are predominantly susceptible to fungal diseases as compared to viral, bacterial, and nematode infections (Das and Rakshit 2016). In contrast to advancements made in genetic enhancement of other primary cereal crops, finger millet has experienced a slower rate of progress. Enhancing the genetic transformation of millets is essential to enhance resistance to both biotic and abiotic challenges (Ceasar and Ignacimuthu 2009).

The worldwide distribution and destructive nature under favorable conditions characterize finger millet blast, induced by the filamentous fungus *Magnaporthe grisea* (anamorph: *Pyricularia grisea*), and ranks as the most devastating disease for finger millet (Yoshida et al. 2016). At every stage of the crop, the finger millet blast exerts its detrimental effects, and results in a complete loss of grain yield, reaching up to 100% (Takan et al. 2011, Senthil et al. 2012). In Eastern Africa and India, this disease was recognized as the foremost obstacle to finger millet production, given the high susceptibility of most genotypes.

The primary challenge in achieving resistance to the pathogen arises from its remarkably unstable genome, resulting in a rapid deterioration of the specific qualities in the developed varieties. The resistance observed in cultivars developed through conventional breeding programs has a brief duration, primarily attributed to the instability of the *Magnaporthe grisea* genome and the rapid evolution of associated genes (Mgonja et al. 2013). Identification of new resistance sources and integrating that resistance into cultivated varieties through conventional breeding methods is both time-consuming and demanding in terms of labor. The evolution of high-throughput sequencing platforms has led to a significant expansion of modern genomic tools. Molecular markers, genome-wide association studies, expressed sequence tags (ESTs), genetic transformations, gene expression profiling, and the recent inclusion of genome editing have demonstrated successful applications across different crops. These tools have been instrumental in the exploration of the genetic foundations of stress tolerance, providing valuable guidance for the development of superior-quality plants (Mbinda and Masaki 2021). Recognizing the significance of finger millet blast disease, it is essential to adopt effective disease control measures. This is particularly crucial in the arid and semiarid regions of Africa and Asia, where finger millet is a staple crop vital for ensuring global food security.

The initial steps in transforming finger millet commenced with the application of the biolistic method. This aimed to assess the efficiency of five gene promoters, including cauliflower mosaic virus 35s (CaMV35S), maize ubiquitin (UqI), rice actin gene promoter ActI, ribulose-1,5-biohosphate carboxylase small subunit gene promoter (RbcS), and *Flaveria trinervia* β-glucuronidase (FtuidA), in expressing the β-glucuronidase (GUS) reporter gene. The development of leaf blast disease-resistant transgenic finger millet involved the integration of the prawn-derived antifungal protein (PIN) gene into the plant's genome (Latha et al. 2005). Following the chemical synthesis of the PIN gene, it was cloned into the plasmid pPin35S. The CaMV35S promoter was utilized to incorporate the construct into the plant genome through the biolistic method. Likewise, the introduction of Chitinase 11 gene (Chi 11) from rice was incorporated into the GPU45 genotype of finger millet using *Agrobacterium*-mediated transformation to confer leaf blast resistance (Ignacimuthu and Ceasar 2012) (Table 5). Transgenic plants, characterized by the overexpression of the foreign gene, demonstrated enhanced resistance to leaf blast in comparison to their non-transgenic counterparts. Nevertheless, there is a lack of available reports regarding transgenic finger millet that exhibits resistance to neck and finger blasts. Consequently, the screening of a variety of potential antifungal genes and the technique of gene pyramiding holds promise for the development of transgenic finger millet resistance to a diverse range of fungal diseases.

With the advent of advanced high-throughput sequencing technologies and omics, there has been a substantial improvement in the comprehension of finger millet blast resistance, and provides invaluable support for breeding programs. The level of resistance in finger millet against *Magnaporthe grisea* is determined by quantitative genetic control (Fu 2015). The homology between EST sequences in the NBS-LRR regions of finger millet and Pi-kh and Pi21 was established by Kumar

Table 5 Finger millet transformation studies

Trait	Promoter/ reporter gene	Functional gene	Transformation method	References
Leaf blast disease	*CaMV35S GUS*	Antifungal protein (PIN) gene of prawn	Biolistic	Latha et al. (2005)
Leaf blast disease	*Uql GUS*	Rice Chitinase gene	*Agrobacterium*-mediated	Ignacimuthu and Ceasar 2012)

et al. in 2016, suggesting a potential role for orthologs of rice blast R-genes in providing resistance in finger millet. The identification of quantitative trait loci (QTLs) linked to NBS-LRR (nucleotide binding site—leucine-rich repeats) sites for finger millet blast resistance through association mapping is strongly emphasized. SSR markers, including NBS-5, NBS-9, NBS-7, and NBS-3, identified from the conserved region of NBS-LRR and EST-SSRs (EST-SSR4, FMBLEST5) are potentially linked to blast resistance. A comparative analysis of NBS-LRR regions of finger millet and rice revealed a high degree of similarity between sequences from resistant genotypes and the Pi genes as well as the NBS-LRR region linked to rice blast. Sequences derived from resistant alleles demonstrate shared regions on the 11th chromosome of rice and corresponding rice ESTs. This suggests that the chromosomal region might play a significant role in conferring resistance to blast disease in finger millet.

The genome-wide association study (GWAS) has emerged as a valuable tool for QTL mining and identification of functional genes from natural populations of crop varieties, particularly in the context of pest and disease resistance in maize (Samayoa et al. 2015) and rice (Wang et al. 2014a; Li et al. 2017, 2019; Feng et al. 2019). A gene encoding an SHR5-receptor-like kinase, homologous to Seita.2G097500 and Seita.2G097400 in foxtail millet, emerged as a crucial regulator of blast resistance. Through a transgenic overexpression approach in rice, it has been confirmed that Seita.2G097500 and Seita.2G097400 are potential blast-resistance genes in foxtail millet (Li et al. 2021). Likewise, in the study conducted by Babu et al. (2014b), five QTLs associated with finger blast resistance were identified in finger millet. These QTLs exhibited a robust correlation with the genic-SSR primer FMBLEST32 and the rice SSR RM262. In a separate investigation, Ramakrishnan et al. (2016) detected two QTLs for leaf blast resistance, showing strong associations with markers UGEP101 and UGEP95 through association mapping.

The emergence of NGS (next-generation sequencing) has introduced novel possibilities for high-marker density genotyping techniques such as genotyping by sequencing (GBS). This enables the exploration of a vast array of single nucleotide polymorphisms (SNPs) for purposes such as diversity analysis, species identification, genome-wide association studies (GWAS), and linkage mapping (Elshire et al. 2011). The integration of GBS and GWAS provides a powerful approach to discover new genetic information crucial for implementing marker-assisted breeding strategies against blast disease (Mbinda and Masaki 2021).

In gene pyramiding, multiple genes are stacked together, leading to the concurrent expression of various genes in a single plant. Despite its efficacy in managing blast and other fungal diseases in rice and other cereal crops such as maize (Zhu et al. 2018) and wheat (Cruppe et al. 2020), this application of this approach is yet to be tested in finger millet. Gene pyramiding has disadvantages such as compromised effectiveness of stacked genes, interactions between genotype and environment, masking of gene expression, impacts on agronomic characters, and the mutation of virulent pathogen genes, which are typically unrelated to each other. The disadvantages mentioned earlier have been resolved through the application of modern biotechnological tools, such as marker-assisted selection, genetic engineering, and genome editing.

In contrast to other cereal crops such as rice, finger millet sequence data is currently quite limited. To overcome this scarcity, the application of comparative genomics becomes crucial for identifying resistant alleles against blast disease. This approach involves matching large sequence datasets from finger millet against the sequences available for rice in diverse databases. Through a comparative analysis of finger millet and rice genomes, it has been established that a significant proportion of chromosomes display are markable collinearity, showing 85% synteny (Srinivasachary et al. 2007). Neck and finger blast resistance QTLs in finger millet are genetically linked to rice genes such as Pi5, Pi21, Pi-d(t) and NBS-LRR. These rice genes can be explored for their potential in conferring blast resistance in finger millet (Babu et al. 2014a).

The progress of genetic engineering in finger millet has been slower compared to several other cereal crops due to its pronounced recalcitrant nature, polyploidy, and genotypic dependence. These factors present significant challenges to the regeneration of plant tissues and, consequently, hinder the overall transgenic production. The most widely utilized and successful method for delivering DNA in transgenic finger millet production is through *Agrobacterium*-mediated transformation (Hema et al. 2014). The creation of blast-resistant transgenics in finger millet has been addressed in two prominent studies. One approach involves incorporating genes from the antimicrobial peptide of prawn (Latha et al. 2005), while the other utilizes rice chitinase (*chi*1) genes (Ignacimuthu and Ceasar 2012).

The CRISPR (clustered regulatory interspaced short palindromic repeats) system, coupled with its associated proteins (Cas), guides the complex to cleave DNA sequences that are complementary to the guide provided. The CRISPR/Cas system is recognized as a revolutionary and innovative tool for plant genome editing due to its simplicity, broad applicability, and cost-effectiveness (Langner et al. 2018). The absence of a complete genome sequence in finger millet poses a barrier for molecular biologists aiming to utilize genome editing tools such as CRISPR-Cas. This limitation prevents the production of cost-effective and resistant genotypes in finger millet without incorporating gene from external sources (Langner et al. 2018). The contributions of various authors with respect to screening of germplasm and/or advanced breeding lines for biotic stresses in small millets is shown in Table 6.

12 Abiotic Stress Tolerance

As we approach the close of the twenty-first century, projections suggest a rise in the global average temperature ranging from 4 to 5 °C. This climatic shift is anticipated to adversely affect the cultivation of major cereal crops such as wheat and rice. In addition to having better nutrition than large cereal crops, small millet crops are naturally resistant to abiotic stresses such as high temperatures, salinity, poor soil fertility, and drought (Goron and Raizada 2015; Singh et al. 2021). Their

Table 6 Resistant sources identification to biotic stresses in small millets

S. no.	Millet crop	Biotic stress	Genetic intervention	Reference
1	Finger millet	Leaf blast	Molecular markers UGEP 95, UGEP 101 were identified	Ramakrishnan et al. (2016)
		Neck blast	Molecular marker UGEP 18 was identified	Babu et al. (2014a, b)
		Finger blast	Molecular markers RM262, FMBLEST32, UGEP81, UGEP53, UGEP24 were identified	Babu et al. (2014a, b)
		Blast resistance	Genetic regions related to blast resistance viz., leucine-rich repeat receptor-like protein kinase family, Kelch repeat-containing family protein, zinc finger protein Sdr4, tryptophan aminotransferase, AP2-like ethylene-responsive transcription factor, chaperone protein DnaJ, hydroxyproline-rich glycoprotein, reductase family protein, F-box domain-containing protein, bromodomain-containing protein, basic helix-loop-helix transcription factor, short-chain dehydrogenase, hexosyltransferase, isoflavone reductase	Dida et al. (2021)
		Aphid	PR 1044, JAN 852, INDAF 5, HR 374, HR 174, PES 176, B7–43, PES 8, RAU 1	Kishore and Jotwani (1980); Murthi and Harinarayana (1989)
		Ear head worm	PR 202, PES 1877, TNAU 294, Indaf 7, TNAU 1877, Indaf 8, PR 177, HR 374, PES 110	Murthi and Harinarayana (1989)
2	Foxtail millet	Blast disease	Chromosomes 2 and 9 had genomic loci associated with blast resistance from GWAS study	Li et al. (2021)
			Cytokinin regulation-related genes present in chromosomes 1, 3, 4, 5, 6, 7, and 11 are expected to role in blast resistance	Wang et al. (2014b)
			Sophisticated lineage structures of *Magnaporthe oryzae* isolates at microgeographic level of green foxtail in Japan through DNA fingerprinting	Yamagashira et al. (2008)
3	Barnyard millet	Grain smut	Identification of resistant sources, namely PRB 47	Kumar (2013)
		Banded leaf and Sheath blight	Identification of resistant sources, namely RBM 12, VL220, TNAU 130	Jain and Anil Kumar Gupta (2010)

(continued)

Table 6 (continued)

S. no.	Millet crop	Biotic stress	Genetic intervention	Reference
		Shoot fly	Resistant genotypes were identified, namely Bhageshwar Local-2, PUNE 2386, ECC 20, VL 24, VL 30, GECH 120, 142, 218	Murthi and Harinarayana (1989)
4	Kodo millet	Head smut	Identification of resistant sources, namely JK13, GPLM 176, RPS 575, 590, 818	Jain (2005); Jain et al. (2013)
		Sheath blight	Resistant sources were identified, namely RPS529, 508, 550, 502, 543,503, 516, 531, 510, 535, 548	Joshi et al. (2014)
		Shoot fly	Resistant sources were identified, namely JK13, RPS917, RPS871, RPS515, RPS628, RPSRPS583, RPS612, RPS938, RPS685, RPS763, RPS806, RPSRPS939, RPS810, RPS934, RPS642, RPS822, RPS941, GPUK 3, RPS970, RPS944	Joshi et al. (2014)
5	Little millet	Head smut	Resistant sources were identified, namely OLM203, GPMR80, GPMR65, Varisukdhara, GPMR82, RLM13, GPMR67, VMLC 296, GPMR105, GPMR70, TNAU98, RLM14, TNAU 89, VMLC 281	Jain (2003); Jain and Tripathi (2007)
		Shoot fly	Resistant sources were identified, namely GPMR164, GPMR274, GPMR236, GPMR243, GPMR110, GPMR213, GPMR584, GPMR683, GPMR569, GPMR189, GPMR241, GPMR98, GPMR163, GPMR324, GPMR670, GPMR598, GPMR192, GPMR96, GPMR583, GPMR161, GPMR596, GPMR95, GPMR190	Gowda et al. (1996)
6	Proso millet	Grain smut	Resistant sources were identified, namely Barnaulskoye 80, I Orenburgskoe9, novskoe Kh86, Gorlinka, MS1316, Khar'kovskoe 86, Veselepodolyankoye 632	Sharma et al. (1993); Krasvin and Usmanova (1988); Ilyin et al. (1993)
		Banded leaf and sheath blight	Identification of resistant sources, namely RAUM 8, GPUP 22, TNAU 137	(Jain and Tiwari (2013))
		Shoot fly	Resistant sources were identified, namely GPMS 136, GPMS 105, GPMS 164, GPMS 108, GPMS 157, GPMS 112, GPMS 125, GPMS 114, GPMS 102, RAU 2, Co 1, GPMS 117, GPMS 122, GPMS 156, BR 6, MS1307, GPMS 115, GPMS 153, GPMS 126, MS1437, GPMS 124, RAU 1, GPMS 135, GPMS 101, GPMS 138, GPMS 123, GPMS 148, GPMS 152, GPMS 155, GPMS 159	Murthi and Harinarayana (1986)

excellent and effective water and nitrogen utilization helps them maintain physiological sustainability, even in adverse environmental conditions (Baltensperger 2002; Saha et al. 2016). The presence of these qualities designates small millets as the ideal smart crop for cultivation amid the changing climate.

12.1 Why Millets Are Climate Resilient?

Millets, in comparison to major cereals, demonstrate enhanced tolerance to environmental stressors, attributed to a variety of morpho-physiological, molecular, and biochemical traits. Essentially, the brief life-cycle of millets contributes to stress evasion, as they can complete their life-cycle in 12–14 weeks, compared to a maximum of 20–24 weeks needed by rice and wheat. Nevertheless, the impact of stress conditions and their outcomes can be counteracted through various characteristics, including reduced stature, diminished leaf area, thickened cell walls, and the ability to establish a robust root system (Li and Brutnell 2011). Moreover, the C4 photosynthetic trait offers substantial benefits to millets. Small millets, with their varied climate-resilient attributes, are predominantly cultivated in arid and semiarid regions.

12.2 Millets as the Model for Stress Biology

The remarkable resilience of millets to a spectrum of abiotic stresses, including drought, light, salinity, and heat, positions them as an attractive platform for examining stress-responsive traits at cellular, molecular, and physiological levels. Numerous investigations into the morpho-physiological and biochemical aspects of millets provide insights into their effective strategies for stress adaptation. Likewise, there is an elevation in biochemical activities, encompassing increased levels of antioxidants, reactive oxygen species, and their scavenging enzymes, along with enhanced catalase and superoxide enzyme activity, and the synthesis of osmolytes and other stress-related proteins has been observed in foxtail millet as a response to abiotic stresses (Lata et al. 2011), little millet (Ajithkumar and Panneerselvam 2014). Furthermore, a multitude of novel genes, alleles, and QTLs have been recently discovered in millets, and their functional characterization has disclosed their significance in providing stress tolerance (Table 7).

12.3 Drought Stress

Drought, a major abiotic stress, significantly influences small millet yield. Contrary to the controlled conditions, drought stress induced leaf withering and rolling, resulting in a decline in chlorophyll and leaf-compatible solute potential as a consequence of more frequent expression of various stress-responsive genes (Parvathi

Table 7 Abiotic stress-related genes in various studies on finger and foxtail millets

Millet	Gene	Nature of study	Reference
Finger millet	Phosphate transporters (Pt)	Cloning and characterization of four Pt genes that exhibit their role in response to Pi stress	Pudake et al. (2017)
	NAC transcription factor	Overexpressing the NAC67 gene in rice, as reported by Rahman et al. in 2016, results in improved tolerance to both salinity and drought stress	Rahman et al. (2016)
	bHLH transcription factor	Elevating the expression of the bHLH57 gene in tobacco provides enhanced tolerance to salinity, oxidative, and drought stress	Ramegowda et al. (2012)
	Dehydrin7	The overexpression of Dehydrin7 gene in tobacco confers increased tolerance to drought stress	Singh et al. 2015
Foxtail millet	Argonaute protein 1 encoding gene	AGO1b has been demonstrated to modulate the stress response in foxtail millet	Liu et al. (2016)
	Late embryogenesis abundant protein (LEA)	The enhanced expression of LEA14 in foxtail millet provides tolerance to drought, salt, and osmotic stress	Wang et al. (2014c)
	ABA-responsive DRE-binding protein (ARDP)	The overexpression of ARDP in foxtail millet enhances their tolerance to drought and salt stress	Li et al. (2014)
	WD-40	Establishing the relationship between WD40 and the dehydration stress-responsive pathway	Mishra et al. (2012)
	Dehydration-responsive element-binding protein 2 (DREB2)	Cloning and characterization of DREB2 demonstrated its role in providing dehydration tolerance	Lata et al. (2011)
	NAC transcription factor	Cloning and characterization of NAC078 revealed its role in conferring tolerance to salinity	Puranik et al. (2011)
	Aldose reductase	Determining the correlation of individual genes with the pathway responsive to salinity stress	Veeranagamallaiah et al. (2007)

et al. 2013). Findings indicated a detrimental influence of drought stress on crop yield, membrane integrity, water potential maintenance, osmotic balance, and photosynthetic efficiency (Ajithkumar and Panneerselvam 2014). In 2013, Parvathi et al. conducted a study involving the generation of transcriptome data from finger millet leaf tissue subjected to drought stress conditions. The finger millet cultivar GPU 28 exhibited the expression of candidate genes as a response to drought stress, as identified in the research. The activation of various genes involved in the drought-stress signaling cascade, including calcineurin B-like interacting protein kinase 31 (CIPK31), serine-threonine protein phosphatase 2A (PP2A), signal recognition particle receptor α (SRPR α), farnesyl pyrophosphate synthase (FPS), among others was revealed through functional annotation and pathway analysis. Moreover, the results suggested that specific basal regulatory genes, such as TATA-binding protein (TBP)-associated factors (TAFs), demonstrate a response to drought conditions. It is clear from these data that basal regulatory process genes are active in finger millet in response to water deficit. Abiotic stress tolerance, particularly in tobacco and rice, was achieved through the overexpression of finger millet TF genes *EcNAC1* (Ramegowda et al. 2012) and *EcNAC67* (Rehman et al. 2016). Hittalmani et al. (2017), through whole genome sequencing, established a connection between 2866 drought-responsive genes and a substantial rise in 19 Pfam domains representing transcription factor (TF) families. In contrast to their wild-type counterparts, transgenic finger millet plants with an inserted bacterial mannitol-1-phosphate dehydrogenase (mtlD) gene showcased enhanced osmotic stress tolerance, as evidenced by chlorophyll retention and a sustained green appearance during drought stress (Hema et al. 2014).

The drought tolerance observed in small millets is characterized by a diversity of agronomic traits. Nitrogen-use efficiency, early flowering, photosynthetic efficiency, and rapid growth are key traits that are essential for small millets in evading the challenges posed by drought conditions. Millets leverage additional traits related to water-use efficiency, ensuring they can prevent drought stress by sustaining water potential within cells. In a similar manner, millets achieve drought tolerance through their exceptional antioxidant properties and the accumulation of osmolytes. For instance, according to Balsamo et al. (2006), the enhancement of tensile strength is linked to bundle sheath lignification. Moreover, little millet demonstrated an increased accumulation of osmolytes in response to drought stress. The nature of the root architecture represents yet another significant morphological trait to endure drought conditions (Fig. 3).

12.4 Salinity Stress

In irrigated agricultural fields, salinity constitutes a prominent issue, leading to adverse effects on both crop yield and nutritional quality. Considering this, the impact of salinity stress on small millets was examined by Rasool et al. (2020) and Shah et al. (2020). Owing to the impact of salt stress, a decline in biomass, photosynthetic efficiency, and water potential was reported. The salt tolerance in finger millet contributes to the sustained production of grains in saline environments. The salt tolerance ability of numerous finger millet genotypes has been explored, highlighting substantial genotype-specific variations in tolerance levels among them. Likewise, the salt tolerance of four foxtail millet genotypes was studied, revealing diverse morphological and anatomical transformations across sensitive and tolerant cultivars. Parvati and Natarajan (2017) reported a considerable decrease in both plant growth and biomass in the seedlings of finger millet genotype GPU 28 subjected to PEG and oxidative stress. The presence of salinity stress led to variations in water content, leaf expansion, plant height, grain weight, length and width of a

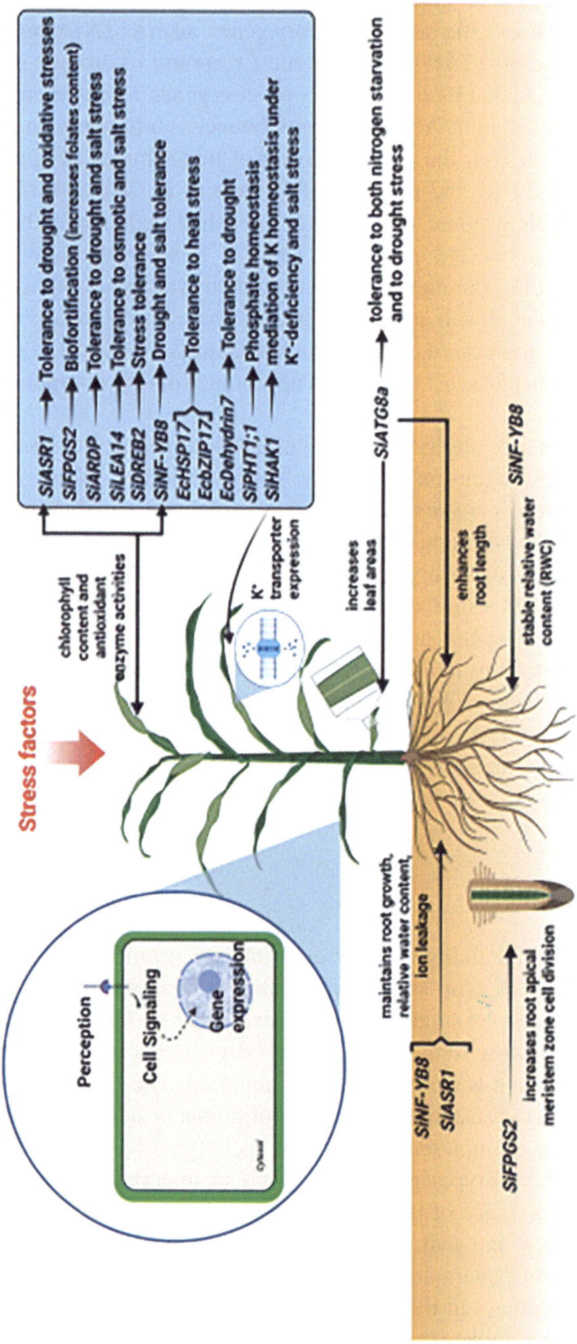

Fig. 3 Physiological functions of various genes in small millets under stress conditions (*Source:* Choudhary et al. 2023)

finger, and flowering (Anjaneyulu et al. 2014). Transgenic finger millet cultivars exhibit a greater degree of salinity tolerance when compared to their wild-type counterparts. Under salinity conditions, the overexpression of *Porteresia coarctata*'s serine-rich protein (*PcSrp*) gene was observed in finger millet, with the plant displaying normal growth and flowering behavior when grown under 250 mM NaCl stress conditions (Mahalakshmi et al. 2006).

12.5 Temperature Stress

Heat tolerance characterizes millets, but they undergo several molecular transformations when exposed to elevated temperatures. For instance, photosynthetic efficiency, respiration, and transpiration are fundamental biological processes acutely responsive to temperature changes, exerting a substantial influence on yield (Ayele et al. 2001). Consequently, the annual surge in global temperature has a considerable impact on crop productivity, posing a threat to food security. Despite being extensively cultivated in regions with a maximum temperature of 36 °C, finger millet cultivars experience a detrimental impact on yield with a further temperature rise (Yogeesh et al. 2016). Findings from the study revealed substantial diversity in the germination rates of different finger millet germplasms in response to elevated temperature. This indicates the necessity for enhanced genotype development through breeding initiatives.

13 Limitations in Small Millets

Small millets have faced neglect from consumers, farmers, and researchers for the past two to three decades. Limited research in small millets is primarily oriented toward varietal development, with only a few varieties released, particularly in the last 5 years, with finger millet being an exception. Browntop millet is still at its infancy. The majority of small millet varieties released are derived from pure line selections originating from land races or germplasm sources. Variability in small millets is limited owing to its self-pollinated nature and natural cross pollination hardly exceeds 1%. Kodo millet flower is cleistogamous and self-pollination is a rule. Moreover, very small size of the florets does not facilitate an easy emasculation. Hot water treatment appears to be promising in finger millet; however, increased temperature may result in the drying of panicles. Gametocides were not as successful as other commercial crops because of its inherent problems. Quest for male sterility in germplasm was in vain for all the small millets except for finger millet, where genetic male sterility was discovered. Germplasm, INFM 95001, and PS 1-IC0598201 are GMS lines. Nevertheless, they exhibited only partial male sterility, and a 100% success rate in crossing was not recorded (Nagaraja et al. 2023). Research on in vitro culture techniques in millets dates back to earlier times (Vasil 1987), but the transformation of millets has not progressed as quickly as that of major cereals. It may be because of the fact that consumption of small millets is very low and mostly at household levels in rural areas. A major drawback in all small millets, except for finger millet, is the lack of accessible primary processing units at convenient locations for the dehulling of grains. If processing mills are provided, small millet consumption will increase, creating a demand for production. This, in turn, attracts funding for small millet research. The international year of millet has sparked initiatives, aimed at promoting millet awareness, integrating them into the public distribution system, declaration of minimum support price for small millets to motivate farmers for their cultivation, and allocating funding for research. Molecular, transgenic, and omics approaches available for the identification of important traits is very limited in small millets. The key constraints that hinder small millet research progress include the inadequate funding in comparison to other cereals, lack of support and encouragement for researchers dedicated to small millets, and comparatively lower priority relative to major crops.

14 Approaches to Utilize Climate Resilient Features of Millets for Crop Improvement

Millets exhibit significant tolerance to abiotic stress under prevailing climatic conditions and have been identified as valuable sources of stress-resistant traits that can contribute to crop improvement. Here is a list of techniques/methodology through which we can harness the climate-resilient traits of millets to improve other cereal crops:

1. Screening of germplasm or breeding lines for various biotic and abiotic stresses and identifying resistant sources. Similarly, germplasm can be screened for various nutrients to identify nutri-dense lines.
2. Understanding the genetics of resistance to biotic or abiotic stress, nutrient density, and agronomic traits.
3. Molecular dwelling through recognizing molecular markers or QTLs linked to stress tolerance or nutrient density through QTL mapping or GWAS.
4. Syntony studies of millets with other cereals, such as maize and rice, where numerous reports on genes or gene families related to stress resistance or nutrient density are available.
5. Landing on gene(s) from linked markers to identify the exact gene(s) responsible for the trait. Allele mining is conducted to drill significant alleles, thereby enriching the genetic potential of crops.
6. The expression of candidate gene(s) can also be targeted through a transcriptomic approach.
7. In the current omics era, integrating various omics, such as transcriptomics, proteomics, and genomics, could be a more effective option for identifying probable candidate genes in millets.
8. Transgressing QTL(s) or genes(s) into improved varieties and potentially implementing gene pyramiding for various economic traits.

All of these paves the way for the identification of genes/QTLs responsible for nutrient density or tolerance mechanism in millets. After identification, the introgression of targeted gene/QTLs can be achieved through molecular breeding, conventional breeding, or transgenic approaches. Additionally, gene correction/deletion of genes with negative effects could be performed through genome editing (Singh et al. 2021).

References

Aggarwal PI, Lydia P, Choudhary P, Singh R, Shukla P, Prasad M (2022) Multi-omics intervention in Setaria to dissect climate-resilient traits: progress and prospects. Front Plant Sci 13:892736. https://doi.org/10.3389/fpls.2022.892736

Ajithkumar IP, Panneerselvam R (2014) ROS scavenging system, osmotic maintenance, pigment and growth status of Panicum sumatrense Roth. under drought stress. Cell Biochem Biophys 68:587–595. https://doi.org/10.1007/s12013-013-9746-x

Anjaneyulu E, Reddy PS, Sunita MS, Kishor PB, Meriga B (2014) Salt tolerance and activity of antioxidative enzymes of transgenic finger millet overexpressing a vacuolar HC-pyrophosphatase gene (SbVPPase) from Sorghum bicolor. J Plant Physiol 171:789–798. https://doi.org/10.1016/j.jplph.2014.02.001

Anonymous (2022) Millets: The future super food for India. Knowledge Paper, The Associated Chambers of Commerce and Industry of India

Ayele M, Blum A, Nguyen HT (2001) Diversity for osmotic adjustment and root depth in TEFF [Eragrostis teff (Zucc) Trotter]. Euphytica 121:237–249. https://doi.org/10.1023/A:1012099914738

Babu BK, Dinesh P, Agrawal PK, Sood S, Chandrashekara C, Bhatt JC, Kumar A (2014a) Comparative genomics and association mapping approaches for blast resistant genes in finger millet using SSRs. PloS One 9(6):1–12. e99182. https://doi.org/10.1371/journal.pone.0099182

Babu BK, Agrawal PK, Pandey D, Jaiswal JP, Kumar A (2014b) Association mapping of agro-morphological characters among the global collection of finger millet genotypes using genomic SSR markers. Mol Biol Rep 41:5287–5297. https://doi.org/10.1007/s11033-014-3400-6

Balsamo RA, Willigen CV, Bauer AM, Farrant J (2006) Drought tolerance of selected Eragrostis species correlates with leaf tensile properties. Ann Bot 97(6):985–991. https://doi.org/10.1093/aob/mcl068

Baltensperger DD (2002) Progress with proso, pearl and other millets. Trends New Crop New Uses:100–103

Bennetzen JL, Schmutz J, Wang H, Percifield R, Hawkins J, Pontaroli AC (2012) Reference genome sequence of the model plant Setaria. Nat Biotechnol 30(6):555–561. https://doi.org/10.1038/nbt.2196

Brhane H, Haileselassie T, Tesfaye K, Ortiz R, Hammenhag C, Abreha KB, Vetukuri RR, Geleta M (2022) Finger millet RNA-seq reveals differential gene expression associated with tolerance to aluminium toxicity and provides novel genomic resources. Front Plant Sci 13:1068383. https://doi.org/10.3389/fpls.2022.1068383

Ceasar SA, Ignacimuthu S (2009) Genetic engineering of millets: current status and future prospects. Biotechnol Lett 31:779–788

Ceasar SA, Maharajan T, Ajeesh Krishna TP, Ramakrishnan M, Victor Roch G, Satish L, Ignacimuthu S (2018) Finger millet [*Eleusine coracana* (L.) Gaertn.] improvement: current status and future interventions of whole genome sequence. Front Plant Sci 9:1054

Choudhary P, Shukla P, Muthamilarasan M (2023) Genetic enhancement of climate-resilient traits in small millets: a review. Heliyon 9:e14502

Cho YI, Chung JW, Lee GA, Ma KH, Dixit A, Gwag JG, Park YJ (2010) Development and characterization of twenty-five new polymorphic microsatellite markers in proso millet (*Panicum miliaceum* L.). Genes Genom 32:267–273

Danquah EY, Hanley SJ, Brookes RC, Aldam C, Karp A (2002) Isolation and characterisation of microsatellites in *Echinochloa* (L.) Beauv. spp. Mol Ecol Notes 2:54–56

Das I, Rakshit S (2016) Millets, their importance, and production constraints. biotic stress resistance in millets. Elsevier, pp 3–19

David RHA, Ramakrishnan M, Maharajan T, BarathiKannan K, Babu GA, Daniel MA (2021) Mining QTL and genes for root traits and biochemical parameters under vegetative drought in South Indian genotypes of finger millet (*Eleusine coracana* (L.) Gaertn) by association mapping and in silico comparative genomics. Biocatal Agric Biotechnol 32:101935. https://doi.org/10.1016/j.bcab.2021.101935

Deepak RGT, Thakur V (2018) Minor millets-their potential health benefits and medicinal properties: a review. Int J Pure Appl Sci 6(1):1677

Dida MM, Oduori CA, Manthi SJ, Avosa MO, Mikwa EO, Ojulong HF, Odeny DA (2021) Novel sources of resistance to blast disease in Finger millet. Crop Sci 61(1):250–262

Dida MM, Srinivasachary, Ramakrishnan S, Bennetzen JL, Gale MD, Devos KM (2007) The genetic map of finger millet, Eleusine coracana. Theor Appl Genet 114:321–332

Elshire RJ, Glaubitz JC, Sun Q, Poland JA, Kawamoto K, Buckler ES, Mitchell SE (2011) A robust, simple genotyping-by-sequencing (GBS) approach for high diversity species. PloS one 6(5):e19379

Francis N, Rajasekaran R, Rajagopalan VR, Bakya SV, Muthurajan R, Kumar AG, Alagarswamy S, Krishnamoorthy I, Thiyagarajan C (2023) Molecular characterization and SNP identification using genotyping-by-sequencing in high yielding mutants of proso millet. Front Plant Sci 14:1108203. https://doi.org/10.3389/fpls.2023.1108203

Feng Z, Kang H, Li M, Zou L, Wang X, Zhao J, Lang W, Zhou N, Li Q, Lan Y, Zhang Y, Chen Z, Liu W, Pan X, Wang G, Zuo S (2019) Identification of new rice cultivars and resistance loci against rice black-streaked dwarf virus disease through genome-wide association study. Rice 12(1):49. https://doi.org/10.1186/s12284-019-0310-1

Fu YB (2015) Understanding crop genetic diversity under modern plant breeding. Theor Appl Genet 128:2131–2142. https://doi.org/10.1007/s00122-015-2585-y

Goron TL, Raizada M (2015) Genetic diversity and genomic resources available for the small millet crops to accelerate a New Green Revolution. Front Plant Sci 6:157. https://doi.org/10.3389/fpls.2015.00157

Gowda K, Gowda KNM, Ramesh S, Jagadish PS, Gowda K (1996) Field evaluation of little millet germplasm for resistance to the shoot fly, *Atherigonapulla Wiede*. (Diptera:Anthomyiidae) Karnataka. J Agric Sci 9(1):155–158

Cruppe G, Cruz CD, Peterson GL, Pedley KF, Asif M, Fritz AK, Calderón L, Cruppe G, Cruz CD, Peterson G, Pedley K, Asif M, Fritz A, Calderon L, Lemes da Silva C, Todd T, Kuhnem P, Singh PK, Singh RP, Braun HJ, NCD B, Valent B (2020) Novel sources of wheat head blast resistance in modern breeding lines and wheat wild relatives. Plant Dis 104(1):35–43. https://doi.org/10.1094/pdis-05-19-0985-re

Guo LL, Qiu J, Ye C, Jin G, Mao L, Zhang H (2017) Echinochloa crus-galli genome analysis provides insight into its adaptation and invasiveness as a weed. Nat Commun 8(1):1–10. https://doi.org/10.1038/s41467-017-01067-5

Gupta P, Rustgi S, Kulwal P (2005) Linkage disequilibrium and association studies in higher plants: present status and future prospects. Plant Mol Biol 57:461–485. https://doi.org/10.1007/s11103-005-0257-z

Hatakeyama M, Aluri S, Balachadran MT, Sivarajan SR, Patrignani A, Grüter S (2018) Multiple hybrid de novo genome assembly of finger millet, an orphan allotetraploid crop. DNA Res 25(1):39–47. https://doi.org/10.1093/dnares/dsx036

Hema R, Vemanna RS, Sreeramulu S, Reddy CP, Senthil Kumar M, Udayakumar M (2014) Stable expression of mtlD gene imparts multiple stress tolerance in finger millet. PloS One 9:e99110. https://doi.org/10.1371/journal.pone.0099110

Hilu K (1994) Evidence from RAPD markers in the evolution of Echinochloa millets (Poaceae). Plant Syst Evol 189:247–257

Hittalmani S, Mahesh HB, Shirke MD, Biradar H, Uday G, Aruna YR (2017) Genome and transcriptome sequence of Finger millet (Eleusine coracana (L.) Gaertn.) provides insights into drought tolerance and nutraceutical properties. BMC Genomics 18(1):1–16. https://doi.org/10.1186/s12864-017-3850-z

Ignacimuthu S, Ceasar SA (2012) Development of transgenic finger millet (*Eleusine coracana* (L.) Gaertn.) resistant to leaf blast disease. J Biosci 37:135–147

Ilyin VA, Zolotukhin EN, Ungenfukht IP, Tikhonov NP, Markin BK (1993) Importance, cultivation and breeding of proso millet in Povolzye province in Russia K.W. In: Riley SC, Gupta A, Seetharam JNM (eds) Advances in small millets. Oxford and IBM Publishers, New Delhi, India, pp 109–116

Indirani K, Devasena M (2021) Review on nutritional profiles and health benefits of little millets –India. Int J Res Eng Sci (IJRES) 9:7

Jain AK, Tripathi SK (2007) Management of grain smut (*Macalpinomycessharmae*) in little millet. Indian Phytopathol 60:467–471

Jain A, Gupta AK (2010) Occurrence of banded leaf and sheath blight on foxtail and barnyard millets in Madhya Pradesh. Annals Plant Prot Sci 18., Society of Plant Protection Sciences(1):268–270

Jain A (2005) Stable sources of resistance for head smut in kodo millet. Indian Phytopathol 58., Springer(1):117

Jain AK (2003) Occurrence of grain smut of little millet caused by Macalpinomycessharmae in Madhya Pradesh. Plant Protect Bull 55:30–32

Jain AK, Tiwari A, Dhingra MR (2013) Resistance against head smut in kodo millet (*Paspalum scrobiculatum* L.) Mysore. J Agric Sci 47(3):645–647

Jain AK, Tiwari A (2013) Evaluation of promising genotypes of proso millet against banded leaf and sheath blight disease caused by Rhizoctonia solani Kuhn. *Mysore J Agric Sci* 47., University of Agricultural Sciences, Bangalore(3):648–650

Jaiswal V, Bandyopadhyay T, Gahlaut V, Gupta S, Dhaka A, Ramchiary N, Prasad M (2019a) Genome-wide association study (GWAS) delineates genomic loci for ten nutritional elements in foxtail millet (*Setaria italica* L.). J Cereal Sci 85:48–55. https://doi.org/10.1016/j.jcs.2018.11.006

Jaiswal V, Gupta S, Gahlaut V, Muthamilarasan M, Bandyopadhyay T, Ramchiary N, Prasad M (2019b) Genome-wide association study of major agronomic traits in foxtail millet (*Setaria italica* L.) using ddRAD sequencing. Sci Rep 9:5020. https://doi.org/10.1038/s41598-019-41602-6

Jayakodi M, Madheswaran M, Adhimoolam K, Perumal S, Manickam D, Kandasamy T, ... Natesan S (2019) Transcriptomes of Indian barnyard millet and barnyardgrass reveal putative genes involved in drought adaptation and micronutrient accumulation. Acta Physiol Plant 41:1–11

Jia G, Huang X, Zhi H, Zhao Y, Zhao Q, Li W, Chai Y, Yang L, Liu K, Lu H, Zhu C, Lu Y, Zhou C, Fan D, Weng Q, Guo Y, Huang T, Zhang L, Lu T, Han B (2013) A haplotype map of genomic variations and genome-wide association studies of agronomic traits in foxtail millet (*Setaria italica*). Nat Genet 45:957–961. https://doi.org/10.1038/ng.2673

Jia X, Zhang Z, Liu Y, Zhang C, Shi Y, Song Y, Wang T, Li Y (2009) Development and genetic mapping of SSR markers in foxtail millet [*Setaria italica* (L.) P. Beauv.]. Theor Appl Genet 118:821–829

Joshi DC, Meena RP, Chandora R (2021) Genetic resources: collection, characterization, conservation, and documentation. In: Millets and pseudo cereals. Woodhead Publishing, pp 19–31

Joshi RP, Jain AK, Chauhan SS (2014) Collection and evaluation of kodo millet land races for agro-morphological traits and biotic stresses in Madhya Pradesh. JNKVV Res J 48(2):162–169

Kishore P, Jotwani MG (1980) Screening of some improved genotypes of the finger millet (Eleusine coracana Gaertn.) for resistance to Sesamiainferens Walker and Myllocerusmaculosus Desbrochers. J Entomol Res 4:221–223

Kumar B (2013) Management of grain smut disease of barnyard millet (Echinochloa frumentacea). Indian Phytopathol 66., Springer(4):403–405

Kumar A, Pathak RK, Gupta SM, Gaur VS, Pandey D (2015) Systems biology for smart crops and agricultural innovation: filling the gaps between genotype and phenotype for complex traits linked with robust agricultural productivity and sustainability. OMICS 19:581–601. https://doi.org/10.1089/omi.2015.0106

Kumar A, Babu BK, Yadav S, Agrawal PK (2016) Allele mining for resistance gene analogs (RGAs) in crop plants: a special emphasis on blast resistance in finger millet (*Eleusine coracana* L.). Indian J Genet Plant Breed 76(1):1. https://doi.org/10.5958/0975-6906.2016.00001.8

Kumari K, Muthamilarasan M, Misra G, Gupta S, Subramanian A, Parida SK, Chattopadhyay D, Prasad M (2013) Development of eSSR-Markers in Setaria italica and their applicability in studying genetic diversity, cross-transferability and comparative mapping in millet and non-millet species. PloS One 8:e67742

Krasavin VD, Usmanova RI (1988) Proso millet variety Orenburgskoe 9 resistant to head smut. SelektsiyaiSemenovodstvo (Moscow), 1. pp 54–55

Langner T, Kamoun S, Belhaj K (2018) CRISPR crops: plant genome editing toward disease resistance. Annu Rev Phytopathol 56(1):479–512. https://doi.org/10.1146/annurev-phyto-080417-050158

Lata C, Bhutty S, Bahadur RP, Majee M, Prasad M (2011) Association of an SNP in a novel DREB2-like gene SiDREB2 with stress tolerance in foxtail millet [*Setaria italica* (L.)]. J Exp Bot 62:3387–3401. https://doi.org/10.1093/jxb/err016

Latha AM, Rao KV, Reddy VD (2005) Production of transgenic plants resistant to leaf blast disease in finger millet (*Eleusine coracana* (L.) Gaertn.). Plant Sci 169:657–667

Li C, Yue J, Wu X, Xu C, Yu J (2014) ABA-responsive DREbinding protein gene from *Setariaitalica*, SiARDP, the target gene of SiAREB, plays a critical role under drought stress. J Exp Bot 65:5415–5427. https://doi.org/10.1093/jxb/eru302

Li P, Brutnell TP (2011) *Setaria viridis* and *Setaria italica*, model genetic systems for the panicoid grasses. J Exp Bot 62:3031–3037

Li C, Wang D, Peng S, Chen Y, Su P, Chen J, Zheng L, Tan X, Liu J, Xiao Y, Kang H, Zhang D, Wang G-L, Liu Y (2019) Genome-wide association mapping of resistance against rice blast strains in South China and identification of a new Pik allele. Rice 12(1):47. https://doi.org/10.1186/s12284-019-0309-7

Li W, Zhu Z, Chern M, Yin J, Yang C, Ran L, Cheng M, He M, Wang K, Wang J, Zhou X, Zhu X, Chen Z, Wang J, Zhao W, Ma B, Qin P, Chen W, Wang Y, Liu J (2017) A natural allele of a transcription factor in rice confers broad-spectrum blast resistance. Cell 170(1):114–126.e15. https://doi.org/10.1016/j.cell.2017.06.008

Li R, Han Y, Zhang Q, Chang G, Han Y, Li X, Zhang B (2020) Transcriptome profiling analysis reveals co-regulation of hormone pathways in foxtail millet during *Sclerospora graminicola* infection. Int J Mol Sci 21(4):1226

Li Z, Jia G, Li X, Li Y, Zhi H, Tang S, Ma J, Zhang S, Li Y, Shang Z, Diao X (2021) Identification of blast-resistance loci through genome-wide association analysis in foxtail millet (Setaria italica (L.) Beauv.). *J Integr Agric* 20. Elsevier BV(8):2056–2064

Liu X, Tang S, Jia G, Schnable JC, Su H, Tang C, Zhi H, Diao X (2016) The C-terminal motif of SiAGO1b is required for the regulation of growth, development and stress responses in foxtail millet (*Setaria italica* (L.) P. Beauv). J Exp Bot 67:3237–3249. https://doi.org/10.1093/jxb/erw135

Mahalakshmi S, Christopher G, Reddy T, Rao K, Reddy V (2006) Isolation of a cDNA clone (PcSrp) encoding serine-rich-protein from *Porteresia coarctata* T. and its expression in yeast and finger millet (*Eleusine coracana* L.) affording salt tolerance. Planta 224:347–359. https://doi.org/10.1007/s00425-005-0218-4

Mbinda W, Masaki H (2021) Breeding strategies and challenges in the improvement of blast disease resistance in finger millet. A current review. Front Plant Sci 11(11). https://doi.org/10.3389/fpls.2020.602882

Mgonja, M., Audi, P., Mgonja, A. P., Manyasa, E. O., &Ojulong, O. (2013). Integrated blast and weed management and microdosing in finger millet: a HOPE project manual for increasing finger millet productivity

Mishra AK, Puranik S, Bahadur RP, Prasad M (2012) The DNA binding activity of an AP2 protein is involved in transcriptional regulation of a stress-responsive gene, SiWD40, in foxtail millet. Genomics 100:252–263. https://doi.org/10.1016/j.ygeno.2012.06.012

Murthi TK, Harinarayana G (1989) Insect pests of small millets and their management in India. In: Seetharam A, Riley KW, Harinarayana G (eds) Small millets in global agriculture, Proceedings of the First International Small Millets Workshop Bangalore, India, October 29–November 2, 1986, pp 255–270

Muthamilarasan M, Prasad M (2021) Small millets for enduring food security amidst pandemics. Trends Plant Sci 26(1):33–40. https://doi.org/10.1016/j.tplants.2020.08.008

Murthi T, Harinarayana G (1986) Insect pests of small millets and their management in India. In: Seetharam A, Riley KW, Harinarayana G (eds) Small millets in global agriculture. Oxford and IBH Publishing, New Delhi, India, pp 255–270

Nagaraja TE, Nandini C, Bhat S, Parveen S (2023) Artificial hybridization techniques in small millets—a review. Front Plant Sci 14:1112117

Nah G, Im JH, Kim JW, Park HR, Yook MJ, Yang TJ, Kim DS (2015) Uncovering the differential molecular basis of adaptive diversity in three Echinochloa leaf transcriptomes. PLoS One 10(8):e0134419. https://doi.org/10.1371/journal.pone.0134419

Ni X, Xia Q, Zhang H, Cheng S, Li H, Fan G (2017) Updated foxtail millet genome assembly and gene mapping of nine key agronomic traits by resequencing a RIL population. GigaScience 6(2):1–8. https://doi.org/10.1093/gigascience/giw005

Pandey G, Mishra G, Kumari K, Gupta S, Parida SK, Chattopadhyay D, Prasad M (2013) Genome-wide development and use of microsatellite markers for large-scale genotyping applications in Foxtail millet [*Setaria italica* (L.)]. DNA Res 20:197–207

Panwar P, Jha AK, Pandey PK, Gupta AK, Kumar A (2011) Functional markers based molecular characterization and cloning of resistance gene analogs encoding NBS-LRR disease resistance proteins in finger millet (*Eleusine coracana*). Mol Biol Rep 38:3427–3436

Parvathi MS, Nataraja KN (2017) Discovery of stress responsive TATA box binding protein associated Factor6 (TAF6) from finger millet (*Eleusine coracana* (L.) Gaertn). J Plant Biol 60:335–342. https://doi.org/10.1007/s12374-016-0574-6

Parvathi M, Nataraja KN, Yashoda B, Ramegowda H, Mamrutha H, Rama N (2013) Expression analysis of stress responsive pathway genes linked to drought hardiness in an adapted crop, finger millet (*Eleusine coracana*). J Plant Biochem Biotechnol 22:193–201. https://doi.org/10.1007/s13562-012-0135-0

Parvathi, M. S., Nataraja, K. N., Reddy, Y. N., Naika, M. B., & Gowda, M. C. (2019) Transcriptome analysis of finger millet (*Eleusine coracana* (L.) Gaertn.) reveals unique drought responsive genes. J Gene 98(2):46

Pendergast TH IV, Qi P, Odeny DA, Dida MM, Devos KM (2022) A high-density linkage map of finger millet provides QTL for blast resistance and other agronomic traits. Plant Genome 15(1):e20175. https://doi.org/10.1002/tpg2.20175

Pramitha J, Ganesan J, Francis N, Rajasekharan R, Thinakaran J (2023) Revitalization of small millets for nutritional and food security by advanced genetics and genomics approaches. Front Genet 13:1007552. https://doi.org/10.3389/fgene.2022.1007552

Pramitha J, Subhalakshmi VKI, Jeeva G, Surendar A (2021) Small millets exploring the hidden potential. Scripown Publications, North West Delhi, India

Pudake RN, Mehta CM, Mohanta TK, Sharma S, Varma A, Sharma AK (2017) Expression of four phosphate transporter genes from Finger millet (*Eleusine coracana* L.) in response to mycorrhizal colonization and Pi stress. Biotechnology 7:17. https://doi.org/10.1007/s13205-017-0609-9

Puranik S, Bahadur RP, Srivastava PS, Prasad M (2011) Molecular cloning and characterization of a membrane associated NAC family gene, SiNAC from foxtail millet [*Setaria italica* (L.) P. Beauv.]. Mol Biotechnol 49:138–150. https://doi.org/10.1007/s12033-011-9385-7

Puranik S, Sahu PP, Beynon S, Srivastava RK, Sehgal D, Ojulong H, Yadav R (2020) Genome-wide association mapping and comparative genomics identifies genomic regions governing grain nutritional traits in finger millet (*Eleusine coracana* L. Gaertn.). Plants People Planet 2:649–662. https://doi.org/10.1002/ppp3.10120

Rahman H, Jagadeeshselvam N, Valarmathi R, Sachin B, Sasikala R, Senthil N, ... Muthurajan R (2014) Transcriptome analysis of salinity responsiveness in contrasting genotypes of finger millet (*Eleusine coracana* L.) through RNA-sequencing. Plant Mol Biol 85:485–503

Rahman H, Ramanathan V, Nallathambi J, Duraialagaraja S, Muthurajan R (2016) Over-expression of a NAC 67 transcription factor from finger millet (*Eleusine coracana* L.) confers tolerance against salinity and drought stress in rice. BMC biotechnology 16:7–20

Rajput SG, Santra DK, Schnable J (2016) Mapping QTLs for morphoagronomic traits in proso millet (*Panicum miliaceum* L.). Mol Breed 36(4):37–18. https://doi.org/10.1007/s11032-016-0460-4

Ramakrishnan M, Antony Ceasar S, Duraipandiyan V, Vinod KK, Kalpana K, Al-Dhabi NA (2016) Tracing QTLs for leaf blast resistance and agronomic performance of finger millet (*Eleusine coracana* (L.) Gaertn.) genotypes through association mapping and in silico comparative genomics analyses. PloS One 11(7):pe0159264. https://doi.org/10.1371/journal.pone.0159264

Ramakrishnan M, Ceasar SA, Vinod KK, Duraipandiyan V, Ajeesh Krishna TP, Upadhyaya HD (2017) Identification of putative QTLs for seedling stage phosphorus starvation response in finger millet (*Eleusine coracana* L. Gaertn.) by association mapping and cross species synteny analysis. PloS One 12(8):e0183261. https://doi.org/10.1371/journal.pone.0183261

Ramegowda V, Senthil-Kumar M, Nataraja KN, Reddy MK, Mysore KS, Udayakumar M (2012) Expression of a finger millet transcription factor. EcNAC1, in tobacco confers abiotic stress-tolerance. PloS One 7:e40397. https://doi.org/10.1371/journal.pone.0040397

Rao DB, Bhaskarachary K, Arlene Christina GD, Sudha Devi G, Vilas AT, Tonapi A (2017) Nutritional and health benefits of millets. ICAR, Indian Institute of Millets Research IIMR, Rajendranagar, Hyderabad, p 112

Rasool A, Shah WH, Tahir I, Alharby H (2020) Exogenous application of selenium (Se) mitigates NaCl stress in proso and foxtail millets by improving their growth, physiology and biochemical parameters. Acta Physiol Plant 42(7):1–13. https://doi.org/10.1007/s11738-020-03109-w

Rehman H, Ramanathan V, Nallathambi J, Duraialagaraja S, Muthurajan R (2016) Over-expression of a NAC 67 transcription factor from finger millet (*Eleusine coracana* L.) confers tolerance against salinity and drought stress in rice. BMC Biotechnol 11:35. https://doi.org/10.1186/s12896-016-0261-1

Roy SK, Kwon SJ, Yu JH, Sarker K, Cho SW, Moon YJ (2017) Comparison of protein profiles of Proso Millet (*Panicum miliaceum*) seeds of various Korean. Korean J Crop Sci 62:40–50

Saha MV, Gowda C, Arya L, Verma M, Bansal KC (2016) Genetic and genomic resources of small millets. Crit Rev Plant Sci 35(1):56–79. https://doi.org/10.1080/07352689.2016.1147907

Samayoa L, Malvar R, Olukolu BA, Holland JB, Butrón A (2015) Genome-wide association study reveals a set of genes associated with resistance to the Mediterranean corn borer (Sesamianonagrioides L.) in a maize diversity panel. BMC Plant Biol 15(1):35. https://doi.org/10.1186/s12870-014-0403-3

Senthil R, Shanmugapackiam S, Raguchander T (2012) Evaluation of biocontrol agents and fungicides for the management of blast disease of finger millet. J Mycol Plant Pathol 42:454–458

Shah WH, Rasool A, Tahir I, Rehman RU (2020) Exogenously applied selenium (Se) mitigates the impact of salt stress in *Setaria italica* L. and *Panicum miliaceum* L. Front Plant Sci 1–13. https://doi.org/10.1007/s13237-020-00326-z

Shankar C, Anjani K (2022) Morpho-molecular genetic diversity analysis of little millet (*Panicum sumatrense*) using yield attributing traits and ISSR markers to evaluate its performance as a summer crop. Environ Ecol 41(3B):1788–1798

Sharma D, Tiwari A, Sood S, Jamra G, Singh NK, Meher PK (2018) Genome wide association mapping of agro-morphological traits among a diverse collection of finger millet (*Eleusine coracana* L.) genotypes using SNP markers. PloS One 13(8):e0199444. https://doi.org/10.1371/journal.pone.0199444

Sharma D, Tiwari A, Sood S, Meher PK, Kumar A (2022) Identification and validation of candidate genes for high calcium content in finger millet [*Eleusine coracana* (L.) Gaertn.] through genome-wide association study. J Cereal Sci 107:103517

Sharma PN, Sugha SK, Panwar KS, Sagwal JC (1993) Performance of proso millet (*Panicum miliaceum*) germplasm in relation to smut (*Sphacelotheca destruens*). Indian. J Agric Sci 63:675

Singh RK, Muthamilarasan M, Prasad M (2021) Biotechnological approaches to dissect climate-resilient traits in millets and their application in crop improvement. J Biotechnol 327:64–73. https://doi.org/10.1016/j.jbiotec.2021.01.002

Singh RK, Singh VK, Raghavendrarao S, Phanindra ML, Venkat Raman K, Solanke AU, Kumar PA, Sharma TR (2015) Expression of finger millet EcDehydrin7 in transgenic tobacco confers tolerance to drought stress. Appl Biochem Biotechnol 177:207–216. https://doi.org/10.1007/s12010-015-1738-4

Srikanth N (2023) Millets-based indigenous ethnomedicinal and nutritional practices: need for scientific validation and value addition. J Drug Res Ayurvedic Sci 8(Suppl 1):S8–S11

Srinivasachary DMM, Gale MD, Devos KM (2007) Comparative analyses reveal high levels of conserved colinearity between the finger millet and rice genomes. Theor Appl Genet 115(4):489–499. https://doi.org/10.1007/s00122-007-0582-5

Takan JP, Chipili J, Muthumeenakshi S, Talbot NJ, Manyasa EO, Bandyopadhyay R, Sere Y, Nutsugah SK, Talhinhas P, Hossain M, Brown AE, Sreenivasaprasad S (2011) *Magnaporthe oryzae* populations adapted to finger millet and rice exhibit distinctive patterns of genetic diversity, sexuality and host interaction. Mol Biotechnol 50:145–158. https://doi.org/10.1007/s12033-011-9429-z

Tiwari A, Sharma D, Sood S, Prakash J, Pratap S, Wasudeo P, Kumar A (2020) Genome-wide association mapping for seed protein content in finger millet (*Eleusine coracana*) global collection through genotyping by sequencing. J Cereal Sci 91:102888. https://doi.org/10.1016/j.jcs.2019.102888

Upadhyaya HD, Vetriventhan M (2018) Underutilized climate-smart nutrient rich small millets for food and nutritional security. Regional expert consultation on underutilized crops for Food and Nutritional Security in Asia and the Pacific—thematic, strategic papers and country status reports.

Vasil IK (1987) Developing cell and tissue culture systems for the improvement of cereal and grass crops. J Plant Physiol 128:193–197

Veeranagamallaiah G, Chandraobulreddy P, Jyothsnakumari G, Sudhakar C (2007) Glutamine synthetase expression and pyrroline-5-carboxylate reductase activity influence proline accumulation in two cultivars of foxtail millet (*Setaria italica* L.) with differential salt sensitivity. Environ Exp Bot 60:239–244. https://doi.org/10.1016/j.envexpbot.2006.10.012

Suresh BV, Muthamilarasan M, Misra G, Prasad M (2013) FmMDb: a versatile database of foxtail millet markers for millets and bioenergy grasses research. PloS One 8:e71418

Vetriventhan M, Azevedo VC, Upadhyaya HD, Nirmalakumari A, Kane-Potaka J, Anitha S, Tonapi VA (2020) Genetic and genomic resources, and breeding for accelerating improvement of small millets: current status and future interventions. Nucleus 63:217–239

Vinoth A, Ravindhran R (2017) Biofortification in millets: a sustainable approach for nutritional security. Front Plant Sci 8:29. https://doi.org/10.3389/fpls.2017.0002

Wang ZM, Devos KM, Liu CJ, Wang RQ, Gale MD (1998) Construction of RFLP-based maps of foxtail millet, Setaria italica (L.) P. Beauv. Theor Appl Genet 96:31–36

Wang C, Yang Y, Yuan X, Xu Q, Feng Y, Yu H, Wang Y, Wei X (2014a) Genome-wide association study of blast resistance in indica rice. BMC Plant Biol 14(1):311. https://doi.org/10.1186/s12870-014-0311-6

Wang Y, Liu H, Xin Q (2014b) Genome-wide analysis and identification of cytokinin oxidase/dehydrogenase (CKX) gene family in foxtail millet (Setaria italica). Crop J 2(4):244–254

Wang M, Li P, Li C, Pan Y, Jiang X, Zhu D, Zhao Q, Yu J (2014c) SiLEA14, a novel atypical LEA protein, confers abiotic stress resistance in foxtail millet. BMC Plant Biol 14:290. https://doi.org/10.1186/s12870-014-0290-7

Yadav CB, Muthamilarasan M, Pandey G, Khan Y, Prasad M (2014) Development of novel microRNA-based genetic markers in foxtail millet for genotyping applications in related grass species. Mol Breed 34:2219. https://doi.org/10.1007/s11032-014-0137-9

Yogeesh LN, Naryanareddy AB, Nanjareddy YA, Gowda MVC (2016) High temperature tolerant genotypes of finger millet (*Eleusine coracana* L.). Nat Environ Pollut Technol 15:1293–1296

Yamagashira A, Iwai C, Moroishi M, Misaka M, Fujita Y, Hirata K, Kusaba M (2008) Population structure of Magnaporthe oryzae isolates from green foxtail in Japan examined by DNA fingerprint analysis. Mycoscience 49., Elsevier BV(6):351–358

Yoshida K, Saunders DGO, Mitsuoka C, Natsume S, Kosugi S, Saitoh H, Inoue Y, Chuma I, Tosa Y, Cano LM, Kamoun S, Terauchi R (2016) Host specialization of the blast fungus Magnaporthe oryzae is associated with dynamic gain and loss of genes linked to transposable elements. BMC Genomics 17(1):370. https://doi.org/10.1186/s12864-016-2690-6

Yue H, Wang L, Liu H, Yue W, Du X, Song W, Nie X (2016) De novo assembly and characterization of the transcriptome of broomcorn millet (*Panicum miliaceum* L.) for gene discovery and marker development. Front Plant Sci 7:1083

Zargar M, Dyussibayeva E, Orazov A, Zeinullina A, Zhirnova I, Yessenbekova G, Rysbekova A (2023) Microsatellite-based genetic diversity analysis and population structure of proso millet (*Panicum miliaceum* L.) in Kazakhstan. Agronomy 13:2514. https://doi.org/10.3390/agronomy13102514

Zhang Y, Gao X, Li J, Gong X, Yang P, Gao J (2019) Comparative analysis of proso millet (*Panicum miliaceum* L.) leaf transcriptomes for insight into drought tolerance mechanisms. BMC Plant Biol 19:1–17

Zhang G, Liu X, Quan Z, Cheng S, Xu X, Pan S (2012) Genome sequence of foxtail millet (*Setaria italica*) provides insights into grass evolution and biofuel potential. Nat Biotechnol 30(6):549–554. https://doi.org/10.1038/nbt.2195

Zhang M, Xu Y, Xiang, J, Zheng B, Yuan Y, Luo D, Fan, J (2021) Comparative evaluation on phenolic profiles, antioxidant properties and α-glucosidase inhibitory effects of different milling fractions of foxtail millet. J Cereal Sci 99:103217

Zou C, Li L, Miki D, Li D, Tang Q, Xiao L (2019) The genome of broomcorn millet. Nat Commun 10:436. https://doi.org/10.1038/s41467-019-08409-5

Zhu X, Zhao J, Hafiz LY-J, Cheng M, Huang J, Cheng W-C, Wang B, Bai C, Wang G, Dong W (2018) Pyramiding of nine transgenes in maize generates high-level resistance against necrotrophic maize pathogens. Theor Appl Genet 131(10):2145–2156. https://doi.org/10.1007/s00122-018-3143-1

Principles of Quality Seed Production in Millets

Vishwanath Rohidas Yalamalle and B. S. Tomar

Abstract

The seed plays a vital role in agriculture and acts as a carrier of the genetic potential of improved varieties. The National Agriculture Research System of India (NARS) has developed several improved millet varieties with higher yields and resistance to diseases and pests. The timely provision of high-quality seeds at affordable rates of these improved varieties to the farmers is crucial for the nutritional and food security of the nation. The present chapter discusses the genetic and agronomic principles of seed production in millets. Genetic principles such as developmental variation, mechanical mixture, mutations, natural crossing, minor genetic variation, selective influence of diseases, and breeding techniques are discussed. Various strategies for maintaining genetic purity, including controlling seed sources, land selection, and implementing isolation measures can be adopted for producing quality seeds. Agronomic principles including the suitable areas for seed production are discussed. The chapter also emphasizes the significance of the seed certification system, outlining standards and procedures for seed quality assurance. The important considerations during hybrid seed productions such as planting ratios and synchronization of flowering between parental lines to achieve synchronized flowering are provided. In conclusion, the chapter highlights the critical role of quality seeds in ensuring food security and provides various principles for quality seed production, certification, and distribution in millets.

Keywords

Millets · Millet seed production · Quality seed · Generation system of seed production · Genetic purity

V. R. Yalamalle (✉) · B. S. Tomar
ICAR-Indian Agricultural Research Institute, New Delhi, India

D. C. Joshi et al. (eds.), *Minor Millets*,
https://doi.org/10.1007/978-981-96-4265-6_9

1 Introduction

Quality seed is crucial for achieving high production and productivity (Yalamalle and Tomar 2019). The investments made in fertilizers, pesticides, and water will fail to yield the expected dividends if the seed is of poor quality. Millets are often produced by small and marginal farmers and often produced under limited resources, instead of seeds farmers use grain for sowing. The major difference between grain and seed is that seed is an outcome of a scientific program, which results in the production of quality seeds; the important characteristics of quality seeds include genetic purity, physical purity, seed germination and vigor, seed moisture, and seed health. From a marketing point of view characteristics such as seed size, weight, and color contribute significantly to overall seed quality. The nation's food security is dependent on the use and supply of quality seeds, as well as improved crop varieties and hybrids. It is of paramount importance that the seeds of the improved genotypes should be multiplied and marketed in such a way that their seed quality attributes, including genetic purity, are maintained.

Every year, several improved varieties of millets and hybrids (in sorghum and pearl millet) are released by the Indian Council of Agricultural Research and State Agricultural Universities for the benefit of the farming community. Supply of quality seed assures the potential of crop production under suitable and favorable agro-climatic conditions. It is the prime duty of personnel in both public and private sectors to ensure that every farmer has access to healthy seeds that are genetically pure, with high seed vigor and germination. The genetic purity of the nucleus seed, breeder seed, and foundation seed under a generation system of multiplication determines the purity of commercial seeds, which is used by the ultimate beneficiary. The adherence to effective certification procedures in quality seed production significantly contributes to the augmentation of food production in our country. To ensure this, the Government of India enacted the Seeds Act, 1966, which establishes minimum standards for seed production and guidelines for testing, certification, and marketing of seeds.

2 Principles of Quality Seed Production in Millets

To produce high-quality seeds, certain principles must be followed, which are divided into two broad categories, namely genetic principles and agronomic principles.

3 Genetic Principles of Seed Production

Genetic purity ensures that the seed being offered for sale is true to type, containing all the genetic qualities that the crop breeder for which the variety is bred and released. Various elements during the production process can have a negative impact on a variety's genetic purity. Kadam (1942) recognized major elements that contribute to variety decline, such as

1. Developmental variation,
2. Mechanical mixture,
3. Mutation,
4. Natural crossing,
5. Minor genetic variation,
6. Selective influence of diseases, and
7. The techniques used by plant breeders.

Genetic deterioration during seed production can be primarily attributed to mechanical mixing, natural crossbreeding, and the selective impact of diseases. Additionally, cultivating seed crops in regions outside their usual habitats may induce developmental variations and contribute to genetic shifts in varieties.

4 Developmental Variation

When a seed crop is cultivated in challenging environmental conditions, such as varying soil fertility, saline or alkaline conditions, distinct photo-periods, diverse elevations, or persistent stress, developmental variations can emerge, leading to differential growth responses. To mitigate or minimize such developmental variations, it is essential to cultivate the variety consistently in areas where it exhibits adaptability or in regions for which it has been officially released. The foundational seeds, nucleus, and breeder seeds, should undergo multiplication exclusively in areas that are adaptable to ensure the preservation of desired characteristics and performance.

5 Mechanical Mixture

The most significant contributors to genetic deterioration during seed production are mechanical mixtures. Mechanical mixtures can occur at various stages, ranging from sowing to harvesting and processing, including:

1. *Contamination through the field*: This can result from self-sown seeds or volunteer plants.
2. *During sowing*: Using the same seed drill for sowing two or more varieties can lead to unintentional mixing.
3. *Not following prescribes isolation distance*: Growing two different varieties adjacent to each other may cause unintended mechanical mixtures.
4. *Threshing floor*: Contamination can occur during the threshing process.
5. *Harvesters or threshers*: The use of equipment for harvesting or threshing that was previously used for another variety can lead to mixing.

6. *Bags or seed bins*: Cross-contamination may happen during storage in shared containers.
7. *Seed processing*: Mixing can occur during the processing of seeds.

6 Mutations

The significance of this issue is relatively low due to the infrequent occurrence of spontaneous mutations. However, if any visible mutations are detected, it is can be addressed by meticulous roughing of plants showing undesirable traits or variations, ensuring the preservation of the desired characteristics and genetic purity of the crop.

7 Natural Crossing

Natural crossing is the most significant contributor to genetic deterioration particularly in cross-pollinated and often cross-pollinated species. The extent of genetic contamination in seed fields is due to natural crossing depending on the breeding system of the species, isolation distance, varietal mass, and pollinating agent. To overcome the problem of natural crossing, isolation distance and border rows have to be maintained as per prescribed standards.

8 Minor Genetic Variation

The significance of this issue is relatively low; however, some minor genetic changes may occur during production cycles due to differences in the environment. Due to these changes, yields may be reduced. To avoid minor genetic variations, periodic testing of the varieties must be done during breeder seed and nucleus seed. Minor genetic variation is a common feature of often cross-pollinated species; therefore, care should be taken during the maintenance of the nucleus and breeder seed.

9 Selective Influence of Disease

The selective impact of diseases on varietal deterioration holds significant importance. Often newly released varieties become susceptible to emerging races of diseases. Similarly, vegetatively propagated stocks experience rapid deterioration when infected by viral, fungal, and bacterial diseases. Therefore, ensuring the production of disease-free seeds or stocks during seed production becomes a crucial factor in maintaining the health and longevity of crop varieties.

10 Techniques of the Plant Breeder

In certain instances, serious instabilities may occur in varieties due to cytogenetic irregularities. Premature release of a variety that has been bred for a particular disease leads to the production of resistant and susceptible plants, which may be an important cause of deterioration not properly assessed in the new varieties before their release. Other factors, such as breakdown in male sterility, certain environmental conditions, and other heritable variations, may considerably lower the genetic purity.

11 Maintain Genetic Purity During Seed Production

11.1 Control of Seed Source

The seed source must be carefully selected based on the generation system. The breeder seed can be obtained from the respective crop institute or through the DAC indent. Foundation seed can be obtained from SAUs, NSC/State Seed Corporations, and ICAR institutes such as ICAR-IARI, New Delhi and IIMR, Hyderabad.

11.2 Land Requirement

It is important to choose a field where the same crop was not grown during the previous season unless it was found to be suitable for producing quality seed.

11.3 Isolation

Isolation is a crucial factor during seed production to minimize the possibility of cross-pollination between different cross-compatible plots or fields. Proper isolation also helps prevent the mixing of crops during harvesting and the spread of pests and diseases from alternative host crops. Millets except sorghum and pearl millet generally require less isolation owing to the self-pollinated nature of these crops. Isolation distance also varies based on the type of seed, such as foundation and certified seeds. The recommended isolation distance for different crops is listed in Table 1.

Apart from distance isolation, there are other two methods of isolation for seed production. The first one is time isolation, where care is taken to ensure the flowering of cross-compatible species is so adjusted that the flowering time does not coincide; this method is not advocated in tillering millets such as pearl millet. The second method is barrier isolation, which involves using a horizontal plastic sheet (8–10 feet in height) between two seed production plots, particularly for wind-pollinated crops.

Table 1 Isolation distance acquirement for important millet crops

Crop	Foundation seed (m)	Certified seed (m)
Sorghum open-pollinated varieties	200	100
Sorghum hybrids	300	200
Pearl millet composites, synthetics, and open-pollinated varieties	400	200
Pearl millet hybrids	1000	200
Barnyard millet	3	3
Common millet	3	3
Finger millet	3	3
Italian millet	3	3
Kodo millet	3	3
Little millet	3	3

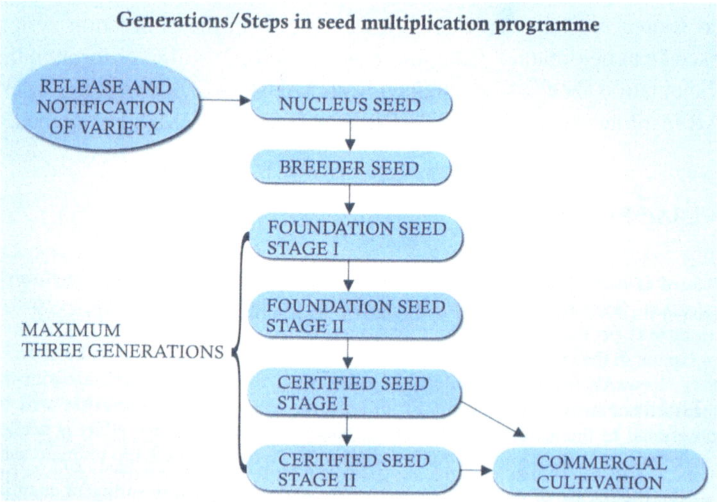

Fig. 1 Generation system of seed production in millets

11.4 Generation System of Seed Multiplication

It is the system of seed multiplication that involves the production of a particular class of seed from a specific class of seed up to a certified seed stage. There are four classes of seeds (Fig. 1):

Nucleus seed: The initial handful of seeds obtained from selected individual plants of a particular variety to purify and maintain that variety by the originating plant breeder.

Breeder seed: The production of nucleus seed is directly controlled by the originating institute or sponsoring breeder; it is multiplied further to obtain foundation seed.

Foundation seed: The progeny of breeders or foundation seed is handled to maintain specific genetic purity and identity. This seed is the source of all other certified seed classes.

Certified seed: It is the progeny of foundation seed or certified seed. The seed production is monitored by seed certification agencies; only those seed lots passing the certification criteria are issued certification tags. The detailed certification process is discussed under seed certification.

12 Rouging

It involves removal of off-types, diseased plants from seed production plots, e.g., in sorghum the heads infected by Kernel smut or Grain smut (*Sphacelotheca sorghi* (Link) Clinton) and Head smut (*Sphacelotheca reiliana* (Kuhn.) Clinton) have to be removed before the filed inspections for seed certification (Trivedi and Gunasekaran 2013). Off-types in millet seed production plots can include other varieties of the same crop, different cross-compatible species, and the same variety that does not meet varietal purity requirements. The recommended stages for rouging are as follows:

(a) *Vegetative stage/before flowering*: volunteer plants, plants that are morphologically different and visually distinguishable are rogued.
(b) *Flowering stage*: pollen shedders, early or late flowering plants with different color and shapes of flowers are removed.
(c) *Maturity and harvest*: Plants with variations in seed color, and non-synchronous maturity are uprooted.

13 Seed Certification

The seed certification system is a legally authorized mechanism designed to ensure quality control throughout seed production and testing processes. It encompasses the verification of seed sources, field inspections of seed crops, supervision of post-harvest stages such as processing and packaging, sampling and analysis of seeds, issuance of certificates, and provision of certification tags, labels, and sealing. Seed certification is applicable to notified varieties and hybrids, and to enforce the provisions of Section 8 of the Seed Act, 1966, seed certification agencies have been established at the state level. Additionally, a Central Seed Certification Board has been instituted to offer advisory services on scientific and operational matters to both the Central Government and State Seed Certification Agencies, as amended in 1972.

The Indian Minimum Seed Certification Standards, 2013 (IMSCS), define the minimum standards necessary for seed certification by a certification agency, which is crucial for assessing seed quality. The standards for sorghum, pearl millet, finger millet, proso millet, foxtail millet, kodo millet, little millet, and barnyard millet are provided in Tables 2 and 3 and are formulated by officially constituted committees or agencies. They are crucial in verifying varietal purity, isolation, seed-borne diseases, weed seeds, and other crop seeds. Furthermore, they assess germination, seed health, and moisture content during field inspections and laboratory tests. Continuous reviews of seed certification standards under the Indian Seeds Act, 1966, are conducted to make necessary amendments based on scientific findings over the years.

Table 2 Prescribed field standards for different millets

Crop	Class of seed	Number of field inspection	Off-types/ pollen shedders (applicable to hybrids) (%)	Specific requirements		
				Heads infected by Kernel smut or Grain smut (%)	Plants infected by Downy mildew (%)	Ergotted ear heads (%)
Sorghum open-pollinated varieties	FS	3	0.05	0.05	–	–
	CS	3	0.10	0.10	–	–
Sorghum hybrids	FS	4	0.05	0.05	–	–
	CS	4	0.10	0.10	–	–
Pearl millet composites, synthetics, and open-pollinated varieties	FS	3	0.05	0.05	0.05	0.02
	CS	3	0.10	0.10	0.10	0.04
Pearl millet hybrids	FS	4	0.05	0.05	0.05	0.02
	CS	4	0.10	0.10	0.10	0.04
Barnyard millet	FS	2	0.05	0.05	–	–
	CS	2	0.10	0.10%	–	–
Common millet	FS	2	0.05	0.05	–	–
	CS	2	0.10	0.10	–	–
Finger millet	FS	2	0.05	0.05	–	–
	CS	2	0.10	0.10	–	–
Italian millet	FS	2	0.05	0.05	–	–
	CS	2	0.10	0.10	–	–
Kodo millet	FS	2	0.05	0.05	–	–
	CS	2	0.10	0.10	–	–
Little millet	FS	2	0.05	0.05	–	–
	CS	2	0.10	0.10	–	–

Table 3 Seed standards for different millets

Crop	Class of seed	Germination (%) (min.)	Pure seed (%) (min.)	Inert matter (%) (max.)	Other crop seeds (/kg) (max.)	Weed seeds (/kg) (max)	Other distinguishable varieties (/kg) (maximum)	Moisture (max.)		Specific requirements
								Normal packing (%)	Moisture impervious (%)	Seeds infected with Ergot and sclerotia
Sorghum open-pollinated varieties and hybrids	FS	75	98.0	2.0	5	5	10	12	8	0.020% (by number)
	CS	75	98.0	2.0	10	10	20	12	8	0.040% (by number)
Pearl millet composites, synthetics, OPV, and hybrids	FS	75	98.0	2.0	10	10	–	12	8	0.020% (by number)
	CS	75	98.0	2.0	20	20	–	12	8	0.040% (by number)
Barnyard millet	FS	75	97.0	3.0	10	10	–	12	8	–
	CS	75	97.0	3.0	20	20	–	12	8	–
Common millet	FS	75	97.0	3.0	10	10	–	12	8	–
	CS	75	97.0	3.0	20	20	–	12	8	–
Finger millet	FS	75	97.0	3.0	10	10	–	12	8	–
	CS	75	97.0	3.0	20	20	–	12	8	–
Italian millet	FS	75	97.0	3.0	10	10	–	12	8	–
	CS	75	97.0	3.0	20	20	–	12	8	–
Kodo millet	FS	75	97.0	3.0	10	10	–	12	8	–
	CS	75	97.0	3.0	20	20	–	12	8	–
Little millet	FS	75	97.0	3.0	10	10	–	12	8	–
	CS	75	97.0	3.0	20	20	–	12	8	–

14 Grow Out Test

It is imperative to confirm the genetic purity of varieties or hybrids before introducing the seed lot to the market, as mandated by the Seed Act, 1966. This confirmation can be achieved through either conducting a Grow Out Test (GOT) or utilizing biochemical and molecular markers in a laboratory setting. GOT involves growing plants alongside standard or reference samples to assess genetic purity based on observable characteristics that are minimally influenced by the environment and highly heritable. GOT is mandatory certification criteria for hybrid seed obtained by the use of chemical hybridizing agents and manual emasculation and pollination. The purity levels in various seed classes are outlined in Table 4.

15 Agronomic Principles of Seed Production

The selection of an appropriate area for seed production is crucial to maximize seed yield and quality while minimizing the impact of disease and pests. The suitability of an area depends on various factors such as temperature, rainfall, soil condition, and insect activity, and it is important to choose an area that is well-suited to the particular variety or hybrid being grown. Careful considerations must be made in selecting an agro-climatic region conducive to the photoperiod and temperature conditions suitable for the crop variety, failure to do so may result in the breakdown of male sterility. Rai et al. (2001) observed higher frequency of pollen shedders in Tift 23A1 (A1 cytoplasm) in pearl millet during the hot dry. The chosen seed plot should be free from volunteer and weed plants, with optimal soil texture, fertility, and resistance to soil-borne diseases and pests. Isolation from neighboring fields is essential to meet certification standards. Proper land preparation enhances germination, stand establishment, and weed control, while the chosen variety should exhibit disease resistance, earliness, grain quality, and suitability to the regional agroclimatic conditions. Seed treatment, appropriate planting time, lower seed rates, mechanical drilling, and precise sowing depth contribute to optimal seed production. Timely rouging, weed and pest control, proper nutrition, and strategic irrigation further ensure the production of high-quality seeds. The provision of honey bees for cross-pollination enhances seed set and yield, emphasizing the comprehensive approach required for successful seed production.

Table 4 Genetic purity requirements for different kinds of seeds

Class of seed	Permissible limit of genetic purity (%) (min)
Foundation seed	99
Certified seed (variety)	98
Certified seed (hybrid)	95
Hybrid by hand emasculation and pollination	90

16 Considerations for Hybrid Seed Production

At present, hybrids are available commercially only in sorghum and pearl millet. Hybrid seeds in these crops are being produced by different methods and systems commercially. In sorghum, genetic male sterility (GMS) and cytoplasmic genetic male sterility (CGMS) system have been reported (Tonapi and Singhal 2003). Owing to the drawbacks of the GMS system, the CGMS system is widely used in sorghum hybrid seed production (Bamji et al. 2023). Like sorghum in pearl millet, both GMS and CGMS system have been identified (Chandra et al. 2022).

17 Planting Ratio

Hybrid seed production relies on effective seed set in the female plants within the designated area. A higher proportion of the female population, organized in rows or blocks, is expected to yield more hybrid seeds. Success depends on the pollen produced and delivered by the male parent, considering its dispersal ability and viability. The tillering nature of millets such as pearl millet can also be utilized for prolonged pollen availability (Govila and Singhal 2003). The planting ratio between female and male is higher when pollination is arranged by hand or by wind. The planting ratio generally followed in hybrid seed production of sorghum and pearl millets is presented (Table 5).

18 Synchronization of Flowering

Achieving proper synchronization of flowering between parental lines is crucial for successful hybrid seed production. Failure of synchronization often leads to poor or no seed set. The timing of flowering is pivotal because the seed parent's seed set relies on viable pollen from the male parent during the receptive period of the stigma. Synchronization involves the simultaneous flowering of both parents, necessitating an understanding of their distinct growth durations and flowering patterns. Sorghum is predominately self-pollinated but protogyny is also reported (Upadhyaya et al. 2008). Pearl millet is a classical example of protogyny (Morgan et al. 1997), where achieving synchronization is very crucial. Knowledge of initiation, peak flowering, and termination aids in planning seeding and sowing, ensuring synchronization. The seed parent, especially in systems such as cytoplasmic male sterility (CMS), protogyny, and protandry, should ideally flower slightly earlier than the male parent, considering the prolonged stigma receptivity of the seed parent. Proper

Table 5 Planting ratio and hybrid seed production in sorghum and pearl millet

Crop	Planting ratio (M:F)
Sorghum	4A:2R
Pearl millet	6A:2R

seeding intervals, staggered sowing, and agronomic interventions such as seed treatment and chemical applications contribute to achieving synchronized flowering, contingent on prior awareness of parental line flowering behavior. Notably, in cases of extended flowering in the seed parent with CMS, protogyny, and protandry, staggered sowing in the pollen parent (3–5 days apart) is recommended to ensure an ample supply of pollen during the seed parent's flowering, ultimately preventing a low seed set. The synchronization of flowering as compiled by Chakrabarty in sorghum and pearl millet is compiled below.

19 Sorghum

1. Adopting staggered sowings is essential, taking into account plant growth, development, and the timing and duration of flowering in the parental lines (Murthy et al. 1994).
2. Apply 500 mg of maleic hydrazide in 1 L of water to the advancing parent 45 days after sowing.
3. Late parent plants can be treated with a 1% urea solution (Kannababu et al. 2002).
4. Consider skipping one irrigation cycle during the advancing flowering stage.
5. Use CCC (chloro chlorine chloride) at 300 ppm to delay flowering.

20 Pearl Millet

1. Staggered sowing is recommended if the difference in flowering period is more than 10–15 days and staggered sowing of early parent line is recommended (Govila and Singhal 2003).
2. Apply a 1% urea solution through spraying on the late parental line.
3. Delays in flowering can be achieved by jerking the early parent.
4. Consider additional nitrogen fertilization or a foliar spray for the late parent (Govila and Singhal 2003).

21 Harvesting of Hybrid Seed

When harvesting hybrid seed production plots and threshing panicles from the A-line rows, utmost care must be taken to prevent seed contamination. Typically, the R-line is harvested initially, and the harvest is removed from the field. Subsequently, the A-line rows are meticulously examined for off-types and other potential admixtures before harvesting.

Seeds play a crucial role in agriculture, carrying the genetic potential of various varieties. Collaborative efforts from AICRP—Sorghum, AICRP—Pearl Millet, and AICRP on Small Millets have led to the development of high-yielding early maturing, pest and disease tollerent varieties. Ensuring timely availability of these varieties as quality seeds, in sufficient quantities and at reasonable prices, is essential.

This can be achieved through systemic seed production, adhering to genetic and agronomic principles, and rigorous seed certification- ultimately contributing to enhanced food security and sustainable agricultural growth worldwide.

References

Bamji R, Davda B, Kanjariya K, Patel K (2023) Heterosis analysis of yield contributing traits in F1 hybrids developed by using CMS background in sorghum (Sorghum bicolor (L.) Moench). Pharma Innov J 12(9):1714–1720

Chandra S, Singh SP, Satyavathi CT, Sankar SM, Singh AM, Bhat JS (2022) Genetics of fertility restoration for the A1 cytoplasmic genic male sterility system in pearl millet (Pennisetum glaucum (L.) R. Br.). Indian J Genet Pl Breed 82(01):65–72. https://doi.org/10.31742/IJGPB.82.1.9

Govila OP, Singhal NC (2003) Pearl millet. In: Singhal NC (ed) Hybrid seed production in field crops (principles and practices). Kalyani Publishers, New Delhi, pp 253–268

Kadam BS (1942) Deterioration of varieties of crops and the task of the plant breeder. Indian J Genet Plant Breed 2:159–172

Kannababu N, Tonapi VA, Rana BS, Rao SS (2002) Influence of different synchronization treatments on floral behaviour of parental lines and hybrid seed set in sorghum. Indian J Plant Physiol 7:362–366

Morgan R, Alvernaz J, Arthur L et al (1997) Genetic characterization and floral development of female sterile and stubby head, two aposporous mutants of pearl millet. Sex Plant Reprod 10:127–135. https://doi.org/10.1007/s004970050079

Murthy DS, Tabo R, Ajayi O (1994) Sorghum hybrid seed production and management. Information bulletin no. 41. ICRISAT Patancheru, India, p 67

Rai K, Anand Kumar K, Andrews D et al (2001) Commercial viability of alternative cytoplasmic-nuclear male-sterility systems in pearl millet. Euphytica 121:107–114. https://doi.org/10.1023/A:1012039720538

Tonapi AV, Singhal NC (2003) Sorghum. In: Singhal NC (ed) Hybrid seed production in field crops (principles and practices). Kalyani Publishers, New Delhi, pp 285–330

Trivedi RK, Gunasekaran M (2013) Indian minimum seed certification standards. The Central Seed Certification Board, Department of Agriculture and Co-operation, Ministry of Agriculture, Government of India

Upadhyaya HD, Reddy VG, Sastry DVSSR (2008) Regeneration guidelines: sorghum. In: Crop specific regeneration guidelines [CD-ROM]. CGIAR System-wide Genetic Resource Programme, Rome, pp 1–8

Yalamalle VR, Tomar BS (2019) Effect of scape regulation on seed yield and quality in onion (Allium cepa). Indian J Agric Sci 89(1):56–59. https://doi.org/10.56093/ijas.v89i1.86130

Genomic Resources and Modern Breeding Approaches to Accelerate Genetic Gain in Minor Millets

Amit Kumar Singh, Monika Singh, Lalit Arya, Badal Singh, and Manjusha Verma

Abstract

Minor millets are an important group of crops with small-sized grains, primarily cultivated in many Asian and African countries. These crops have been an integral part of human diets since ancient times; however, their acreage declined globally due to their low productivity as compared to cereals such as wheat and rice. There is a renewed interest in these crops owing to their abundance of vital nutrients, minerals, and bioactive compounds essential for human well-being, coupled with their resilience to varying climate conditions. Therefore, the need of the hour is to enhance their productivity, a goal achievable through focused and systematic genetics and breeding research. Advances in the field of genomics, which can provide a much deeper understanding of various traits, can significantly accelerate genetic improvement in these crops. In recent years, genomic resources in the form of simple sequence repeat (SSRs), single nucleotide polymorphism (SNPs), and other sequence-based markers have been developed in the majority of the minor millets. Further, advanced sequencing technologies have also led to the construction of reference genomic assemblies in most of the millet crops. Given the widespread availability of genomic resources in these crops, the emphasis should now shift toward mapping and deploying crucial genes associated with increased yield and conferring tolerance to various biotic and abiotic traits. Furthermore, a range of advanced genomic techniques, such as genome editing, genomic selection, and haplotype breeding, offer the potential to accelerate the development of millet varieties with high yield and resilience to climate, and these should be integrated into the millet genetic enhancement initiative.

A. K. Singh (✉) · M. Singh · L. Arya · B. Singh · M. Verma
Division of Genomic Resources, ICAR-National Bureau of Plant Genetic Resources, New Delhi, India
e-mail: amit.singh5@icar.gov.in

D. C. Joshi et al. (eds.), *Minor Millets*,
https://doi.org/10.1007/978-981-96-4265-6_10

Keywords

Minor millets · Genomic resources · Breeding approaches · Genetic gain · Simple sequence repeat (SSR) · Single nucleotide polymorphism (SNP) · Genome editing

1 Introduction

Minor millet crops are members of the Poaceae family, which are characterized by their small seeded coarse grains. These crops are predominately cultivated in Asia and Africa where they are primarily grown on the marginal lands as a constituent of subsistence agriculture. Once the staple crop in many regions, minor millets faced a decline in prominence with the advent of the Green Revolution, which encouraged the widespread cultivation of wheat and rice. Notably, in the past few years, they have acquired renewed importance due to their climate resilience and better nutritional properties (Ceasar and Maharajan 2022). However, their productivity across cultivation areas is hindered by a lack of high-yielding varieties and appropriate agro-techniques as a consequence of insufficient attention to genetic and breeding research and lack of funding support. Trait improvement in millets, such as any other crop, requires knowledge of the genetic diversity in their gene pool as well as the genes/alleles controlling various useful traits. Genomic resources have an important role in conducting these studies. Unfortunately, most minor millets except for foxtail millet are considered orphan concerning the availability of genomic information. In the past few years, the emergence of genomic technologies, particularly next-generation sequencing (NGS), has facilitated the generation of genomic resources, including genome sequences, sequence-based markers, and information related to gene expression in these crops (Goron and Raizada 2015; Vetriventhan et al. 2020). Molecular markers are valuable genomic tools for revealing details about the extent and pattern of distribution of genetic diversity within the gene pool of minor millets. Several investigations have already documented analyses of genetic diversity in the extensive germplasm collection of foxtail millet (Wang et al. 2012) and finger millet (Ramakrishnan et al. 2016; Backiyalakshmi et al. 2021). Moreover, molecular markers can aid in the identification of genes and alleles associated with valuable traits essential for the genetic enhancement of these crops via quantitative trait locus (QTL) mapping. Several QTL mapping investigations have been carried out for diverse beneficial traits such as yield, as well as resistance or tolerance to biotic and abiotic stresses in small millet crops (Liu et al. 2020; He et al. 2021; Pendergast IV et al. 2022). Further, the integration of genomic information, comprising sequence variation and gene expression patterns, with phenotypic data is pivotal for a comprehensive understanding of trait variations and their underlying genetic mechanisms. Therefore, there is a crucial need to prioritize the development of genomic tools, such as molecular markers, and leverage them for the identification of candidate genes associated with various economically significant traits. In recent years, many advanced genetics and genomic tools have been also developed. Approaches such as speed breeding (SB), genome-wide association studies (GWAS), genome editing and genomic selection (GS) have gained immense popularity among researchers for the precise genetic dissection and improvement of desirable traits in crops. The application of these approaches is still in its early stages for minor millet crops, with only a few studies reporting on genome editing and genomic selection in some of the millets. This chapter delves into the progress achieved in the generation of genomic tools as well their application in trait mapping and genomics-assisted breeding of millet.

2 Genomic Resources in Minor Millets

2.1 Marker Resources

2.1.1 Marker Resources in Foxtail Millet

Molecular markers play a crucial role in genetic and genomic investigations of crop plants, encompassing areas such as genetic diversity, DNA profiling, population structure studies, QTL mapping, and gene tagging. In the context of small millets, an extensive array of molecular markers has been created and employed in the molecular characterization of foxtail millet, followed by finger millet and other small millets. The effective application of simple sequence repeat (SSR) markers in molecular studies is notable due to their codominant inheritance, substantial allelic variation, and high abundance. Numerous studies have been conducted to develop SSR and other markers specifically for foxtail millet (Table 1).

Additionally, a large number of Single nucleotide polymorphisms (SNPs), insertion/deletion polymorphisms, and structural variations (SV) have been also identified and can be used in genomics-assisted genetic studies of foxtail millet (Bai et al. 2013; Zhao Qing Ying et al. 2018). Yadav et al. (2014) developed 176 microRNA (miRNA)-based molecular markers and mapped them to the foxtail millet genome and 66 markers showed 100% amplification and ~ 67% transferability in millets and non-millet species. Transposable elements (TE)-based markers were also developed with 30 of them found polymorphic in a diverse panel of foxtail millet genotypes (Yadav et al. 2015) (Foxtail millet Transposable Elements-based Marker Database; http://59.163.192.83/ltrdb/index.html).

2.1.2 Marker Resources in Finger Millet

In finger millet, genomic SSRs have been developed by Dida et al. (2007), Gimode et al. (2016), Lee et al. (2017), Bheema Lingeswara Reddy and Sivaramakrishnan (2017); Hittalmani et al. (2017); and Kumar et al. (2021). Expressed sequence tag (EST)/genic SSRs have been developed by Arya et al. (2009); Panwar et al. (2011); Naga et al. (2012); Nirgude et al. (2014), and Brhane et al. (2021). Further, 80 polymorphic SNP markers with an average of Polymorphism Information Content (PIC) of 0.29 and 5226 bi-allelic SNPs with an average PIC of 0.23 were developed by Gimode et al. (2016) and Brhane et al. (2022a, b), respectively.

Table 1 Marker resources in foxtail millet

Marker type	Genotypes/ accessions used	Utilization	References
26 SSR markers from 1213 EST sequences	12 foxtail millet cultivars and one accession of *Setaria viridis*	Mean number of alleles per locus: 2.5 Cross-transferable markers: 10	Jia et al. (2007)
100 SSR markers	40 foxtail millet accessions	Alleles generated: 228 Mean number of alleles per locus: 6.16	Jia et al. (2009)
45 SSR markers	12 foxtail millet populations	Mean number of alleles per locus: 3.73 Average expected heterozygosity: 0.587 Cross-transferable markers: 24	Lin et al. (2011)
327 EST-SSR marker mapped on 9 chromosome 40 EST-SSR markers used for diversity analysis	41 accessions	Diversity range: 0.02–0.65 Average transferability rate: 87.8%	Kumari et al. (2013)
64 SSRs 39 SSRs used for diversity analysis	41 accessions	Average number of alleles per locus: 2.7 PIC: 0.31	Gupta et al. (2013a, b)
15,573 mapped SSR markers	8 accessions	Out of 159 markers tested, 67% were polymorphic Cross-species transferability to switchgrass: ~79% and ~93% to pearl millet	Pandey et al. (2013a, b)
733 SSR	28 diverse genotypes	Number of alleles: 2 to 16 Average PIC: 0.67	Zhang et al. (2014)
98 novel intron length polymorphic (ILP)		Average number of alleles per locus: 2.6 Mean Nei's average gene diversity: 0.171	Gupta et al. (2011)
4049 ILP markers identified, 440 of them used for diversity analysis	89 cultivated and 8 accessions of wild species	Average number of alleles per locus: 3.73 Average PIC: 0.20	Muthamilarasan et al. (2014)

2.1.3 Marker Resources in Proso Millet

In proso millet, genomic/genic SSRs have been developed through enriched genomic library construction, cross-transferability from switchgrass, RNA sequencing, and NGS (Cho et al. 2010; Rajput et al. 2014; Hou et al. 2017; Liu et al. 2017; Jiang et al. 2018), and an average number of alleles per locus ranging from 2.67 to 4.4 was reported.

2.1.4 Marker Resources in Barnyard Millet

In barnyard millet, not many reports are available for molecular marker development. Most of the markers have been developed through cross-transferability or through transcriptome-based resources (Kumari et al. 2013; Arya et al. 2014; Babu and Chauhan 2017; Babu et al. 2018, Kalyana Babu et al. 2018; Manimekalai et al. 2018; Jayakodi et al. 2019; Renganathan et al. 2021). The only report available with respect to genomic SSR marker development is by Meniya et al. (2023), wherein a total of 46,157 SSRs have been identified and 15 validated SSR loci showed an average PIC of 0.28. And with respect to SNPs marker identification, there is one report by Wallace et al. (2015). So, there is a dire need to work on the generation of genomic resources in barnyard millet for its improvement and enhanced utilization.

2.1.5 Marker Resources in Kodo Millet

In kodo millet, molecular marker development is still in its infancy. No reports other than Arya et al. (2014) and ICAR (2023) are available on SSR marker development in kodo millet, wherein SSR markers have been developed through cross-transferability and from RNA seq data. To name a few others are Dof domains and *Dof* gene-based markers by Kushwaha et al. (2015) and another one is on neutral and functional markers-based diversity analysis by Yadav et al. (2016). And with respect to SNP marker development, a total of 3461 SNP markers were used for population structure analysis in 165 accessions of kodo millet (Johnson et al. 2019).

2.1.6 Marker Resources in Little Millet

In little millet, not much work has been done pertaining to the generation of genomic resources. Natesan et al. (2020) developed two polymorphic SSRs from 25 maize SSR markers and one polymorphic SSRs from 25 pearl millet SSRs. Desai et al. (2021) reported 4443 genic-SSR motifs and out of 50 primers validated, 48% SSRs showed polymorphism. Further, genic SSRs have been developed and used for the characterization of 96 accessions of little millet (ICAR 2023).

2.1.7 Marker Resources in Fonio

In fonio, Barnaud et al. (2012) developed 21 nuclear microsatellite primers and 2–22 alleles per locus were generated in 36 West Africa accessions of *Digitaria exilis*. The expected heterozygosity of 0.03–0.93 was revealed based on the markers developed. Ngom et al. (2018) developed SSR markers through cross-species amplification in *D. exilis* Stapf and in related *Digitaria* species. Olodo et al. (2019) designed 126 new putative primer pairs in *D. exilis* and 24 primer pairs were transferred to *D. longiflora* with 96% transferability and to *D. iburua* 71% transferability. Ibrahim Bio Yerima et al. (2021) developed 688 DArTseq-derived SNP markers and used to characterize 59 fonio individuals from West Africa.

2.1.8 Marker Resources in tef

Marker resources in tef are very limited. Zeid et al. (2010) developed SSR markers from 812 EST-derived markers of other grass species and were used for the development of the genetic map and QTL analysis, and further Zeid et al. (2012) used 68 of these markers to characterize 37 tef accessions and PIC value ranged from 0.05 to 0.86.

2.2 Genomes and Pangenomes

2.2.1 Genomes

A complete high-quality reference genome serves as a valuable resource to study and understand genetic mechanisms controlling different agronomic traits, stress tolerance, climate resilience, genome evolution, and so on. In small millets, partial, draft, or reference genomes have been assembled in finger millet, foxtail millet, proso millet, barnyard millet, little millet, fonio, and teff and details have been highlighted in Table 2.

2.2.2 Pan-Genomes

Reference genome of a particular species contains only a subset of the genome sequence information of that species. When we sequence and assemble genomes of multiple accessions/varieties of that species and add novel genomic regions/sequences to the reference genome, the resulting comprehensive reference is called a pan-genome. Presence/absence variation (PAVs) is common among the accessions/varieties. Genes common to all accessions are called as 'core' and others as 'dispensable'. Decreasing genome sequencing costs and requirement of more diversity to reveal causative loci for a particular trait has led to the emergence of pan-genome assembly. Iterative mapping approach is most commonly used (Ruperao et al. 2021) for assembly of pan-genomes. The availability of pan-genomes in major crops has facilitated discovery of genes (associated with flowering time, organ size, biomass, seed composition, glucosinolate metabolism, disease resistance etc.) not present in reference genomes and presence/absence variations among the accessions. In case of small millets, pan-genomes have been assembled in foxtail millet and proso millet.

He et al. (2023) reported the graph-based assembly of *Setaria* pan-genome using 110 representative genomes (40 landraces, 35 each wild, and modern cultivated accessions). These accessions were chosen based on geographic distribution and phylogenetic relationships, breeding and/or research utility, and subgroup distribution. The pan-genome comprised of 73,528 gene families (23.8% core, 42.9% soft core, 29.4% dispensable, and 3.9% private genes) and 202,884 nonredundant structural variants. A 366-bp presence/absence promoter variant in yield gene *SiGW3*, important for crop improvement, was identified. Further, 68 traits/13 environment experiments led to the identification of potential genes useful for genome editing, genomic selection, and marker-assisted breeding.

Pan-genome was also developed in broomcorn millet or proso millet (*Panicum miliaceum* L.) (Chen et al. 2023) to improve production and quality, as well as ensure food security. Chen analyzed 516 broomcorn millet genomes to decipher the diversity, genetic variations, and population structure among them. Further, a

Table 2 Details of whole genome sequencing in small millets

Crop	Binomial name	Cultivar	Genome size	Number of chromosomes/ scaffolds/ contigs)	% assembled of estimated genome size	Genes/protein coding genes	Sequencing technology	References
Finger millet	*E. coracana* subsp. *coracana*	ML 365	~1.2 Gb (1196 Mb) 1453 Mb based on flow cytometry	Scaffolds (525,627)	Assembled 82% of total genome estimated by flow cytometry	85,243	Illumina NextSeq; Illumina HiSeq; SOLiD	Hittalmani et al. (2017)
Finger millet	*E. coracana* subsp. *coracana*	PR 202	~1.2 Gb (1189 Mb) 1.5 Gb based on flow cytometry	Scaffolds (1897)	Assembled 78.2% of total genome estimated by flow cytometry	62,348	Illumina MiSeq; NextSeq 500; PacBIo RSII; Bionano Irys R system	Hatakeyama et al. (2018)
Finger millet	*E. coracana* subsp. *coracana*	KNE 796-S	1.1 Gb(1129 Mb)	Chromosome/ scaffolds (18/532)	27% smaller than the most recent estimated sizes	48,836	Illumina X10 and NovoSeq6000 platforms, and the PACBIO SEQUEL I platform	Devos et al. (2023)
Foxtail millet	*S. italica*	Yugu1	400 Mb	Chromosome/ scaffolds (9/336)	Assembled 80% of the predicted genome size of ~510 Mb (http://data.kew.org/cvalues/)	24,000–29,000	ABI3730xl capillary sequencing machines	Bennetzen et al. (2012)
Foxtail millet	*S. italica*	Zhang gu	~423 Mb (490 Mb by cytogenetic method)	Chromosome/ scaffolds (9/613)	Assembled 86% of total genome estimated by cytogenetic method	38,801	Whole genome shotgun combined with NGS	Zhang et al. (2012)

(continued)

Table 2 (continued)

Crop	Binomial name	Cultivar	Genome size	Number of chromosomes/scaffolds/contigs)	% assembled of estimated genome size	Genes/protein coding genes	Sequencing technology	References
Foxtail millet	*S. italica*	TT8	477.5 Mb	Chromosome/scaffolds (9/2688)	Assembled 97% of total genome estimated genome size of 490 Mb	44,292 predicted gene models	454; Illumina HiSeq; Illumina MiSeq	Tsai et al. (2016)
Foxtail millet	*S. italica*	Huagu 11	408.37 Mb	Chromosome/scaffolds (9/173)	Assembled 90% of total genome estimated genome size of 456 Mb	41,984 protein coding genes	SMRT, hi-C and NGS	Wang et al. (2021a, b)
Proso millet	*P. milliaceum*	Landrace (accession number 00000390)	854.7 Mb ~923 Mb based on a K-mer analysis	Chromosome/scaffolds (18/1309)	Assembled 92.6% of total genome estimated genome size of 923 Mb	55,930	PacBio; Illumina HiSeq	Zou et al. (2019)
Little millet	*P. sumatrense*	Lit 171		Partial genome assembly				NCBI
Barnyard millet	*E. colona* var. *frumentacea*	PI463783	1.13 Gb (1.18 Gb based on a K-mer analysis)	Draft genome assembly (chromosome level)	Assembled 95.7% of total genome estimated genome size of 1.18 Gb	93,697	Illumina HiSeq2500; PacBio sequel II	Wu et al. (2022)

Barnyard millet	*E. frumentacea*	CO(KV)2		Partial genome assembly 11,39,481 contigs			Ion S5 next generation sequencer	Meniya et al. (2023)
Fonio	*D. exilis*	CM05836	716.471 Mb 893 Mb based on flow cytometry	Chromosome level assembly 18/8457 scaffolds	Assembled 80.23% of total genome estimated genome size of 893 Mb	57,021	Illumina 10X genomics; hi-C Bionano	Abrouk et al. (2020)
Fonio	*D. exilis*	Niatia	~761 Mb	3329 contigs		67,855	PacBio	Wang et al. (2021a, b)
Teff	*E. tef*	Tsedey (DZ-Cr-37	672 Mbp 772 Mbp by flow cytometry	14,057 contigs	Assembled 87% of the genome size estimated from flow cytometry	28,113 proteins	Illumina HiSeq2000 454-FLX technology	Cannarozzi et al. (2014)
Teff	*E. tef*	Dabbi	576 Mb 622 Mb based on flow cytometry	Chromosome level assembly 10/1344 contigs	Assembled 92.6% of total genome estimated genome size of 622 Mb	68,255	PacBio	VanBuren et al. (2020)

graph-based pangenome of broomcorn millet was developed based on de novo genome assemblies (long-read) of 32 representative accessions. A total of 139 loci related with 31 key agronomic and domestication traits were identified. Gene *longmi038370* controlling panicle shape was identified. Structural variations, particularly PAVs, were found highly linked with transposable elements and contributed to 68.3% of total PAVs. The resources generated can be utilized for genomics-assisted breeding programs in proso millet.

2.3 Linkage Maps

Genetic maps play a crucial role in identifying QTLs, gene mapping, cloning, and employing marker-assisted breeding strategies to enhance genetic improvements in any crop. Initially, in small millets, these maps were developed using restriction fragment length polymorphism (RFLP) and later with amplified fragment length polymorphism (AFLP) and SSR markers. However, for precise candidate gene identification, the need for high-resolution maps and fine-mapping techniques becomes imperative. With the advent of genotyping technologies, SNPs and DArT markers have been used to construct high-density genetic maps in some millets. Progress made in the construction of linkage maps is discussed in the following sections.

2.3.1 Finger Millet

Dida et al. (2007) employed an F2 population resulting from the inter-subspecific cross of cv. Okhale-1 (*E. coracana* subsp. *coracana*) and acc. MD-20 (*E. coracana* subsp. *africana*) to create the first genetic map in finger millet utilizing RFLP, AFLP, and EST-SSR markers. This map encompassed all 18 chromosomes, albeit partially, spanning 721 cM on the A genome and 787 cM on the B genome. Among the 82 SSRs developed during the study, 31 could be successfully mapped. Subsequently, Qi et al. (2018) established the initial high-density genetic map in finger millet by crossing the cultivated accession Okhale-1 with the wild accession MD-20, employing the UGbS-Flex pipeline. The linkage map of MD-20 × Okhale-1 revealed notable marker intervals on 9A and 9B, potentially attributed to the presence of a reciprocal 9A–9B translocation in one of the mapping parents.

Furthermore, Pendergast IV et al. (2022) developed a genetic map with high marker density (5422 markers, averaging 301 markers per linkage group) for finger millet's F2:3 population from cross ACC 100007 × GBK 030647. This map, spanning 18 linkage groups and 2938 cM, with 1466 cM on the A genome and 1472 cM on the B genome, was created through genotyping-by-sequencing (GBS) and the UGbS-Flex pipeline. Comparative genomic analyses were carried out on foxtail millet, rice, and sorghum, unveiling conserved acrocentric homoeologous chromosomes (such as 4A and 4B in finger millet) present across all these species. The study identified QTLs associated with plant height, panicle number, flowering date, and blast incidence and severity. The genomic regions and genes pinpointed hold significant relevance for blast resistance and various agronomic traits.

2.3.2 Foxtail Millet

Wang et al. (1998) developed an RFLP-based map (comprising nine linkage groups with a span of 964 cM) containing 160 loci in the Longgu 25 × Pagoda Flower Green cross. Segregation data revealed the presence of a gamete fertility-affecting gene on chromosome VIII in the cross. Sato et al. (2013) created a genetic map (13 linkage groups) and located the gene (*stb1*) responsible for the "spikelet-tipped bristles" (stb) trait in foxtail millet on chromosome 2. Eventually, they constructed nine linkage groups (totaling 1287.5 cM) based on SSR markers, pinpointing stb1 precisely on chromosome 2. Fang et al. (2016) conducted genotyping on 167 individuals from an F2 population resulting from the Yugu1 × Longgu7 cross to develop a high-density genetic map. The map comprised 1035 loci, spanning 1318.8 cM, with an average distance of 1.27 cM between neighboring markers. In their study, they identified 29 QTLs that accounted for 7.0–14.3% of phenotypic variation across 11 traits. Notably, QTLs for peduncle length originated from Longgu7, while for other traits, QTLs were derived from Yugu1, with the exception of qLMS6.1. In a separate study, Wang et al. (2017a, b) employed the RAD-seq approach to establish a high-density genetic map covering 1648.8 cM, with an average marker distance of 0.17 cM in foxtail millet. They used 9968 SNP markers to genotype 124 F2 progenies of Hongmiaozhangu and Changnong35, identifying 11 major QTLs related to eight agronomic traits. Fukunaga et al. (2022) constructed a linkage map using flexible ddRAD-seq with parents (Taiwan and Yugu1) and RILs, detecting major QTLs for spikelet-tipped bristles (stb) (a homeobox (HOX14) gene on chromosome II), leaf sheath colors (C gene on chromosome IV), and days to heading (DTH) on chromosome II. Guo et al. (2023) created a high-density genetic linkage map (spanning 1227.383 cM with an average interval of 0.879 cM between adjacent bin markers) based on a 217 F7 recombinant inbred line (RIL) population of the Hongjiugu × Yugu18 cross. They identified a total of 39 QTLs related to four hull color traits, with six major QTLs on chromosomes 1 and 9.

2.3.3 Proso Millet

Rajput et al. (2016) established the first genetic linkage map of proso millet by utilizing 93 recombinant inbred lines derived from the cross between "Huntsman" and "Minsum." A total of 18 major linkage groups and 84 minor linkage groups were delineated, incorporating 833 GBS-SNP markers. Notably, 519 markers remained ungrouped. Within the 18 major linkage groups, 117 SNP markers were distributed, spanning a cumulative distance of 2137 cM with an average inter-marker distance of 18 cM. The length of markers ranged from 54.6 to 236 cM, and the number of markers varied from 4 to 12 in each major linkage group. Furthermore, the study identified 18 QTLs associated with eight morpho-agronomic traits. These QTLs explained phenotypic variances ranging from 13.2% to 34.7% each and were detected on 14 linkage groups. This comprehensive work is anticipated to contribute to genomics-assisted advancements in proso millet improvement.

2.3.4 Tef

Bai et al. (1999) employed 211 AFLP loci in their study, constructing a linkage map spanning 2149 cM with 25 linkage groups. The research focused on 85 F8 RILs resulting from the intraspecific cross Kaye Murri × Fesho. Zhang et al. (2001) detailed a genetic linkage map, encompassing 1489 cM and 149 marker loci distributed across 20 linkage groups with an average interval of 9.99 cM between markers. This map was derived from an investigation involving 159 tef cDNA and 162 heterologous probes in 116 F8 RILs from the cross between cv Kaye Murri (*E.* tef) × accession 30–5 (*Eragrostis pilosa*). Chanyalew et al. (2005) reported a linkage map covering 2112.3 cM and 20 linkage groups, utilizing ISSR, AFLP, rice EST-SSR, and tef-specific EST-SSR markers in their study with 124 F8 RILs from the interspecific cross *E. tef* (DZ-01-2785) × *E. pilosa* (30–5). Yu et al. (2006) employed 94 F9 RILs from the cross between cv Kaye Murri (*E. tef*) × accession 30–5 (*Eragrostis pilosa*). They utilized RFLP, EST-SSR, SNP/INDEL, IFLP, and ISSR markers, resulting in a genetic map with 121 markers from 156 loci organized into 21 linkage groups, covering 2081.5 cM with a mean density of 12.3 cM per locus. This map is valuable for comparative analyses with other cereal crops and for further genetic studies in tef. Zeid et al. (2010) utilized 151 F9 RILs from Kaye Murri × *E. pilosa* (acc. 30–5), employing IFLP, RFLP, EST-SSR, and ISSR markers, generating a genetic map covering 1277 cM and consisting of 30 linkage groups in teff.

3 Molecular Level Diversity Characterization in Minor Millet Germplasm

The genetic improvement program in millets requires knowledge of the extent and pattern of genetic diversity available in the germplasm collection of these crops. International research institutes such as the ICRISAT for small millet, foxtail millet, barnyard millet, and Kodo millet have preserved genetic resources in small millet germplasm collections as core collections, constituted mainly utilizing morphological characterization data. NBPGR, the Ethiopian Biodiversity Institute, ICARDA, USDA, the N.I. Vavilov Russian Scientific Research Institute, the Ustymivika Experimental Research Station, and the Kenya Agricultural Livestock Research Station are some more genetic reservoirs for small millets (Upadhyaya et al. 2008; Joshi et al. 2021). The diversity of these collections has been characterized using morphological parameters. Most millets display extensive morphological variations, encompassing differences in seed size, shape, and color (Upadhyaya et al. 2009). Due to the challenges faced by breeders in utilizing the vast millet germplasm collection, smaller core and mini-core collections have been established. These collections serve as excellent gene pools with diverse germplasm for studying population structure, genetic diversity, and identifying new genes and alleles (Upadhyaya et al. 2012). In recent years, there has been a realization that morphological parameters alone are insufficient for obtaining an accurate estimate of diversity. Consequently, there is a burgeoning need for diversity assessment through molecular markers. This shift reflects an understanding that molecular markers offer a more precise means of gauging genetic variation, contributing to a more comprehensive evaluation of biodiversity. Until recently, very limited reports were published with respect to molecular studies in millets, which could be mainly attributed to the limited availability of genomic resources in these crops (Gupta et al. 1999; Kumar 1999; Gupta and Varshney 2000; Varshney et al. 2005). Initially, studies in millets used RFLP, AFLP, RAPD, and other random-based markers for marker-based diversity analysis. The increased adoption of direct sequencing technologies, driven by a reduction in costs, has made high-throughput GBS feasible, regardless of the available genetic information for the crop. This advancement, as highlighted by Varshney et al. (2010) and Ganal et al. (2012), enables large-scale genotyping of plant accessions and the simultaneous discovery of markers such as SNPs, as demonstrated by Elshire et al. (2011) and Mir et al. (2013). In the context of genetic relationship studies in small millets, both abstract and specific molecular markers have been utilized (M'Ribu and Hilu 1996; d'Ennequin et al. 2000; Karam et al. 2004; Dida et al. 2008; Cho et al. 2010; Panwar et al. 2010a; Lin Heng Sheng et al. 2012; Kumari et al. 2013; Vetriventhan and Upadhyaya 2018; Babu et al. 2018). Molecular markers have played a pivotal role in characterizing germplasm, as exemplified in studies conducted by Hu et al. (2009), Panwar et al. (2010b), Gupta et al. (2011), and Vetriventhan et al. (2012). SSR markers and intron-length polymorphic markers derived from the foxtail millet genome were employed for characterizing germplasm, constructing linkage maps, and conducting comparative genetic analyses in other millets (Pandey et al. 2013a, b; Muthamilarasan et al. 2014). Similarly, the utilization of cross-species transferable molecular markers has been instrumental in exploring diversity and relationships in small millets (Jia et al. 2007; Arya et al. 2009; Reddy and Sivaramakrishnan 2017; Kumari et al. 2013). Numerous studies have identified molecular markers associated with various traits in finger millet and foxtail millet (Lata et al. 2021; Lata and Prasad 2011; Lata and Prasad 2013a, b). Genetic diversity was evaluated among elite finger millet genotypes using SSR markers (Manjappa et al. 2018), and a similar approach was employed for proso millet (Zeinullina et al. 2023).

The rapid advancements in millet genomics are contributing significantly to the enhancement of genomic resources in small millets in various ways. The abundant genetic diversity found in small millets germplasm serves as a valuable reservoir of genes and alleles. These can be employed in marker-assisted breeding programs to achieve genetic improvement in these crops.

4 Bioinformatic Resources

Bioinformatics resources contribute to the integration and visualization of data from omics platforms and provide information on the genetic variation of germplasm resources that can provide mainstream requirements and support the millet researchers. In this context, there have been several efforts to create bioinformatic resources and databases to facilitate the investigators. The rapid advancements in high-throughput NGS technologies have led to an abundance of bioinformatics data in several small millet crops. Following the unveiling of the foxtail millet whole genome sequence, extensive molecular marker resources were created to enhance foxtail millet improvement efforts. The Foxtail millet Marker Database (FmMDb; http://www.nipgr.res.in/foxtail.html), introduced by Bonthala et al. (2013), serves as a comprehensive online repository for retrieving, visualizing, and managing extensive foxtail millet marker datasets. It offers valuable marker information to researchers and breeders dedicated to advancing crop improvement and developing elite cultivars of millet and bioenergy grass species. This database contributes significantly to global food security initiatives. Furthermore, the Foxtail millet Transcription Factors Database (FmTFDb) was launched, encompassing 2297 putative TFs in 55 families, along with sequence features, chromosomal locations, tissue-specific gene expression data, gene ontology (GO) assignments, and phylogeny. Released by Bonthala et al. 2014, FmTFDb provides public access for visualizing individual members of a TF family through various query interfaces and analysis tools, including BLAST searches, annotation query interfaces, and tools for identifying enriched GO terms and visualizing physical maps. Similarly, the Multi-omics Database (MDSi) includes the genome of Xiaomi with 161,844 annotations, 34,436 protein-coding genes, and expression information in 29 different tissues. It also encompasses whole-genome resequencing (WGS) data of 398 germplasms, consisting of 360 foxtail millets and 38 green foxtails, along with corresponding metabolic data (Li et al. 2023). MDSi facilitates the interactive search and comparison of SNPs and Indels in these germplasms through implemented tools. Recently, the Milletdb database has been developed that data for six millets and their one related species genomes, graph-based pan-genomics of pearl millet, and stress-related multi-omics data (Sun et al. 2023). This enables Milletdb to be the most complete millet multi-omics database available. It archives the findings from genome-wide association studies (GWAS) on 20 traits related to yield, observed in three different environmental conditions (field with no stress, early drought, and late drought) over a span of 2 years. This repository empowers users to pinpoint stress-related genes crucial for enhancing yield. Milletdb streamlines the functional genomics analysis of millets by offering a suite of 20 tools for analysis and retrieval. Additionally, a comprehensive transcriptome database, LMTdb, was conceived to unravel the genetic signatures of the relatively unknown little millet after completing its transcriptome sequencing (Shekhar et al. 2023). This resource encompasses information on transcriptome sequences, functional annotations, microsatellite markers, differentially expressed genes (DEGs), and pathway details.

5 Trait Mapping in Minor Millets

Genetic improvement requires knowledge on gene and genomic regions controlling various desirable traits. Since the last decade, DNA marker-based methods, such as biparental QTL mapping and genome-wide association studies (GWAS) aiming at effective genomics-assisted breeding, have been applied for crop improvement (Abberton et al. 2016; Yabe and Iwata, 2020). Biparental mapping populations represent a set of individuals derived from the crosses between two diverse parents with respect to phenotype and traits to be shared, for the identification of QTLs. The availability of versatile DNA markers and the advancement of statistical models and techniques have led to the widespread use of this strategy (Kearsey and Farquhar 1998). SNP markers are being widely adopted for QTL mapping. SNP markers can be identified using DNA array technologies, NGS technologies such as RAD-Seq or GBS (Baird et al. 2008; Elshire et al. 2011; Babiker et al. 2015; Erath et al. 2017). Since the development of NGS technology, GWAS has emerged as a preferred technique for the identification of markers associated with desirable traits. GWAS, also known as linkage disequilibrium mapping, can be performed using natural genetic resources or breeding populations and high-density markers (Hamblin et al. 2011).

Minor millets except for foxtail millet and finger millet have largely remained unexplored for the genes and QTLs due nonavailability of genomic resources and little attention of researchers on this group of crops. Initial mapping studies used random markers such as RAPD, ISSR, and so on to generate linkage maps in millets. However, recently the development of linkage maps and trait mapping in millets has picked up as a result of NGS-based genotyping approaches such as GBS, resequencing, and which can aid in the discovery of SNPs with minimum cost per sample. Breeding methods along with new genomic and molecular techniques offer an avenue for the identification of candidate genes or markers associated with agronomically important traits or for stress tolerance. Candidate genes for increasing yield, stress tolerance, and other important agronomic features have been identified in minor millets (Gupta et al. 2013a, b; Singh et al. 2014; Li et al. 2014; Wang et al. 2014; Choudhary et al. 2023; Maharajan et al. 2023). In the following section, we have described the progress made in the direction of the development of linkage maps and mapping key traits in minor millets.

Molecular investigations into minor millets have yielded insights into phenotypic variations, characteristics associated with climate resilience, genome-wide SNP markers, and the heritability of significant traits and markers contributing to genetic diversity. This marker data has not been fully utilized for the development of improved varieties and crop improvement programs in minor millets. However, the advancement in genomics and bioinformatics allows the application of upcoming tools for improvement in these crops. The trait mapping employing different approaches in minor millets has been summarized in Table 3.

6 Functional Genomics for Identification of Key Genes

6.1 Transcriptomics

Transcriptomics is an effective approach for discovering gene/pathways controlling plant responses for biotic and abiotic stresses as well plant growth and development. Some important transcriptomics studies conducted in minor millets are summarized in Table 4.

Table 3 Trait mapping in minor millets

Crop	Plant material	Approach used	Trait	Key findings	References
Mapping for quality/agronomically important traits					
Tef	In eight locations in Ethiopia, *Eragrostis tef*, cv. Kaye Murri, and *E. pilosa* (30–5) evaluated for 22 morphological and yield-related and traits during 1999 and 2000	Composite interval mapping	Grain yield and yield components	99 QTLs on 15 linkage groups (LGs) identified for 19 traits, 12 QTLs for grain yield on 9 LGs, and more than 5 QTLs for different traits on 7 LGs. Ten QTLs constituting the largest cluster found to be located on LG8; eight of which were related to yield or yield components. Two of the yield QTLs showed consistency in various agro-ecological zones	Yu et al. (2007)
Tef	151 F_9 RILs derived from crossing *E. tef* with its wild relative *E. pilosa*	Linkage map constructed using SSR markers	Grain yield and lodging resistance	The linkage map comprised 30 LGs. Using 11 environment conditions and marker data, QTLs governing lodging, grain yield, and 15 other associated traits identified	Zeid et al. (2011)
Tef	42 tef lines and 40 *Eragrostis* species assessed using SNP data produced by GBS	>1000 SNPs identified and high-quality SNPs employed to investigate phylogenetic relationships, genetic diversity, population structure, sequence similarity, and divergence		In the tef reference genome, *E. aethiopica*, *E. ferruginea*, *E. lehmanniana*, *E. lugens*, *E. obtusa*, and *E. pilosa* shared 92% of the sequences. Wild *Eragrostis* species found to diverse than the tef cultivars	Girma et al. (2018)
Finger millet	113 accessions assessed for 14 agro-morphological parameters at two locations in India (Almora and Pantnagar)	SNPs employed to find markers associated with grain yield and its component traits using association mapping. Single locus single trait (SLST), multi-locus mixed model (MLMM) and multi-trait mixed model (MTMM) used to study marker trait associations (MTAs)	Grain yield and its component traits	20, 36, and 53 MTAs for five traits were identified using SLST, MLMM, and MTMM, respectively. Among these, nine MTAs were found common	Sharma et al. (2018)

Kodo millet, little millet, and proso millet	190 accessions each of kodo millet, little millet, and proso millet examined for population structure and genome-wide marker datasets	3461, 2245, and 1882 SNPs identified for kodo, little, and proso millet, respectively, using GBS		Eight putative subpopulations each of little millet and proso millet and seven of kodo millet found by population genetic analysis	Johnson et al. (2019)
Foxtail millet	142 geographically diverse genotypes evaluated for ten agronomic traits using GWAS	10 K SNPs (generated using GBS-ddRAD approach)	Agronomic traits	Number of SNPs on each chromosome ranged from 844 (chromosome 5) to 2153 (chromosome 8). The linkage disequilibrium (LD) decay ranged from 76 kb (chromosome 6) to 357 kb (chromosome 4). 81 MTAs for ten traits identified. For flag leaf width, grain yield, and thousand-grain weight (TGW), high confidence MTAs found	Jaiswal et al. (2019)
Fonio millet	183 cultivated and wild accessions studied	Chromosome-scale reference assembly along with the deep re-sequencing	Agronomic traits	Two genes found to be linked with seed size and shattering	Abrouk et al. (2020)
Finger millet	113 genotypes studied in order to understand the complex genetic architecture of seed protein content (SPC) and associated traits such as days to maturity (DM) and grain yield (GY) in two locations (Almora and Pantnagar)	Genome-wide association mapping through GBS	Seed traits	Significant genotypic variations found in both the locations for SPC, DM, and GY In finger millet, five SNPs found to be linked to the SPC and GY and three with DM. Aspartyl protease considered as a promising candidate gene for SPC content in finger millet	Tiwari et al. (2020)

(continued)

Table 3 (continued)

Crop	Plant material	Approach used	Trait	Key findings	References
Prosomillet	88 varieties/landraces evaluated for genomic and phenotypic diversity for seed traits and to identify MTAs	2412 high-quality SNPs, (generated by sequencing of restriction-site associated DNA fragments) used for diversity studies and GWAS	Agronomic and seed traits	Ten MTAs for seed morphology traits and 13 for major agronomic and seed traits identified and the collection found diverse for seed traits	Boukail et al. (2021)
Foxtail millet	333 RILs derived by crossing Ai 88 and Liaogu 1 (an elite cultivar) studied for plant height fora period of 4 years in 13 different environmental conditions	QTL mapping using an ultra-high density bin map with 3744 marker bins	Plant height	Mean plant height ranged from 89.5 to 149.9 cm. 26 QTLs identified for plant height and 13 of these repeatedly detected under multiple environment conditions. *Seita.1G242300* encoding gibberellin 2-oxidase-8 detected to be expressed in nine environments. Six genes for gibberellic acid biosynthesis/signaling, and 15 genes for F-box domain proteins also identified	He et al. (2021)
Prosomillet	200 families of mutant population (M2 generation) derived from mutation in proso millet variety ATL1, studied to identify SNPs and Indels induced by mutations	25,901, 30,335, and 31,488 SNPs, respectively, observed in the wild type, high-yielding ATLhy mutants and low-yielding ATL_ly mutants by GBS	Yield	84 and 171 functional SNPs identified in ATLhy and ATL_lymutants, respectively. Two functional SNPs in ATL_hy and one in ATL_ly found to be related to E3 ubiquitin-protein ligase (*UPL7*) gene. Two SNPs of ATL_ly encoding starch synthase found to be involved in starch biosynthetic pathway	Francis et al. (2023)

Mapping for abiotic stress resistant and climate resilient traits					
Prosomillet	Association analysis on 143 genotypes and 15 agronomic traits using mixed linear model (MLM)	514 polymorphic AFLP markers generated by 11 primer combinations	Salt tolerance	Under salt stress, M14/E10–14, M14/E10–64 markers (for seed yield), and M14/E10–64 marker (for forage yield) showed stable association	Yazdizadeh et al. (2020)
Finger millet	186 eco-geographically diverse genotypes evaluated for 13 agronomic parameters and reaction to blast	2977 high-quality SNPs generated by GBS	Agronomic traits, Neck blast resistance	132 common MTAs found for agronomic parameters using the GWAS general linear and mixed models. 26 MTAs showed homology with candidate genes involved in plant growth, development, and photosynthesis. Four common MTAs for neck blast resistance found	Sood et al. (2023)
Finger millet	An $F_{2:3}$ population derived by crossing ACC 100007 [*E. coracana* (L.) Gaertn. subsp. *Coracana*] with GBK 030647 [*E. coracana* (L.) Gaertn. subsp. *Africana*], genotyped by GBS	Phenotypic data on morphology, yield, and blast resistance traits studied in $F_{2:3}$ population in a field trial in Kenya	Agronomic traits and blast resistance to *Magnaporthe oryzae*	An 18-linkage-group genetic map with 5422 markers allowed comparative genomic analyses with foxtail millet, rice, and sorghum. Significant QTLs identified for plant height, flowering date, panicle number, and blast incidence and severity	Pendergast IV et al. (2022)

Table 4 Transcriptomics studies in millets

Crop	Trait	Significant achievement	Reference
Kodo millet	Dehydration stress	De novo RNA-sequencing analysis conducted on the dehydration-sensitive kodo millet cultivar "CO3" provided a comprehensive insight into the mechanisms governing the response to dehydration. The study unveiled a multitude of molecular and metabolic pathways intricately regulated by diverse hormones and stress-responsive genes under dehydrated conditions	Suresh et al. (2022)
Finger millet	Drought	Transcriptome analysis of finger millet variety ML-365 subjected to both low moisture stress and non-stress conditions uncovered a set of candidate genes induced by drought. These identified genes hold significant promise for utilization in breeding programs aimed at enhancing drought tolerance in crops	Hittalmani et al. (2017)
	Drought	Transcriptome sequencing of the finger millet variety GPU-24 subjected to drought revealed novel stress-responsive regulators, pathways, and mechanisms that contribute to the inherent drought-hardy nature of this crop	Parvathi et al. (2019)
	Drought	Transcriptome analysis of two distinct groups of finger millet genotypes, consisting of three aluminum (Al)-tolerant and three Al-susceptible varieties, showed crucial insights into the molecular basis of Al tolerance. Differentially expressed genes in Al-tolerant genotypes included a wide range of activities and processes, including catalytic functions, binding mechanisms, transport activities, responses to stimuli, and involvement in metabolic processes such as glyoxylate and dicarboxylate metabolism	Brhane et al. (2022b)
	Salinity	Transcriptome sequencing of leaf and shoot tissue of salt stress susceptible and tolerant finger millet genotypes. Salinity stress inhibited photosynthetic capacity and photosynthesis-related genes in the susceptible genotype CO 12. Numerous genes related to cell growth and differentiation were identified as upregulated in both genotypes, with a more pronounced increase observed in the tolerant genotype	Rahman et al. (2014)
Little millet	Drought	Transcriptome sequencing was used to explore the mechanism of drought tolerance in little millet genotype OLM-203/Tarini. Genes associated with polyamine metabolism were differentially expressed and contributed to enhanced level of drought tolerance in this genotype	Narayanrao et al. (2023)

(continued)

Table 4 (continued)

Crop	Trait	Significant achievement	Reference
Barnyard millet	Grain Fe and Zn content	RNA-sequencing of low and high grain Fe content genotypes during spike development revealed key pathways/genes-associated Fe accumulation. The spike emergence was found to be more critical for Fe accumulation than milking stage	Padhiyar et al. (2023)
Teff	Low calcium stress	Members of Ca^{2+} channels, Ca^{2+}-transporters, Ca^{2+}-binding proteins, and Ca^{2+}-dependent protein kinases found to be differentially regulated. Several Fe/metal transporters, as well as oligo-peptide transporters, found to be differentially regulated between shoot and root in response to Ca^{2+} treatment	Ligaba-Osena et al. (2022)

6.2 Proteomics

Proteomics deals with the study of the entire set of proteins produced in an organism. While the transcriptomics approach provides information at the transcript level, alterations in gene expression do not always relate to corresponding changes at the proteome level. Moreover, several transcripts undergo posttranscriptional and posttranslational modifications, which cannot be understood with transcriptomic analysis. Therefore, it is extremely important to characterize the gene expression pattern at the protein level to have a detailed understanding of molecular mechanisms governing a trait in an organism. Proteome analysis can complement transcriptomics and genomics to provide a comprehensive understanding of mechanisms involved in plant growth and differentiation or tolerance to biotic and abiotic stresses. There are a number of transcriptomics studies conducted in minor millets for various traits.

7 Modern Breeding Approaches

7.1 Marker-Assisted Breeding

The genes/QTLs identified through various marker-based mapping approaches can be precisely incorporated into agronomically superior parent by selecting plants using the markers linked to specific traits. Commonly employed molecular breeding methods are marker-assisted selection (MAS), marker-assisted backcrossing (MABC), and marker-assisted recurrent selection. MABC has found widespread application in the introgression of genes in cereals such as rice and wheat. This approach can significantly contribute to millets improvement by accelerating the genetic gain and the development of varieties resilient to various biotic and abiotic stresses. MABC approach is primarily utilized for transferring traits governed by major genes and has also been shown to expedite the backcrossing process by reducing the number of backcrosses required to regain the recurrent parent phenotype. Furthermore, MABC approach allows background selection using markers polymorphic between parents and thus aid in minimizing the donor's segment. The approach greatly outperforms conventional breeding approach in terms of precision and efficiency. Additionally, using this approach it is possible to efficiently pyramid two or even more disease genes for resistance or any other major gene-controlled trait into a superior genetic background. Following this approach, a number of genes for disease resistance and abiotic stress tolerance have been transferred into rice and wheat. For example, *Sub1* gene for submergence tolerance (Septiningsih et al. 2009), *Piz*, *Pib*, *Pita*, and *Pik* genes for blast resistance (Zampieri et al. 2023), and *Xa1*, *Xa13*, and *Xa21* for bacterial blight resistance (Pandey et al. 2013a, b). However, there is no report of marker enable introgression of genes in any minor millet crop, which is primarily due to lack of genes and markers for any important target trait.

7.2 Genomic Selection

The utilization of marker-assisted selection has shown limited effectiveness in enhancing the genetic characteristics of quantitative traits governed by numerous QTLs, each exerting a small effect. The complex nature of quantitative traits, such as yield, poses significant challenges in identifying all relevant genes/QTLs, estimating their impacts, and successfully integrating them into superior agronomic contexts. Genomic selection (GS), an alternative approach to MAS, is demonstrated to be effective for the improvement of quantitative traits in crops including cereals such as maize (Zhang et al. 2017), rice (Wang et al. (2017b)), wheat (Dreisigacker et al. 2023), and orphan crops such as pigeon pea (Bohra et al. 2020). In GS approach, the effect of each marker is estimated, and the cumulative sum of these marker effects is employed in calculating the genomic-estimated breeding values (GEBV) for each individual. For the practical purpose, the GS approach has two populations: test population and breeding population. The GS model is developed in the training population, the individuals of which are genotyped and also phenotyped for the target trait. The model thus developed is used to predict GEBV solely based on the genotyping data. The approach has great potential in selection of the suitable donors as well as elite genotypes for the target trait during generation advancement. Consequently, GS presents a valuable opportunity to explore and enhance and boosting the genetic gain per selection in a breeding scheme per unit timeline, thereby increasing breeding programs' speed and efficacy (Xu et al. 2017). For example, in oil palm, it takes around 19 years to complete a selection cycle; however, intervention through GS has helped reduce it to 6 years.

The productivity of millets remains low due to the lack of strong systematic breeding programs for their genetic improvement. In fact, there is limited or no information about the genes/markers or genetic basis of the majority of traits including productivity-related traits in minor millets. Moreover, biparental mapping studies could also not be performed in these crops because of their small floret sizes, which made mapping population generation very difficult. In such a scenario, GS can be very useful in accelerating genetic gain in small millets as this allows for the simultaneous identification of genomic regions and the selection of superior lines with high-yield associated traits (Srivastava et al. 2020). There have been some reports of genomic prediction of traits in major millets including bajra (Liang et al. 2018) and maize (Millet et al. 2019). Advances in SNP genotyping methodologies and their ever-falling cost have made it easier to perform large-scale genotyping even non-model species such as millets, making it possible to perform genomic prediction in these crops. The donor germplasm lines for a trait can be easily selected based on the genomic prediction in a small set of training populations in order to accelerate their utilization in systematic breeding. Recently, He et al. (2023) evaluated different marker panels of SNPs and structural variants in foxtail millet for the genomic prediction of 68 agronomic and quality traits under different environments. They showed that prediction accuracy of 0.70 for 97% of the studied phenotypes. Additionally, they computed GEBVs for approximately 1.04 million haplotype combinations, examining their correlations with quality and various yield component traits. The results indicated that the GEBVs for yield and grain quality traits could potentially be enhanced by as much as 50% and 49%, respectively.

7.3 Haplotype-Based Breeding

Haplotype-based breeding has emerged as a powerful genomic-assisted breeding technique for the improvement of complex traits and has been attempted in many crops including rice (Abbai et al. 2019), pigeon pea (Sinha et al. 2020), and so on. Haplotypes are defined as the combination of alleles (variants) that are located on the same genomic region/chromosome and have a tendency to inherit together. The haplotypes are generally defined by the SNPs that are in strong linkage disequilibrium (LD). At the individual level, there could be a maximum of two haplotypes for a genomic segment; however, many haplotypes may be present for the same genomic region in the population of a species. Therefore, sequencing/genotyping of a large number of germplasm lines or individuals of the natural population of a species can facilitate the discovery of haplotypes that can be used to perform GWAS to find more reliable associations for the targeted trait. The haplotypes thus identified can be subsequently validated and explored for selecting targeted train in the breeding population. Even in GS, the haplotypes have been found more efficient as compared to when the individual SNPs are used (Matias et al. 2017). As discussed in the previous section, biparental mapping did not make much progress in millets due to the difficulty in developing a mapping population. In this scenario, haplotype GWAS can prove very useful in minor millet crops because with ever falling sequencing cost, it is feasible to sequence a diverse panel of individuals that can be simultaneously phenotyped for the targeted trait to find associated haplotype/candidate genomic regions. The identified haplotype can be directly used to select the superior genotype in the breeding program. Moreover, targeted sequencing followed by haplo-pheno analysis can be used to identify superior haplotypes for specific genes linked to grain yield, quality, and abiotic stress tolerance (Sinha et al. 2020). The advantage of haplotype-based breeding is that accession that possesses superior haplotype is taken forward in trait improvement ensuring efficient selection of genotypes. Therefore, the approach if applied in the improvement of millets, it can fasten the development of improved varieties for various desirable traits.

7.4 Genome Editing

Genome editing (GE) has emerged as a powerful technique for genome modification in crops for trait discovery. It has allowed precise mutation in the target gene and has helped incorporation of key target traits in major crops during the past 10 years (Mohanta et al. 2017; Zhang et al. 2019). This technology has revolutionized the field of plant science because, in addition to being a useful tool for improving traits, it can be very well applied to gain insight into how genes function. GE approach is fast and more acceptable as compared to transgenic approach, which makes it a very important tool for the genetic improvement of crops and it can be very well extended to millets for introducing key traits. A number of genome editing technologies are available including zinc finger nucleases, transcription activator-like effector nucleases (TALENs), and CRISPR-Cas9 (Mohanta et al. 2017). Among these CRISPR–CAS9 system is very popular with researchers and comprises two components, CAS9 (an endonuclease) and single guide RNA (sgRNA), which has the sequence complementary to the targeted region of the genome. The sgRNA enables recognition of the target region with protospacer-adjacent motif (PAM) to facilitate precise cut by CAS9 in the target. Genome editing has great potential to accelerate trait improvement in millets. One of the major constraints for less preference for millet by the farmers is their low productivity as compared to major cereals. GE can be used to decipher the genetic basis of various productivity-related traits in these crops. The key putative candidate genes for yield-related traits identified by the genetic mapping and transcriptomics studies could be taken up for validation by GE. Furthermore, GE can help in better understanding the mechanisms underlying rich nutrient profiles and climate resilience in millets, thereby facilitating the transfer of these traits to major staple cereals. Despite the importance of these crops, genomic research in general and genome editing in particular, is lagging behind in millets as these are primarily grown in the world's poor countries in Asia and Africa that do not have sufficient financial resources to undertake advanced research in these crops. The standard genome editing procedure in any crop requires its reference scale genome sequence as well as a suitable transformation protocol. Foxtail millet (Bennetzen et al. 2012) was the first millet species for which a reference scale genome was published. However, since then, significant advancements in sequencing technologies have led to the generation of the reference scale genome assemblies of various species including proso millet (Zou et al. 2019), fonio millet (Abrouk et al. 2020), teff (VanBuren et al. 2020), and Indian barnyard millet (Wu et al. 2022). The reference scale and fully annotated whole genome sequence of the millet crop would not only help identify and precisely select the target gene but also facilitate the designing of suitable gRNA. Furthermore, the availability of reference scale sequence can help editing efficiency and the type of editing (insertion and deletion) introduced through resequencing of edited lines. Besides, generating the reference scale genome sequences, the focus should be also given to the development of efficient transformation protocols without that GE cannot be successfully applied in millet crops. Among the minor millets, GE was first applied in the foxtail millet as it was the first millet crop with a reference scale (Lin et al. 2018; Zhang et al. 2021; Cheng et al. 2021; Liang et al. 2022). In foxtail, an efficient plant transformation was not available, so a protoplast transfection approach was used in the first CRISPR/CAS9-mediated GE effort in this crop. This study was primarily meant for screening suitable guide RNA and CAS9 for the generation of stable transformants through protoplast transfection. The phytoene desaturase (*PDS*) gene was targeted for standardization, resulting in the production of some edited foxtail millet lines that were unstable. In another CRISPR study, with the objective to get mutant inducer lines, *MTL* gene was disrupted, which showed mutagenesis efficiency of 26% (Cheng et al. 2021). Recently, Liang et al. (2022) reported development of an efficient GE protocol using CRISPR/CAS9 and base editor. First, they used single- and multi-target knout systems to select the efficient sgRNA for the genes *SiFMBP*, *SiDof4*, *SiBADH2*, *SiGBSS1*, and *SiIPK1* in the foxtail millet protoplast. Then, *Agrobacterium*-mediated transformation protocol was used for the CRISPR/CAS9 editing that enabled recovery of homozygous mutant lines for all the targeted genes. Additionally, using this transformation protocol, single base editing was efficiently and successfully applied using cytosine and adenine base editors to target the *SiALS* and *SiACC* genes. The single base edited lines for *SiALS* gene were herbicide resistant. Recently, Beyene et al. (2022) made the first GE effort in teff and introduced knock-out mutation in semi-dwarfing (SD1) gene homolog of rice by applying the CRISPR/CAS9 system. The edited teff lines had short heights with reduced internodal distance and were resistant to lodging. The lodging is a major problem in teff, the edited lines can be very useful in enhancing the productivity of this crop in Ethiopia where it is considered a major staple food crop.

7.5 Speed Breeding

Speed breeding (SB) protocols shorten breeding cycle, i.e., growing one generation seed to seed and thus very helpful in hastening both selection efficiency and basic research (Watson et al. 2018). The technique has been standardized in many crops. Around six generations per year is possible in crops such as chickpea, wheat, barley, and so on under speed breeding facilities. Moreover, SB can be also integrated with marker-assisted selection or genome editing to fast-forward the genetic gain (Varshney et al. 2021). Rapid genetic gain has been realized in wheat by integrating GS with SB (Voss-Fels et al. 2019). Considering the advantages associated with SB approach, if it is standardized and adopted in millets, it has the potential to accelerate genetic gain in these important group of orphan crops.

8 Conclusions and Future Prospects

Over the past two decades, significant progress has been made in the generation of genomic resources for the model minor millet crop, foxtail millet, as well as for finger millet. For other minor millets also, notable progress has been achieved with respect to genomic resource generation, a trend expected to accelerate in the coming years with the availability of complete genome sequences for these millets. The genomic resources thus generated have been mainly employed for diversity analysis and, to a certain extent, the mapping of valuable traits. Functional genomics technologies, including proteomics, transcriptomics, and metabolomics, have been also implemented in some minor millets and contributed toward understanding the genetic and molecular mechanisms governing various useful traits. Some attempts have also been made to use modern genomic tools such as genome editing and genomic prediction for trait improvement in these crops. Overall, it can be said that the platform for genomics-assisted research is already laid in the minor millets. The need of the hour is to suitably combine various genomic technologies for focused research on genetic dissection of various desirable traits, importantly productivity-related traits. Identifying genes/alleles controlling these traits is crucial for millet improvement. Further, incorporating genomics-assisted breeding techniques, such as marker-assisted selection (MAS), genomic selection, and genome editing, as well as the implementation of speed breeding, can significantly contribute to accelerating genetic gains in millet improvement.

References

Abbai R, Singh VK, Nachimuthu VV, Sinha P, Selvaraj R, Vipparla AK, Singh AK, Singh UM, Varshney RK, Kumar A (2019) Haplotype analysis of key genes governing grain yield and quality traits across 3K RG panel reveals scope for the development of tailor-made rice with enhanced genetic gains. Plant Biotechnol J 17(8):1612–1622. https://doi.org/10.1111/pbi.13087

Abberton MJ, Batley A, Bentley J, Bryant H, Cai J, Cockram AC, de Oliveira LJ, Cseke H, Dempewolf C, Pace D et al (2016) Global agricultural intensification during climate change: a role for genomics. Plant Biotechnol J 14:1095–1098

Abrouk M, Ahmed HI, Cubry P, Šimoníková D, Cauet S, Pailles Y, Bettgenhaeuser J, Gapa L, Scarcelli N, Couderc M, Zekraoui L, Kathiresan N, Čížková J, Hřibová E, Doležel J, Arribat S, Bergès H, Wieringa JJ, Gueye M, Kane NA, Krattinger SG (2020) Fonio millet genome unlocks African orphan crop diversity for agriculture in a changing climate. Nat Commun 11(1):4488. https://doi.org/10.1038/s41467-020-18329-4

Arya L, Verma M, Gupta VK, Karihaloo JL (2009) Development of EST SSRs in finger millet (Eleusine coracana ssp coracana) and their transferability to pearl millet (Pennisetum glaucum). J Plant Biochem Biotechnol 18:97–100

Arya L, Chauhan D, Yadav Y, Verma M (2014) Transferability of simple sequence repeat (SSR) markers developed in finger millet, and pearl millet to kodo millet and barnyard millet. In: Johri

AK, Mishra GC (eds) Innovative approach in stem cell research. Cancer Biology and Applied Biotechnology, Excellent Publishing House, New Delhi, pp 60–64

Babiker EM, Gordon TC, Jackson EW, Chao S, Harrison SA, Carson ML, Obert DE, Bonman JM (2015) Quantitative trait loci form two genotypes of oat (Avena sativa) conditioning resistance to Puccinia coronata. Phytopathology 105:239–245

Babu B, Chauhan R (2017) *In-silico* identification of EST based microsatellite markers and SNPs, and comparative genomic analysis of ESTs in barnyard millet for their omics applications. Curr Agric Res J 5:279–287. https://doi.org/10.12944/CARJ.5.3.03

Babu B, Rashmi C, Sood S (2018) Cross transferability of finger millet and maize genomic SSR markers for genetic diversity and population structure analysis of barnyard millet. Indian J Genet Plant Breeding 78:364–372. https://doi.org/10.31742/IJGPB.78.3.5

Backiyalakshmi C, Vetriventhan M, Deshpande S et al (2021) Genome-wide assessment of population structure and genetic diversity of the global finger Millet germplasm panel conserved at the ICRISAT Genebank. Front Plant Sci 12:692463. Published 2021 Aug 20. https://doi.org/10.3389/fpls.2021.692463

Bai G, Tefera H, Ayele M, Nguyen HT (1999) A genetic linkage map of tef [Eragrostis tef (Zucc.) Trotter] based on amplified fragment length polymorphism. Theor Appl Genet 99(3–4):599–604. https://doi.org/10.1007/s001220051274

Bai H, Cao Y, Quan J, Dong L, Li Z, Zhu Y, Zhu L, Dong Z, Li D (2013) Identifying the genome-wide sequence variations and developing new molecular markers for genetics research by re-sequencing a landrace cultivar of foxtail millet. PLoS One 8(9):e73514. https://doi.org/10.1371/journal.pone.0073514

Baird NA, Etter PD, Atwood TS, Currey MC, Shiver AL, Lewis ZA, Selker EU, Cresko WA, Johnson EA (2008) Rapid SNP discovery and genetic mapping using sequenced RAD markers. PLoS One 3:e3376

Barnaud A, Vignes H, Risterucci AM, Noyer JL, Pham JL, Blay C, Buiron M, Vigouroux Y, Billot C (2012) Development of nuclear microsatellite markers for the fonio, Digitaria exilis (Poaceae), an understudied West African cereal. Am J Bot 99(3):e105–e107. https://doi.org/10.3732/ajb.1100423

Bennetzen JL, Schmutz J, Wang H, Percifield R, Hawkins J, Pontaroli AC, Estep M, Feng L, Vaughn JN, Grimwood J, Jenkins J, Barry K, Lindquist E, Hellsten U, Deshpande S, Wang X, Wu X, Mitros T, Triplett J, Yang X, Devos KM (2012) Reference genome sequence of the model plant Setaria. Nat Biotechnol 30(6):555–561. https://doi.org/10.1038/nbt.2196

Beyene G, Chauhan RD, Villmer J, Husic N, Wang N, Gebre E, Girma D, Chanyalew S, Assefa K, Tabor G, Gehan M, McGrone M, Yang M, Lenderts B, Schwartz C, Gao H, Gordon-Kamm W, Taylor NJ, MacKenzie DJ (2022) CRISPR/Cas9-mediated tetra-allelic mutation of the 'green revolution' SEMIDWARF-1 (SD-1) gene confers lodging resistance in tef (Eragrostis tef). Plant Biotechnol J 20(9):1716–1729. https://doi.org/10.1111/pbi.13842

Bheema Lingeswara Reddy IN, Sivaramakrishnan S (2017) Identification of SSR markers which could differentiate blast disease resistance accessions in finger millet (Eleusine coracana (L.) Gaertn.). J Crop Sci Biotechnol 20:37–44

Bohra A, Saxena KB, Varshney RK, Saxena RK (2020) Genomics-assisted breeding for pigeon pea improvement. Theor Appl Genet 133(5):1721–1737. https://doi.org/10.1007/s00122-020-03563-7

Bonthala VS, Muthamilarasan M, Misra G, Prasad M (2013) FmMDb: a versatile database of foxtail millet markers for millets and bioenergy grasses research. PLoS One 8(8):e71418. https://doi.org/10.1371/journal.pone.0071418

Bonthala VS, Muthamilarasan M, Roy R, Prasad M (2014) FmTFDb: a foxtail millet transcription factors database for expediting functional genomics in millets. Mol Biol Rep 41(10):6343–6348. https://doi.org/10.1007/s11033-014-3574-y

Boukail S, Macharia M, Miculan M et al (2021) Genome wide association study of agronomic and seed traits in a world collection of proso millet (*Panicum miliaceum* L.). BMC Plant Biol 21:330. https://doi.org/10.1186/s12870-021-03111-5

Brhane H, Haileselassie T, Tesfaye K, Hammenhag C, Ortiz R, Abreha KB, Geleta M (2021) Novel expressed sequence tag-derived and other genomic simple sequence repeat markers revealed genetic diversity in Ethiopian finger Millet landrace populations and cultivars. Front Plant Sci 12:735610. https://doi.org/10.3389/fpls.2021.735610

Brhane H, Haileselassie T, Tesfaye K, Ortiz R, Hammenhag C, Abreha KB, Vetukuri RR, Geleta M (2022a) Finger millet RNA-seq reveals differential gene expression associated with tolerance to aluminum toxicity and provides novel genomic resources. Front Plant Sci 13:1068383. https://doi.org/10.3389/fpls.2022.1068383

Brhane H, Haileselassie T, Tesfaye K, Ortiz R, Hammenhag C, Abreha KB, Geleta M (2022b) Novel GBS-based SNP markers for finger Millet and their use in genetic diversity analyses. Front Genet 13:848627. https://doi.org/10.3389/fgene.2022.848627

Cannarozzi G, Plaza-Wüthrich S, Esfeld K, Larti S, Wilson YS, Girma D, de Castro E, Chanyalew S, Blösch R, Farinelli L, Lyons E, Schneider M, Falquet L, Kuhlemeier C, Assefa K, Tadele Z (2014) Genome and transcriptome sequencing identifies breeding targets in the orphan crop tef (Eragrostis tef). BMC Genomics 15(1):581. https://doi.org/10.1186/1471-2164-15-581

Ceasar SA, Maharajan T (2022) The role of millets in attaining united Nation's sustainable developmental goals. Plants People Planet 4:345–349

Chanyalew S, Singh H, Tefera H, Sorrels M (2005) Molecular genetic map and QTL analysis of agronomic traits based on a *Eragrostis tef* x *Eragrostis pilosa* recombinant inbred population. J Genet Breed 59:53–66

Chen J, Liu Y, Liu M, Guo W, Wang Y, He Q, Chen W, Liao Y, Zhang W, Gao Y, Dong K, Ren R, Yang T, Zhang L, Qi M, Li Z, Zhao M, Wang H, Wang J, Qiao Z, Diao X (2023) Pangenome analysis reveals genomic variations associated with domestication traits in broomcorn millet. Nat Genet 55(12):2243–2254. https://doi.org/10.1038/s41588-023-01571-z-z

Cheng Z, Sun Y, Yang S, Zhi H, Yin T, Ma X, Zhang H, Diao X, Guo Y, Li X, Wu C, Sui Y (2021) Establishing in planta haploid inducer line by edited SiMTL in foxtail millet (Setaria italica). Plant Biotechnol J 19(6):1089–1091

Cho YI, Chung JW, Lee GA, Ma KH, Dixit A, Gwag JG, Park YG (2010) Development and characterization of twenty-five new polymorphic microsatellite markers in proso millet (*Panicum miliaceum* L.). Genes Genomics 32:267–273

Choudhary P, Shukla P, Muthamilarasan M (2023) Genetic enhancement of climate-resilient traits in small millets: a review. Heliyon 9:e14502. https://doi.org/10.1016/j.heliyon.2023.e14502

d'Ennequin MLT, Panaud O, Toupance B, Sarr A (2000) Assessment of genetic relationships between *Setaria italic* and its wild relative *S. Viridis* using AFLP markers. Theor Appl Genet 100:1061–1006

Desai H, Hamid R, Ghorbanzadeh Z, Bhut N, Padhiyar SM, Kheni J, Tomar RS (2021) Genic microsatellite marker characterization and development in little millet (Panicum sumatrense) using transcriptome sequencing. Sci Rep 11(1):20620. https://doi.org/10.1038/s41598-021-00100-4

Devos KM, Qi P, Bahri BA et al (2023) Genome analyses reveal population structure and a purple stigma color gene candidate in finger millet. Nat Commun 14:3694. https://doi.org/10.1038/s41467-023-38915-6

Dida MM, Srinivasachary, Ramakrishnan S, Bennetzen JL, Gale MD, Devos KM (2007) The genetic map of finger millet, Eleusine coracana. Theor Appl Genet 114(2):321–332. https://doi.org/10.1007/s00122-006-0435-7

Dida MM, Wanyera N, Harrison Dunn ML, Bennetzen JL, Devos KM (2008) Population structure and diversity in finger millet (*Eleusine coracana*) germplasm. Trop Plant Biol 1:131–141

Dreisigacker S, Pérez-Rodríguez P, Crespo-Herrera L, Bentley AR, Crossa J (2023) Results from rapid-cycle recurrent genomic selection in spring bread wheat. G3 (Bethesda, Md.) 13(4):jkad025. https://doi.org/10.1093/g3journal/jkad025

Elshire RJ, Glaubitz JC, Sun Q, Poland JA, Kawamoto K, Buckler ES, Mitchell SE (2011) A robust, simple genotyping-by-sequencing (GBS) approach for high diversity species. PLoS One 6:e19379

Erath W, Bauer E, Fowler DB, Gordillo A, Korzun V, Ponomareva M, Schmidt M, Schmiedchen B, Wilde P, Schön CC (2017) Exploring new alleles for frost tolerance in winter rye. Theor Appl Genet 130:2151–2164

Fang X, Dong K, Wang X, Liu T, He J, Ren R, Zhang L, Liu R, Liu X, Li M, Huang M (2016) A high density genetic map and QTL for agronomic and yield traits in foxtail millet [Setaria italica (L.) P. Beauv.]. BMC genomics 17:1–2

Francis N, Rajasekaran R, Rajagopalan VR, Bakya SV, Muthurajan R, Kumar AG, Alagarswamy S, Krishnamoorthy I, Thiyagarajan C (2023) Molecular characterization and SNP identification using genotyping-by-sequencing in high yielding mutants of proso millet. Front Plant Sci 14:1108203

Fukunaga K, Abe A, Mukainari Y, Komori K, Tanaka K, Fujihara A, Yaegashi H, Kobayashi M, Ito K, Ohsako T, Kawase M (2022) Recombinant inbred lines and next-generation sequencing enable rapid identification of candidate genes involved in morphological and agronomic traits in foxtail millet. Sci Rep 12(1):218

Ganal MW, Polley A, Graner EM, Plieske J, Wieseke R, Luerssen H, Durstewitz G (2012) Large SNP arrays for genotyping in crop plants. J Biosci 37(5):821–828. https://doi.org/10.1007/s12038-012-9225-3

Gimode D, Odeny DA, de Villiers EP, Wanyonyi S, Dida MM, Mneney EE et al (2016) Identification of SNP and SSR markers in finger millet using next generation sequencing technologies. PLoS One 11(7):e0159437

Girma D, Cannarozzi G, Weichert A, Tadele Z (2018) Genotyping by sequencing reasserts the close relationship between tef and its putative wild eragrostis progenitors. Diversity 10(2):17. https://doi.org/10.3390/d10020017

Goron TL, Raizada MN (2015) Genetic diversity and genomic resources available for the small millet crops to accelerate a new green revolution. Front Plant Sci 24(6):157

Guo S, Chai S, Guo Y, Shi X, Han F, Qu T, Xing L, Yang Q, Gao J, Gao X, Feng B (2023) Mapping of major QTL and candidate gene analysis for hull colour in foxtail millet (Setaria italica (L.) P. Beauv). BMC Genomics 24(1):458

Gupta PK, Varshney RK (2000) The development and use of microsatellite markers for genetic analysis and plant breeding with emphasis on bread wheat. Euphytica 113(3):163–185

Gupta S, Kumari K, Das J, Lata C, Puranik S, Prasad M (2011) Development and utilization of novel intron length polymorphic markers in foxtail millet (Setaria italica (L.) P. Beauv.). Genome 54(7):586–602

Gupta AK, Gaur VS, Gupta S, Kumar A (2013a) Nitrate signals determine the sensing of nitrogen through differential expression of genes involved in nitrogen uptake and assimilation in finger millet. Funct Integ Genom 13(2):179–190

Gupta S, Kumari K, Muthamilarasan M, Subramanian A, Prasad M (2013b) Development and utilization of novel SSR s in foxtail millet [S etaria italica (L.) P. Beauv.]. Plant Breed 132(4):367–374

Hamblin MT, Buckler ES, Jannink JL (2011) Population genetics of genomics-based crop improvement methods. Trends Genet 27(3):98–106

Hatakeyama M, Aluri S, Balachadran MT, Sivarajan SR, Patrignani A, Grüter S, Poveda L, Shimizu-Inatsugi R, Baeten J, Francoijs KJ, Nataraja KN (2018) Multiple hybrid de novo genome assembly of finger millet, an orphan allotetraploid crop. DNA Res 25(1):39–47

He Q, Zhi H, Tang S, Xing L, Wang S, Wang H, Zhang A, Li Y, Gao M, Zhang H, Chen G (2021) QTL mapping for foxtail millet plant height in multi-environment using an ultra-high density bin map. Theor Appl Genet 134:557–572

He Q, Tang S, Zhi H, Chen J, Zhang J, Liang H, Alam O, Li H, Zhang H, Xing L, Li X (2023) A graph-based genome and pan-genome variation of the model plant Setaria. Nat Genet 55(7):1232–1242

Hittalmani S, Mahesh HB, Shirke MD, Biradar H, Uday G, Aruna YR, Lohithaswa HC, Mohanrao A (2017) Genome and transcriptome sequence of finger millet (Eleusine coracana (L.) Gaertn.) provides insights into drought tolerance and nutraceutical properties. BMC Genomics 18:1–6

Hou S, Sun Z, Li Y, Wang Y, Ling H, Xing G, Han Y, Li H (2017) Transcriptomic analysis, genic SSR development, and genetic diversity of proso millet (Panicum miliaceum; Poaceae). Appl Plant Sci 5(7):1600137

Hu X, Wang J, Lu P, Zhang H (2009) Assessment of genetic diversity in broomcorn millet (Panicum miliaceum L.) using SSR markers. J Genet Genomics 36(8):491–500

Ibrahim Bio Yerima AR, Issoufou KA, Adje CA, Mamadou A, Oselebe H, Gueye MC, Billot C, Achigan-Dako EG (2021) Genome-wide scanning enabled SNP discovery, linkage disequilibrium patterns and population structure in a panel of fonio (Digitaria exilis [Kippist] Stapf) germplasm. Front Sustain Food Syst 5:699549

ICAR-NBPGR (2023) Annual report 2022. ICAR-National Bureau of Plant Genetic Resources, New Delhi, p 218

Jaiswal V, Gupta S, Gahlaut V, Muthamilarasan M, Bandyopadhyay T, Ramchiary N, Prasad M (2019) Genome-wide association study of major agronomic traits in foxtail millet (Setaria italica L.) using ddRAD sequencing. Sci Rep 9(1):5020

Jayakodi M, Madheswaran M, Adhimoolam K, Perumal S, Manickam D, Kandasamy T, Yang TJ, Natesan S (2019) Transcriptomes of Indian barnyard millet and barnyard grass reveal putative genes involved in drought adaptation and micronutrient accumulation. Acta Physiol Plant 41:1–1

Jia XP, Shi YS, Song YC, Wang GY, Wang TY, Li Y (2007) Development of EST-SSR in foxtail millet (Setaria italica). Genet Resour Crop Evol 54:233–236

Jia X, Zhang Z, Liu Y, Zhang C, Shi Y, Song Y, Wang T, Li Y (2009) Development and genetic mapping of SSR markers in foxtail millet [Setaria italica (L.) P. Beauv.]. Theor Appl Genet 118(4):821–829

Jiang Y, Li H, Zhang J, Xiang J, Cheng R, Liu G (2018) Whole genomic EST-SSR development based on high-throughput transcript sequencing in proso millet (*Panicum miliaceum*). Int J Agric Biol 20:617–620

Johnson M, Deshpande S, Vetriventhan M, Upadhyaya HD, Wallace JG (2019) Genome-wide population structure analyses of three minor millets: Kodo millet, little millet, and proso millet. Plant Genome 12(3):190021

Joshi DC, Meena RP, Chandora R (2021) Genetic resources: Collection, characterization, conservation, and documentation," in Millets and Pseudo Cereals (India: Woodhead Publishing Series in Food Science, Technology and Nutrition), pp. 19–31. https://doi.org/10.1016/B978-0-12-820089-6.00003-3

Kalyana Babu B, Sood S, Kumar D, Joshi A, Pattanayak A, Kant L, Upadhyaya HD (2018) Cross-genera transferability of rice and finger millet genomic SSRs to barnyard millet (Echinochloa spp.). 3. Biotech 8(2):95

Karam D, Westra P, Nissen SJ, Ward SM, Figueiredo JE (2004) Genetic diversity among proso millet (Panicum miliaceum) biotypes assessed by AFLP technique. Planta Daninha 22:167–174

Kearsey MJ, Farquhar AG (1998) QTL analysis in plants; where are we now? Heredity 80(2):137–142

Kumar LS (1999) DNA markers in plant improvement: an overview. Biotechnol Adv 17(2–3):143–182

Kumar SM, Madhura JN, Sheshashayee MS, Hittalamani S, Udayakumar M (2021) Development of microsatellite markers in finger millet (*Eleusine coracana* Geartn.). Indian J Genet 81(3):462–465. https://doi.org/10.31742/IJGPB.81.3.14

Kumari K, Muthamilarasan M, Misra G, Gupta S, Alagesan S, Parida S et al (2013) Development of eSSR-markers in Setaria italica and their applicability in studying genetic diversity, cross-transferability and comparative mapping in millet and non-millet species. PLoS One 8:e67742. https://doi.org/10.1371/journal.pone.0067742

Kushwaha H, Jillo KW, Singh VK, Kumar A, Yadav D (2015) Assessment of genetic diversity among cereals and millets based on PCR amplification using Dof (DNA binding with one finger) transcription factor gene-specific primers. Plant Syst Evol 301:833–840

Lata C, Prasad M (2011) Role of DREBs in regulation of abiotic stress responses in plants. J Exp Bot 62(14):4731–4748. https://doi.org/10.1093/jxb/err210

Lata C, Prasad M (2013a) Validation of an allele-specific marker associated with dehydration stress tolerance in a core set of foxtail millet accessions. Plant Breed 132(5):496–499

Lata C, Prasad M (2013b) Association of an allele-specific marker with dehydration stress tolerance in foxtail millet suggests SiDREB2 to be an important QTL. J Plant Biochem Biotechnol 23:119–122

Lata C, Bhutty S, Bahadur RP, Majee M, Prasad M (2021) Association of an SNP in a novel DREB2-like gene SiDREB2 with stress tolerance in foxtail millet [Setaria italica (L.)]. J Exp Bot 62(10):3387–3401

Lee KJ, Yoon MS, Shin MJ et al (2017) Development of SSR markers and their use in studying genetic diversity and population of finger millet (Eleusine coracana L. Gaertn.). Plant Breed Biotechnol 5:183–191. https://doi.org/10.9787/PBB.2017.5.3.183

Li C, Yue J, Wu X, Xu C, Yu J (2014) An ABA-responsive DRE-binding protein gene from Setaria italica, SiARDP, the target gene of SiAREB, plays a critical role under drought stress. J Exp Bot 65(18):5415–5427

Li X, Hou S, Feng M, Xia R, Li J, Tang S, Han Y, Gao J, Wang X (2023) MDSi: multi-omics database for Setaria italica. BMC Plant Biol 23(1):223

Liang Z, Gupta SK, Yeh CT, Zhang Y, Ngu DW, Kumar R, Patil HT, Mungra KD, Yadav DV, Rathore A, Srivastava RK (2018) Phenotypic data from inbred parents can improve genomic prediction in pearl millet hybrids. G3 8(7):2513–2522

Liang Z, Wu Y, Ma L, Guo Y, Ran Y (2022) Efficient genome editing in Setaria italica using CRISPR/Cas9 and base editors. Front Plant Sci 12:815946

Ligaba-Osena A, Salehin M, Numan M, Wang X, Choi SC, Jima D, Bobay LM, Guo W (2022) Genome-wide transcriptome analysis of the orphan crop tef (Eragrostis tef (Zucc.) Trotter) under long-term low calcium stress. Sci Rep 12(1):19552

Lin Heng Sheng LH, Liao GwoIng LG, Chiang ChihYun CC, Kuoh ChangSheng KC, Chang Song Bin CS (2012) Genetic diversity in the foxtail millet (Setaria italica) germplasm as determined by agronomic traits and microsatellite markers. AJCS 6(2):342–349

Lin HS, Chiang CY, Chang SB, Kuoh CS (2011) Development of simple sequence repeats (SSR) markers in Setaria italica (Poaceae) and cross-amplification in related species. Int J Mol Sci 12(11):7835–7845

Lin CS, Hsu CT, Yang LH, Lee LY, Fu JY, Cheng QW, Wu FH, Hsiao HC, Zhang Y, Zhang R, Chang W (2018) Application of protoplast technology to CRISPR/Cas9 mutagenesis: from single-cell mutation detection to mutant plant regeneration. Plant Biotechnol J 16(7):1295–1310

Liu MX, Xu Y, Yang TY, Qiao ZJ, Wang RY, Wang YY et al (2017) Development of species-specific microsatellite markers for broomcorn millet (*Panicum miliaceum* L.) via high-throughput sequencing. Adv Crop Sci Tech 5:297

Liu T, He J, Dong K, Wang X, Wang W, Yang P, Ren R, Zhang L, Zhang Z, Yang T (2020) QTL mapping of yield component traits on bin map generated from resequencing a RIL population of foxtail millet (Setaria italica). BMC Genomics 21:1–3

M'Ribu HK, Hilu KW (1996) Application of random amplified polymorphic DNA to study genetic diversity in Paspalum scrobiculatum L. (Kodo millet, Poaceae). Genet Resour Crop Evol 43:203–210

Maharajan T, Ajeesh Krishna TP, Rakkammal K, Ramakrishnan M, Ceasar SA, Ramesh M, Ignacimuthu S (2023) Identification of QTL associated with agro-morphological and phosphorus content traits in finger millet under differential phosphorus supply via linkage mapping. Agriculture 13(2):262

Manimekalai M, Dhasarathan M, Karthikeyan A, Murukarthick J, Renganathan VG, Thangaraj K, Vellaikumar S, Vanniarajan C, Senthil N (2018) Genetic diversity in the barnyard millet (Echinochola frumentacea) germplasms revealed by morphological traits and simple sequence repeat markers. Curr Plant Biol 14:71–78

Manjappa, Deshpande S, Rangaiah S, Gowda MVC (2018) Assessment of molecular diversity in an elite set of finger millet (*Eleusine coracana* (L.) Gaertn.) genotypes using SSR markers. Electron J Plant Breed 9 (2):564–576. https://doi.org/10.5958/0975-928X.2018.00069.8

Matias FI, Galli G, Correia Granato IS, Fritsche-Neto R (2017) Genomic prediction of autogamous and allogamous plants by SNPs and haplotypes. Crop Sci 57(6):2951–2958

Meniya VH, Padhiyar SM, Desai H, Kheni JK, Tomar RS (2023) Development and validation of microsatellite markers for barnyard millet obtained by partial genome assembly. Ann Arid Zone 62(1):83–89. https://doi.org/10.59512/aaz.2023.62.1.9

Millet EJ, Kruijer W, Coupel-Ledru A, Alvarez Prado S, Cabrera-Bosquet L, Lacube S, Charcosset A, Welcker C, van Eeuwijk F, Tardieu F (2019) Genomic prediction of maize yield across European environmental conditions. Nat Genet 51(6):952–956

Mir RR, Hiremath PJ, Riera-Lizarazu O, Varshney RK (2013) Evolving molecular marker technologies in plants: From RFLPs to GBS. In: Diagnostics in plant breeding. Springer Science+ Business, New York, pp. 229–247. ISBN 978-94-007-5686-1; 978-94-007-5687-8 (eBook)

Mohanta TK, Bashir T, Hashem A, Abd Allah EF, Bae H (2017) Genome editing tools in plants. Genes 8(12):399

Muthamilarasan M, Venkata Suresh B, Pandey G, Kumari K, Parida SK, Prasad M (2014) Development of 5123 intron-length polymorphic markers for large-scale genotyping applications in foxtail millet. DNA Res. 21(1):41–52

Naga BL, Mangamoori LN, Subramanyam S (2012) Identification and characterization of EST-SSRs in finger millet (Eleusine coracana (L.) Gaertn.). J Crop Sci Biotechnol 15:9–16

Narayanrao DR, Tomar RS, Sm P, Jasminkumar K, Ashish G, Chauhan NM, Singh SC, Upadhye V, Kuddus M, Kamble L, Hajare ST (2023) De novo transcriptome sequencing of drought tolerance–associated genes in little millet (Panicum sumatrense L.). Funct Integr Genomics 23(4):303

Natesan S, Venugopalan G, Selvamani SB, Angamuthu N (2020) Characterization of little millet (Panicum sumatrense) varieties using morphological descriptors and SSR based DNA fingerprinting. J Phytology 12:29–34

Ngom A, Gueye MC, Gueye M, Billot C, Calatayud C, Diop BM, Kane NA, Piquet M, Vigouroux Y, Zekraoui L, Mbaye MS (2018) Cross-species amplification of microsatellite loci developed Digitaria exilis Stapf in related Digitaria species. J Appl Biosci 129:12982–12995

Nirgude M, Babu BK, Shambhavi Y, Singh UM, Upadhyaya HD, Kumar A (2014) Development and molecular characterization of genic molecular markers for grain protein and calcium content in finger millet (Eleusine coracana (L.) Gaertn.). Mol Biol Rep 41:1189–1200

Olodo KF, Gueye MC, Calatayud C, Diop BM, Kane NA, Ngom A, Ntui VO, Barreto MM, Uyoh EA, Abraham S, Vigouroux Y (2019) EST-SSR development for Digitaria exilis and its relatives D. iburua and D. longiflora from transcriptome sequences. Plant Genetic Resour 17(3):280–284

Padhiyar SM, Kheni J, Bhatt SB, Desai H, Tomar RS (2023) Comparative transcriptome profiling of high and low grain-iron containing Indian barnyard millet (Echinochloa frumentacea L.) genotypes during different stages of grain development

Pandey G, Misra G, Kumari K, Gupta SA, Parida SK, Chattopadhyay DE, Prasad MA (2013a) Genome-wide development and use of microsatellite markers for large-scale genotyping applications in foxtail millet [Setaria italica (L.)]. DNA Res 20(2):197–207

Pandey MK, Shobha Rani N, Sundaram RM, Laha GS, Madhav MS, Srinivasa Rao K, Sudharshan I, Hari Y, Varaprasad GS, Subba Rao LV, Suneetha K (2013b) Improvement of two traditional basmati rice varieties for bacterial blight resistance and plant stature through morphological and marker-assisted selection. Mol Breed 31:239–246

Panwar P, Nath M, Yadav VK, Kumar A (2010a) Comparative evaluation of genetic diversity using RAPD, SSR and cytochrome P450 gene based markers with respect to calcium content in finger millet (Eleusine coracana L. Gaertn.). J Genet 89:121–133

Panwar P, Saini RK, Sharma N, Yadav D, Kumar A (2010b) Efficiency of RAPD, SSR and cytochrome P 450 gene based markers in accessing genetic variability amongst finger millet (Eleusine coracana) accessions. Mol Biol Rep 37:4075–4082

Panwar P, Jha AK, Pandey PK, Gupta AK, Kumar A (2011) Functional markers based molecular characterization and cloning of resistance gene analogs encoding NBS-LRR disease resistance proteins in finger millet (Eleusine coracana). Mol Biol Rep 38:3427–3436

Parvathi MS, Nataraja KN, Reddy YN, Naika MB, Gowda MC (2019) Transcriptome analysis of finger millet (Eleusine coracana (L.) Gaertn.) reveals unique drought responsive genes. J Genet 98(2):46

Pendergast TH IV, Qi P, Odeny DA, Dida MM, Devos KM (2022) A high-density linkage map of finger millet provides QTL for blast resistance and other agronomic traits. Plant Genome 15(1):e20175

Qi P, Gimode D, Saha D, Schröder S, Chakraborty D, Wang X, Dida MM, Malmberg RL, Devos KM (2018) UGbS-flex, a novel bioinformatics pipeline for imputation-free SNP discovery in polyploids without a reference genome: finger millet as a case study. BMC Plant Biol 18:1–9

Rahman H, Jagadeeshselvam N, Valarmathi R, Sachin B, Sasikala R, Senthil N, Sudhakar D, Robin S, Muthurajan R (2014) Transcriptome analysis of salinity responsiveness in contrasting genotypes of finger millet (*Eleusine coracana* L.) through RNA-sequencing. Plant Mol Biol 85(4–5):485–503. https://doi.org/10.1007/s11103-014-0199-4

Rajput SG, Plyler-Harveson T, Santra DK (2014) Development and characterization of SSR markers in proso millet based on switchgrass genomics. Am J Plant Sci 5:175

Rajput SG, Santra DK, Schnable J (2016) Mapping QTLs for morpho-agronomic traits in proso millet (Panicum miliaceum L.). Mol Breed 36:1–8

Ramakrishnan M, Antony Ceasar S, Duraipandiyan V, Al-Dhabi NA, Ignacimuthu S (2016) Assessment of genetic diversity, population structure and relationships in Indian and non-Indian genotypes of finger millet (Eleusine coracana (L.) Gaertn) using genomic SSR markers. Springerplus 5:1–1

Reddy INBL, Sivaramakrishnan S (2017) Identification of SSR markers which could differentiate blast disease resistance accessions in finger millet (*Eleusine coracana* (L.) Gaertn.). J Crop Sci Biotechnol 20:37–44. https://doi.org/10.1007/s12892-016-0091-2

Renganathan VG, Vanniarajan C, Senthil N, Nirmalakumari A, Karthikeyan A, Veni K, Ramalingam J (2021) Genetics and molecular markers for anthocyanin pigmentation in barnyard millet (Echinochloa frumentacea (Roxb.) Link). Plant Breed 140(2):246–253

Ruperao P, Thirunavukkarasu N, Gandham P, Selvanayagam S, Govindaraj M, Nebie B, Manyasa E, Gupta R, Das RR, Odeny DA, Gandhi H (2021) Sorghum pan-genome explores the functional utility for genomic-assisted breeding to accelerate the genetic gain. Front Plant Sci 12:666342

Sato K, Mukainari Y, Naito K, Fukunaga K (2013) Construction of a foxtail millet linkage map and mapping of spikelet-tipped bristles 1 (stb1) by using transposon display markers and simple sequence repeat markers with genome sequence information. Mol Breed 31:675–684

Septiningsih EM, Pamplona AM, Sanchez DL, Neeraja CN, Vergara GV, Heuer S, Ismail AM, Mackill DJ (2009) Development of submergence-tolerant rice cultivars: the Sub1 locus and beyond. Ann Bot 103(2):151–160

Sharma D, Tiwari A, Sood S, Jamra G, Singh NK, Meher PK, Kumar A (2018) Genome wide association mapping of agro-morphological traits among a diverse collection of finger millet (Eleusine coracana L.) genotypes using SNP markers. PLoS One 13(8):e0199444

Shekhar S, Prasad AS, Banjare K, Kaushik A, Mannade AK, Dubey M, Patil A, Premi V, Vishwakarma AK, Sao A, Saxena RR (2023) LMTdb: a comprehensive transcriptome database for climate-resilient, nutritionally rich little millet (Panicum sumatrense). Front Plant Sci 14:1106104

Singh RK, Phanindra ML, Singh VK, Raghavendrarao S, Solanke AU, Kumar PA (2014) Isolation and characterization of drought responsive EcDehydrin7 gene from finger millet (Eleusine coracana (L.) Gaertn.). Indian J Genetics Plant Breeding 74(04):456–462

Sinha P, Singh VK, Saxena RK, Khan AW, Abbai R, Chitikineni A, Desai A, Molla J, Upadhyaya HD, Kumar A, Varshney RK (2020) Superior haplotypes for haplotype-based breeding for drought tolerance in pigeon pea (Cajanus cajan L.). Plant Biotechnol J 18(12):2482–2490

Sood S, Joshi DC, Rajashekara H, Tiwari A, Bhinda MS, Kumar A, Kant L, Pattanayak A (2023) Deciphering the genomic regions governing major agronomic traits and blast resistance using genome wide association mapping in finger millet. Gene 854:147115

Srivastava RK, Singh RB, Pujarula VL, Bollam S, Pusuluri M, Chellapilla TS, Yadav RS, Gupta R (2020) Genome-wide association studies and genomic selection in pearl millet: advances and prospects. Front Genet 10:1389

Sun M, Yan H, Zhang A, Jin Y, Lin C, Luo L, Wu B, Fan Y, Tian S, Cao X, Wang Z (2023) Milletdb: a multi-omics database to accelerate the research of functional genomics and molecular breeding of millets. Plant Biotechnol J 11:2348–2357

Suresh BV, Choudhary P, Aggarwal PR, Rana S, Singh RK, Ravikesavan R, Prasad M, Muthamilarasan M (2022) De novo transcriptome analysis identifies key genes involved in dehydration stress response in kodo millet (Paspalum scrobiculatum L.). Genomics 114(3):110347

Tiwari A, Sharma D, Sood S, Jaiswal JP, Pachauri SP, Ramteke PW, Kumar A (2020) Genome-wide association mapping for seed protein content in finger millet (Eleusine coracana) global collection through genotyping by sequencing. J Cereal Sci 91:102888

Tsai KJ, Lu MY, Yang KJ, Li M, Teng Y, Chen S, Ku MS, Li WH (2016) Assembling the Setaria italica L. Beauv. genome into nine chromosomes and insights into regions affecting growth and drought tolerance. Sci Rep 6(1):35076

Upadhyaya HD, Gowda CL, Reddy VG, Singh S (2008) Diversity of small millets germplasm in genebank at ICRISAT. University of Southampton, Southampton

Upadhyaya HD, Pundir RP, Dwivedi SL, Gowda CL (2009) Mini core collections for efficient utilization of plant genetic resources in crop improvement programs. In: Information bulletin no. 78. International Crops Research Institute for the Semi-Arid Tropics 2009

Upadhyaya HD, Dronavalli N, Gowda CL, Singh S (2012) Mini core collections for enhanced utilization of genetic resources in crop improvement. Indian J Plant Genet Resour 25(1):111–124

VanBuren R, Man Wai C, Wang X, Pardo J, Yocca AE, Wang H, Chaluvadi SR, Han G, Bryant D, Edger PP, Messing J (2020) Exceptional subgenome stability and functional divergence in the allotetraploid Ethiopian cereal teff. Nat Commun 11(1):884

Varshney RK, Graner A, Sorrells ME (2005) Genic microsatellite markers in plants: features and applications. Trends Biotechnol 23(1):48–55

Varshney RK, Glaszmann JC, Leung H, Ribaut JM (2010) More genomic resources for less-studied crops. Trends Biotechnol 28(9):452–460

Varshney RK, Bohra A, Roorkiwal M, Barmukh R, Cowling WA, Chitikineni A, Lam HM, Hickey LT, Croser JS, Bayer PE, Edwards D (2021) Fast-forward breeding for a food-secure world. Trends Genet 37(12):1124–1136

Vetriventhan M, Upadhyaya HD (2018) Diversity and trait-specific sources for productivity and nutritional traits in the global proso millet (Panicum miliaceum L.) germplasm collection. Crop J 6(5):451–463

Vetriventhan M, Upadhyaya HD, Anandakumar CR, Senthilvel S, Parzies HK, Bharathi A, Varshney RK, Gowda CL (2012) Assessing genetic diversity, allelic richness and genetic relationship among races in ICRISAT foxtail millet core collection. Plant Genetic Resources 10(3):214–223

Vetriventhan M, Azevedo VC, Upadhyaya HD, Nirmalakumari A, Kane-Potaka J, Anitha S, Ceasar SA, Muthamilarasan M, Bhat BV, Hariprasanna K, Bellundagi A (2020) Genetic and genomic resources, and breeding for accelerating improvement of small millets: current status and future interventions. Nucleus 63:217–239

Voss-Fels KP, Herzog E, Dreisigacker S, Sukumaran S, Watson A, Frisch M, Hayes B, Hickey LT (2019) "SpeedGS" to accelerate genetic gain in spring wheat. In: Applications of genetic and genomic research in cereals. Woodhead Publishing, pp 303–327

Wallace JG, Upadhyaya HD, Vetriventhan M, Buckler ES, Tom Hash C, Ramu P (2015) The genetic makeup of a global barnyard millet germplasm collection. Plant Genome 8(1):plantgenome2014-10

Wang ZM, Devos KM, Liu CJ, Wang RQ, Gale MD (1998) Construction of RFLP-based maps of foxtail millet, Setaria italica (L.) P. Beauv Theor Appl Genet 96:31–36

Wang C, Jia G, Zhi H, Niu Z, Chai Y, Li W, Wang Y, Li H, Lu P, Zhao B, Diao X (2012) Genetic diversity and population structure of Chinese foxtail millet [Setaria italica (L.) Beauv.] landraces. G3 2(7):769–777

Wang M, Li P, Li C, Pan Y, Jiang X, Zhu D, Zhao Q, Yu J (2014) SiLEA14, a novel atypical LEA protein, confers abiotic stress resistance in foxtail millet. BMC Plant Biol 14:1–6

Wang J, Wang Z, Du X, Yang H, Han F, Han Y, Yuan F, Zhang L, Peng S, Guo E (2017a) A high-density genetic map and QTL analysis of agronomic traits in foxtail millet [Setaria italica (L.) P. Beauv.] using RAD-seq. PLoS One 12(6):e0179717

Wang X, Li L, Yang Z, Zheng X, Yu S, Xu C, Hu Z (2017b) Predicting rice hybrid performance using univariate and multivariate GBLUP models based on North Carolina mating design II. Heredity 118(3):302–310

Wang J, Li S, Lan L, Xie M, Cheng S, Gan X, Huang G, Du G, Yu K, Ni X, Liu B (2021a) De novo genome assembly of a foxtail millet cultivar Huagu11 uncovered the genetic difference to the cultivar Yugu1, and the genetic mechanism of imazethapyr tolerance. BMC Plant Biol 21(1):271

Wang X, Chen S, Ma X, Yssel AE, Chaluvadi SR, Johnson MS, Gangashetty P, Hamidou F, Sanogo MD, Zwaenepoel A, Wallace J (2021b) Genome sequence and genetic diversity analysis of an under-domesticated orphan crop, white fonio (Digitaria exilis). GigaScience 10(3):giab013

Watson A, Ghosh S, Williams MJ, Cuddy WS, Simmonds J, Rey MD, Asyraf Md Hatta M, Hinchliffe A, Steed A, Reynolds D, Adamski NM (2018) Speed breeding is a powerful tool to accelerate crop research and breeding. Nature plants 4(1):23–29

Wu D, Shen E, Jiang B, Feng Y, Tang W, Lao S, Jia L, Lin HY, Xie L, Weng X, Dong C (2022) Genomic insights into the evolution of Echinochloa species as weed and orphan crop. Nat Commun 13(1):689

Xu Y, Li P, Zou C, Lu Y, Xie C, Zhang X, Prasanna BM, Olsen MS (2017) Enhancing genetic gain in the era of molecular breeding. J Exp Bot 68(11):2641–2666

Yabe S, Iwata H (2020) Genomics-assisted breeding in minor and pseudo-cereals. Breed Sci 70(1):19–31. https://doi.org/10.1270/jsbbs.19100

Yadav CB, Muthamilarasan M, Pandey G, Khan Y, Prasad M (2014) Development of novel microRNA-based genetic markers in foxtail millet for genotyping applications in related grass species. Mol Breed 34:2219–2224

Yadav CB, Bonthala VS, Muthamilarasan M, Pandey G, Khan Y, Prasad M (2015) Genome-wide development of transposable elements-based markers in foxtail millet and construction of an integrated database. DNA Res 22(1):79–90

Yadav Y, Lavanya GR, Pandey S, Verma M, Ram C, Arya L (2016) Neutral and functional marker based genetic diversity in kodo millet (Paspalum scrobiculatum L.). Acta Physiol Plant 38:1–2

Yazdizadeh M, Fahmideh L, Mohammadi-Nejad G, Solouki M, Nakhoda B (2020) Association analysis between agronomic traits and AFLP markers in a wide germplasm of proso millet (Panicum miliaceum L.) under normal and salinity stress conditions. BMC Plant Biol 20:1–8

Yu JK, Kantety RV, Graznak E et al (2006) A genetic linkage map for tef [Eragrostis tef (Zucc.) Trotter]. Theor Appl Genet 113:1093–1102. https://doi.org/10.1007/s00122-006-0369-0

Yu JK, Graznak E, Breseghello F, Tefera H, Sorrells ME (2007) QTL mapping of agronomic traits in tef [Eragrostis tef (Zucc) trotter]. BMC Plant Biol 7:1–3

Zampieri E, Volante A, Marè C, Orasen G, Desiderio F, Biselli C, Canella M, Carmagnola L, Milazzo J, Adreit H, Tharreau D (2023) Marker-assisted pyramiding of blast-resistance genes in a japonica elite rice cultivar through forward and background selection. Plan Theory 12(4):757

Zeid M, Yu JK, Goldowitz I, Denton ME, Costich DE, Jayasuriya CT, Saha M, Elshire R, Benscher D, Breseghello F, Munkvold J (2010) Cross-amplification of EST-derived markers among 16 grass species. Field Crop Res 118(1):28–35

Zeid M, Belay G, Mulkey S, Poland J, Sorrells ME (2011) QTL mapping for yield and lodging resistance in an enhanced SSR-based map for tef. Theor Appl Genet 122:77–93

Zeid M, Assefa K, Haddis A, Chanyalew S, Sorrells ME (2012) Genetic diversity in tef (Eragrostis tef) germplasm using SSR markers. Field Crop Res 127:64–70

Zeinullina A, Rysbekova A, Dyussibayeva E, Zhirnova I, Zhanbyrshina N, Zhunusbayeva Z (2023) Application of sodium azide for chemical induced mutagenesis of proso millet culture (*Panicum miliaceum* L.). Bulgarian J Agric Sci 29(4):623–631

Zhang D, Ayele M, Tefera H, Nguyen HT (2001) RFLP linkage map of the Ethiopian cereal tef [Eragrostis tef (Zucc) trotter]. Theor Appl Genet 102:957–964
Zhang G, Liu X, Quan Z, Cheng S, Xu X, Pan S, Xie M, Zeng P, Yue Z, Wang W, Tao Y (2012) Genome sequence of foxtail millet (Setaria italica) provides insights into grass evolution and biofuel potential. Nat Biotechnol 30(6):549–554
Zhang S, Tang C, Zhao Q, Li J, Yang L, Qie L, Fan X, Li L, Zhang N, Zhao M, Liu X (2014) Development of highly polymorphic simple sequence repeat markers using genome-wide microsatellite variant analysis in foxtail millet [Setaria italica (L.) P. Beauv.]. BMC Genomics 15:1–9
Zhang X, Pérez-Rodríguez P, Burgueño J, Olsen M, Buckler E, Atlin G, Prasanna BM, Vargas M, San Vicente F, Crossa J (2017) Rapid cycling genomic selection in a multiparental tropical maize population. G3 7(7):2315–2326
Zhang Y, Malzahn AA, Sretenovic S, Qi Y (2019) The emerging and uncultivated potential of CRISPR technology in plant science. Nature Plants 5(8):778–794
Zhang W, Zhi H, Tang S, Zhang H, Sui Y, Jia G, Wu C, Diao X (2021) Identification of no pollen 1 provides a candidate gene for heterosis utilization in foxtail millet (Setaria italica L.). Crop J 9(6):1309–1319
Zhao Qing Ying ZQ, Zhang Rui Juan ZR, Wang Rui Liang WR, Gao Jian Hua GJ, Han Yuan Huai HY, Yang Zhi Rong YZ, Wang Xing Chun WX (2018) Genome-wide identification of molecular markers based on genomic re-sequencing of foxtail millet elite cultivar Jingu 21. Acta Agronomica Sinica 44:686
Zou C, Li L, Miki D, Li D, Tang Q, Xiao L, Rajput S, Deng P, Peng LI, Jia W, Huang R (2019) The genome of broomcorn millet. Nat Commun 10(1):436

Genome Sequencing Efforts in Minor Millets: Current Knowledge and Emerging Insights

Roshan Kumar Singh, Anurag Panchal, Mehanathan Muthamilarasan, and Manoj Prasad

Abstract

Small millets (or minor millets) include finger millet (*Eleusine coracana*), foxtail millet (*Setaria italica*), proso millet (*Panicum miliaceum*), barnyard millet (*Echinochloa crus-galli*), kodo millet (*Paspalum scrobiculatum*), little millet (*Panicum sumatrense*), teff (*Eragrostis tef*), fonio (*Digitaria exilis*), job's tears (*Coix lacryma-jobi*), guinea millet (*Brachiaria deflexa*), and browntop millet (*Urochloa ramosa*). These millets are highly nutritious and climate-resilient but marginally cultivated for the production and consumption of particular communities. Though called "poor men's crops," minor millets possess the potential to ensure food and nutritional security amid the threat of global climate change. Thus, scope exists to improve the agronomic traits of these minor millets for commercial cultivation; however, lack of genomic resources remains a bottleneck to this advancement. Genome sequencing not only provides an opportunity to decode the genes encoded by the genome, but also provides avenue for the development of genomic resources. The success of genome

R. K. Singh · A. Panchal
National Institute of Plant Genome Research, Aruna Asaf Ali Marg, New Delhi, India

M. Muthamilarasan
Department of Plant Sciences, School of Life Sciences, University of Hyderabad, Hyderabad, Telangana, India

M. Prasad (✉)
National Institute of Plant Genome Research, New Delhi, India

Department of Genetics, University of Delhi South Campus, New Delhi, India
e-mail: manoj_prasad@nipgr.ac.in; manojprasad@south.du.ac.in

D. C. Joshi et al. (eds.), *Minor Millets*,
https://doi.org/10.1007/978-981-96-4265-6_11

sequencing for resource development and further implementation of these resources have been proven in other crop plants. Among minor millets, genomes of a few species have been sequenced, including finger millet, foxtail millet, proso millet, barnyard millet, teff, fonio, and job's tears. However, the genomes of kodo millet, little millet, guinea millet, and browntop millet remains to be sequenced. In this context, the chapter summarizes the outcomes of sequencing efforts and the application of genome sequence information in accelerating genomics studies in minor millets. The chapter also enumerates the status of transcriptome sequencing and its application in dissecting the genes underlying important traits.

Keywords

Minor millets · Genome sequencing · Transcriptome · Climate-resilience · Crop improvement

1 Introduction

Millets belong to the family Poaceaeare, one of the ancient crops of human civilization and commonly recognized for their nutritional potential and climate resilience. These crops are largely cultivated under marginal/low-fertile soils with limited external application of pesticides and chemical fertilizers. Almost all millets are rain-fed crops and are cultivated in arid–semiarid regions that receive low rainfall and thus provide an alternate source for sustainable agriculture and food security amid global climate change (Bandyopadhyay et al. 2023). Millets are prevalently known as nutri-cereals as their grains are rich in several essential micro- and macronutrients required by the human body. The nutrient content in terms of carbohydrates, proteins, minerals, dietary fibers, antioxidants, and vitamins in millets are significantly manifold higher than that of rice, wheat, and maize (Dhaka et al. 2021). The higher proportion of resistant starch in small millets are making them a suitable food for person with type 2 diabetes mellitus (Rajasekaran et al. 2004; Muthamilarasan et al. 2016). The gluten-free attributes of millets provide an alternative food option for the people with celiac disease. Higher lignocellulosic biomass in most of the millets including foxtail-, little-, and pearl millet can be exploited for the generation of biofuels from lignocellulosic secondary cell-wall biomass. Mainstreaming millet cultivation and consumption can accelerate achieving the goal of zero hunger by the year 2030.

On the basis of grain size and area of cultivation millets are broadly classified into major and minor millets. Major millets comprise of sorghum (*Sorghum bicolor*; jowar) and pearl millet (*Pennisetum glaucum*; bajra) while finger millet (*Eleusine coracana*; ragi), foxtail millet (*Setaria italica*; kangni), little millet (*Panicum sumatrance*; kutki), kodo millet (*Paspalum scrobiculatum*; kedrav), barnyard millet (*Echinochloa frumentacea*; sawan/jhangora), and broomcorn millet (*Panicum miliaceum*) are considered among minor millets. Both major and minor millets have shorter lifecycle (3–4 months), inbreeding nature, C4 photosynthesis, excellent water use efficiency (WUE), can produce large number of seeds, infected by limited number of pathogens compared to major crops, and can adapt in extreme environmental conditions (Singh et al. 2021). The water requirement for millets cultivation is more than 50% less as compared to rice and wheat and thus they are considered drought-tolerant crops. Minor millets host several unique sets of genes that make them withstand adverse environmental conditions and nutritionally rich crops. However, these crops are being underutilized and research neglected for a long time. The availability of genome sequence information is prerequisite to investigate the molecular regulation behind stress response, nutritional superiority, and other essential agronomic traits. The minor millets stand far behind in terms of genomic resource development and gene characterization when compared with major crops including rice, wheat, or maize. Rice was the first cereal to be sequenced at whole genome level at the early 2000s (Yu et al. 2002; Goff et al. 2002). From then, it took almost 10 years to release a millet genome sequence. Foxtail millet was the first millet whose genome was sequenced by the two independent groups, Joint Genome Institute (JGI) of the Department of Energy, USA and Beijing Genome Initiative, China, in the year 2012 (Bennetzen et al. 2012; Zhang et al. 2012). Following foxtail millets, genomes of finger-, pearl-, and broomcorn millets were also sequenced in due course of time (Table 1; Fig. 1). The genome sequence information from these crops has led to the discovery of molecular markers and several novel alleles that are being utilized to further crop improvement. Here, we have discussed the genome sequencing efforts in minor millets and analyzed the sequence information in order to comprehend the genomic attributes of these crops.

2 Genome Sequencing and Genetic Architecture of Minor Millets

2.1 Foxtail Millet (*Setaria italica*)

Foxtail millet is primarily grown for food in tropical semiarid parts of Africa and Asia and as forage for Europe, Australia, and Northa America. It has been chosen as the first millet crop for the genome sequencing due to its small sized genome (~430 Mb), self-pollinating nature, and relatively low content of repetitive DNA (Lata et al. 2013). The release of sequence of *Setaria* genome has become as one of the milestones in millet genomics that prompted the genomic research in other millets as well. The foxtail millet accession "Zhang gu" was used for sequencing by the Beijing Genome Initiative group and identified as total of 38,801 genes while "Yugu1" was sequenced by the Joint Genome Institute (JGI) of the Department of Energy, USA, and predicted the presence of predicted the presence of 35,500–36,000 protein-coding genes in foxtail millets. Genes for non-coding RNAs were also

Table 1 Status of genome sequencing in minor millets

Common name	Organism name	Release year	Size (Mb)	Ploidy level	Sequence level	BioSample ID	BioProject ID	Reference
Job's tears	*Coix lacryma-jobi* var. *lacryma-jobi*	2019	1731.46	Tetraploid	Chromosome	SAMN11865572	PRJNA544872	Liu et al. (2019)
	Coix lacryma-jobi var. *ma-yuen*	2019	1280.02		Scaffold	SAMN12817652	PRJNA573577	–
Fonio	*Digitaria exilis*	2020	760.651	Tetraploid	Contig	SAMN15456850	PRJNA640067	–
		2020	716.471		Chromosome	SAMEA6542037	PRJEB36539	Abrouk et al. (2020)
		2021	716.471		Chromosome	SAMEA6797793	PRJEB36539	
Barnyard millet	*Echinochloa crus-galli*	2021	1340.74	Hexaploid	Chromosome	SAMN20340778	PRJNA748695	–
		2020	1340.77		Scaffold	SAMN16274019	PRJNA665921	Ye et al. (2020)
		2017	1486.61		Scaffold	SAMEA104207156	PRJEB22102	–
Finger millet	*Eleusine coracana* subsp. *coracana*	2018	1229.37	Tetraploid	Scaffold	SAMD00076255	PRJDB5606	Hatakeyama et al. (2018a, b)
		2017	1195.99		Scaffold	SAMN04849255	PRJNA318349	–
Teff	*Eragrostis tef*	2020	575.08	Tetraploid	Chromosome	SAMN28611010	PRJNA841677	VanBuren et al. (2020)
		2015	607.32		Scaffold	SAMN02872936	PRJNA253673	–
Proso millet	*Panicum miliaceum*	2022	862.02	Tetraploid	Chromosome	SAMN30451036	PRJNA872304	–
		2018	848.352		Chromosome	SAMN08335224	PRJNA429322	–
		2018	854.793		Chromosome	SAMN08389585	PRJNA431363	Zou et al. (2019)
Foxtail millet	*Setaria italica*	2016	477.542	Diploid	Chromosome	SAMN04534922	PRJNA314430	–
		2012	405.868		Chromosome	SAMN02981383	PRJNA32913	Bennetzen et al. (2012)

–: Not available (not published)

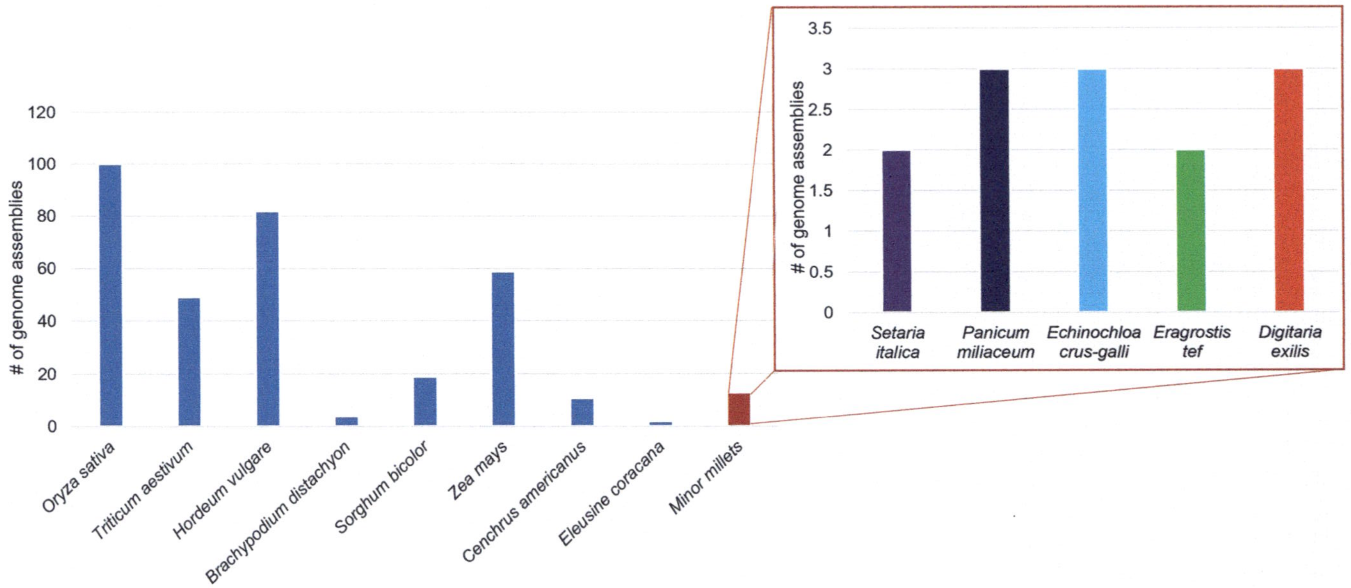

Fig. 1 Proportion of genome assemblies deposited in NCBI genome database showing a comparison between major cereals, major millets, and minor millets (inner box)

identified, which showed the presence of a total of 159 miRNA encoding genes, 704 tRNA genes, 382 genes for small nuclear RNA transcripts, and 99 rRNA genes (18S, 28S, 5.8S, and 5S) in foxtail millet genome (Zhang et al. 2012). Approximately 40% of *Setaria* genome is estimated to be composed of transposable elements that are nearly similar to that of rice (~40%), but relatively lower than maize and wheat (more than 80%) (Bennetzen et al. 2012). In-depth analysis of *Setaria* genome has also identified six gene-clusters associated with the water-deficit conditions and possess a higher number of drought-responsive genes than drought susceptible crops such as rice and wheat (Muthamilarasan and Prasad 2017). The identified genes in this clusters are possible candidates for the functional characterization and could be introgressed to susceptible crops to make them drought tolerant.

Foxtail millet genome when compared with genomes of other grasses including rice, maize, sorghum, and Brachypodium, maximum collinearity was found between foxtail millet-maize (86.7%), followed by foxtail millet-sorghum (71.2%), and foxtail millet-rice (71.8%) genomes (Muthamilarasan et al. 2017). This supports that foxtail millet is evolutionary more closure to wheat and sorghum than rice. In addition to synteny maps at genome level, several comparative maps were constructed between foxtail millet and other cereals genome at molecular maker and gene family levels (Pandey et al. 2013; Yadav et al. 2014, 2015; Muthamilarasan et al. 2014). Unlike the comparative maps at genome levels, the majority of these analyses revealed the higher synteny percentage between foxtail millet and sorghum followed by maize and rice. The marker-based comparative maps are valuable in the map-based mining of novel alleles/genes, controlling improved agronomic traits from foxtail millet using marker-based genotyping data of other related cereal crops.

2.2 Finger Millet (*Eleusine coracana*)

Finger millet is another majorly cultivated millet crops in rural regimes of southern India and Africa. Finger millet plant body is slender, highly tillering terminated by the characteristic finger like inflorescences. Their grains can be stored for longer period of time as they are resistant to insect pests, making the crop an important source of food during famine. The whole sequencing of tetraploid genome of finger millet ML-365 cultivar yielded 1196 Mb, covering approximately 82% of total estimated genome size of ~1.2 Gb (Hittalmani et al. 2017). The combination of Illumina and SOLiD sequencing technologies was used in case of finger millet. Genome analysis estimated a total number of 85,243 genes in finger millet, whereas one-half of the genome (~49.92%) is repetitive in nature. Among repeat sequence, the retroelements constitute around 35.56% and long terminal repeat (LTR) elements were the major components (33.26%) among retroelements. Comparative mapping revealed higher collinearity between the genome of finger millet with that of foxtail millet and rice as compared to other cereal species. The average GC content of finger millet genome was calculated to ~44.76%. Approximately 6500 transposable element-related genes were identified.

Simple sequence repeats (SSRs) play a substantial role in defining evolutionary relationship among genomes. In total, 114,083 SSRs were identified across the genome of ML-365 of which 66,805 (58.56%) were di-, 40,578 (35.57%) tri-, 2179 (1.91%) tetra-, 3010 (2.64%) penta-, and 1511 (1.32%) were hexa-repeat types (Gimode et al. 2016). The abundance of AG/CT as di-, AAG/CTT as tri-, AAAT/ATTT as tetra-, AAAAG/CTTTT as penta-, and AACACC/GGTGTT as hexa-repeats were found to be hinger among others. These SSR markers could be utilized in the construction of linkage maps, diversity studies, association studies, QTL mapping, and marker-assisted selection in molecular breeding programs. The Pfam analysis of finger millet transcriptome identified more than 2800 drought responsive genes distributed across 19 Pfam domains (Hittalmani et al. 2017). The gene families such as NAD-dependent epimerase/dehydratase, protein kinases, U-box containing proteins, universal stress proteins, DCPS domains containing genes were major among them. The study also identified 146 key C_4-specific enzymes encoding genes that include 31 NAD-MDH, 21 NADP-ME, 34 CA, 27 PEPC, 7 PPDK-RP, 17 PPDK, and 9 PPCK. The occurrence of multiple copies of these genes indicates possible gene duplication during the evolution of C_4 photosynthesis.

2.3 Broomcorn Millet (*Panicum miliaceum*)

Broomcorn millet (or proso millet) is one of the highly drought-tolerant cereal crops broadly cultivated in semiarid regions of Asis, Africa, and Europe. Its domestication was reported to be stared from Northern China and it is one among the oldest cultivated crop plants (Barton et al. 2009). Broomcorn millet has superior WUE than all the other cultivated cereal crops (Washburn et al. 2015). It is commonly used as summer rotation crop or in dryland farming in drought prone areas. The genome of a Northern Chinese landrace of broomcorn millet was sequenced and assembled through PacBio sequencing platform (Zou et al. 2019). It is an allotetraploid species with 36 chromosomes ($2n = 4\times = 36$) and the total size of the genome was estimated to be ~923 Mb. Phylogenetic analyses revealed that the two ancestral genomes of broomcorn millet diverged ~5.6 million years ago. The genome of broomcorn millet has 58.2% repeat content of which 92.1% consists of transposable elements only. A total of 112,158 SSR were also identified with a mean occurrence frequency of 22.5 per Mb. Genome analysis estimated approximately 55,930 protein-coding and 339 microRNA (miRNA) genes in broomcorn millet. In addition to that 1640 ribosomal RNAs, 1420 transfer RNAs, and 2302 small nuclear RNAs, encoding genes were also identified. Around 74.5% of broomcorn millet gene families were shared among other grasses including maize, foxtail millet, rice, and, sorghum, whereas only 4.2% of the gene families were unique to the broomcorn millet. More than 73% of the gene models identified in broomcorn millet contained conserved domains listed in the Pfam database.

Phylogenetic analysis exploiting single copy homologous genes indicates its close evolutionary relationship between foxtail millet followed by maize and sorghum. A large number of syntenic regions (604) were identified in broomcorn millet by comparing it to the genome of foxtail millet with an average size of 1.31 Mb, whereas minimum synteny blocks were detected when compared to sorghum. All the genes involved in C_4 photosynthesis (except for dicarboxylate transport 2 and mitochondrial pyruvate carrier) present in broomcorn millet were found to be in almost two-fold higher copy number than the foxtail millet. It was estimated that broomcorn- and foxtail millet diverged from a common ancestor approximately 18 million years ago.

2.4 Barnyard Millet (*Echinochloa crus-galli*)

First draft of barnyard millet genome was reported by Guo et al. (2017) and an improved version was reported by Ye et al. (2020). The genomic DNA of *E. crus-galli* line STB08 was sequenced using Illumina HiSeq 2000 system and Pacbio RS II system to generate 240.3 Gb data that represented 90.7% of genome (Guo et al. 2017). Also, RNA-seq was performed for gene annotation, and the study predicted 108,771 genes encoded by barnyard millet genome. Also, the analysis predicted that long terminal retrotransposons accounted for 21.9% of the genome. To further improve the genome assembly, Ye et al. (2020) sequenced the genomes of *E. oryzicola*, a tetraploid progenitor of barnyard millet, and a diploid species, *E. haploclada*on PacBio Sequel platform. Further, genome annotation, subgenome construction, divergence time estimation, gene family analysis, and selection pressure analysis were performed.

2.5 Teff (*Eragrostis tef*)

The variety, Tsedey (DZ-Cr-37) of teff, was sequenced using 454-FLX pyrosequencing and Illumina platforms to construct genome assembly (Cannarozzi et al. 2014). Parallelly, transcriptome assemblies were also prepared for integrating the genome data for gene prediction. The project achieved an assembly of 672 Mb, representing 87% of the genome. Further, the sequence information was used to mine SSR markers and the genome-wide microsatellite markers were used to study important agronomic and disease resistance traits (Cannarozzi et al. 2014). VanBuren et al. (2020) improved the genomic assembly through long-read sequencing (PacBio) and long-range high-throughput chromatin capture (Hi-C). The genome of teff landrace, Dabbi, was sequenced to achieve 85X coverage of 622 Mb genome. Also, pseudomolecules were constructed using high-density SNP-based genetic map followed by analysis of repetitive elements.

2.6 Fonio (*Digitaria exilis*)

Fonio millet genome was released in 2020 (Abrouk et al. 2020), wherein the DNA of reference accession, CM05836, was sequenced. The Illumina Hi-Seq 2500 and Illumina NovaSeq 6000 platforms were used for sequencing followed by the use of Hi-C to construct chromosome-scale assembly. In addition, the DNA of 641 *D. exilis* accessions along with a few *D. longiflora* accessions were sequenced for variant calling to identify SNPs. This data was further used to analyze genetic diversity and population structure. The authors also analyzed the association of climate, geography, and social factors with the population genetics structure, leading to demographic history reconstruction (Abrouk et al. 2020).

2.7 Job's Tears (*Coix lacryma-jobi*)

The genome of an elite coix cultivar, Beijing (BJ), was sequenced by Liu et al. (2019) on the Illumina HiSeq 2000, PacBio RS II, and Sequel platforms. To ensure the accuracy of the genome, BioNano and Hi-C libraries were constructed followed by FISH (fluorescence in situ hybridization). The analysis generated a genome of 1.73 Gb, comprising of 44,485 predicted protein-coding genes. Further, BJ was crossed to GXG to develop F_2 population, which was resequenced to develop GBS library and construction of genetic map.

3 Status of Transcriptome Sequencing in Minor Millets

A focused study of differentially expressed genes to know the transcriptional dynamics in order to identify key genes involved in developmental change and stress biology can be referred to as transcriptome sequencing analysis. This kind of omics analysis is helpful in giving the first insight into the changes in the gene expression at different developmental stage or the changes occurring in the plant at molecular level during its encounter to any kind of environmental change. Ultimately, this up- and downregulation of genes at different stages leads to the activation or repression of the downstream network or module. One such study provides the clues for the key genes involved in transcriptional dynamics during five different stages of grain filling in foxtail millet (Wang et al. 2020). In total, 11,399 genes including 902 transcription factors were found, which were differentially expressed and were involved in different pathways such as starch biosynthesis, polyamine metabolism, hormone signaling, and cell wall invertases.

One recent attempt was aimed to delineate the pathways associated with nitrogen fertilization during starch synthesis in proso millet (Wang et al. 2024). Co-expression analysis using waxy and non-waxy proso millet varieties revealed the regulation of starch synthesis and quality, which is controlled by nitrogen fertilization at S2 and S3 of grain filling stages. Nitrogen fertilization negatively affected the starch biosynthesis in grains but an increase in its degradation to maltose, dextrins, and then to glucose was observed. Foxtail millet, a naturally abiotic stress-tolerant crop plant, is affected by a biotrophic oomycete *Sclerospora graminicola.* A comparative transcriptome profiling was done in two contrasting cultivars of foxtail millet G1 and JG1 (Wang et al. 2022). Different leaf stages were used for RNA isolation that were 3, 5, and 7 leaf stages. A total of 473 and 1433 genes were found to be upregulated in resistant and susceptible cultivar, respectively. Their pathway analysis revealed highly enriched reads of genes belonging to glutathione metabolism, phenylalanine catabolism, cutin, suberin, and wax biosynthesis genes. This also goes along with the primary defense acquired by plant cell when attacked by foreign pathogen. Some defense-related genes were also found to be upregulated in resistant cultivars such as leucine-rich protein kinases, peroxidases, laccases, and some auxin response factors.

With this, the possible role of many transcription factor (TF) families have also been investigated in different abiotic stresses. For example, one study on NAC TFs in broomcorn millet was done to know the potential role of this family members during drought stress (Shan et al. 2020). Transcriptome analysis revealed the differential upregulation of NAC family members during drought stress, which was even confirmed by qRT-PCR of some selected genes. A number of genes were precisely expressed in seeds *PmNAC092*, *PmNAC051*, and *PmNAC005*, hinting their role in seed development and maturation. Foxtail millet is a plant that is often cultivated with Chinese chestnut. High tannins in foxtail millet can prevent the attack of red spiders on Chinese chestnut plants. To understand the molecular basis of high tannins in leaves, transcriptome analysis was done, and 335 genes were found to be differentially expressed (Li et al. 2022). Together, TFs were also predicted by PlantTFDB. Out of them 22.61% genes were involved in metabolic pathways, 14.78% in some secondary metabolite synthesis, 3.48% in phenylpropanoid biosynthesis pathway. TFs such as bHLH were found to be most upregulated and one bHLH TF *TRANSPARENT TESTA* had already been reported for its role in tannin accumulation in faba bean.

4 Application of Genomic Information in Gene Discoveries in Minor Millets

Through the genome-wide identification of gene families has been thoroughly carried out to analyze and compare the gene structure and function from other species (Puranik et al. 2013; Singh et al. 2016). This analysis generally helps in searching the genes in the genome and then analyzing them for their role in specific developmental stage and role in response to different types of environmental cues. Availability of genomic resource of a crop plant helps in NGS-based allele discovery, and finding superior genes to be introgressed in crop varieties to enhance the yield and productivity of crop plants. Most of the reports present mainly explains the role of minor millet genes in imparting abiotic stress tolerance. For example, overexpression of *SiMYB30* in rice confer low nitrogen stress in transgenic rice (Zhang et al. 2023); *SiMYB19* overexpression rice resulted in salinity stress tolerance in rice (Xu et al. 2022); *SiHSP21.9* overexpression in rice mitigates heat stress tolerance and maintains chloroplast proteome integrity (Singh et al. 2022); *SiATG8a* confers phosphorus starvation in transgenic wheat (He et al. 2023); and *SiNRX1* overexpression in *Arabidopsis* resulted in enhanced drought and salt stress demonstrated by higher survival rate and better growth (Zhang et al. 2022). The availability of genome sequence information has also enabled the application of genome editing in millets. Sequence polymorphism or insertion/deletion of short sequences responsible for altered agronomic traits can be identified and edited for desired phenotypes. Till date, CRISPR/Cas-mediated genome editing has been applied only to foxtail millet and is yet to be employed to other minor millets due to lack of standard reference genomes and their functional annotations, molecular characterization of their genes, and efficient genetic transformation method.

5 Conclusions

Minor millets are naturally tolerant to several environmental stresses with superior nutritional values. Recent advancements in biotechnological research and availability of genomic information of foxtail-, finger-, and broomcorn millets has enabled the exploration of these resources for crop improvement. However, the pace of molecular research for understanding the trait superiority in these crops are lagging behind in comparison to other cereals including rice, maize, and wheat. The sequenced minor millet genomes share high collinearity among themselves and then with neighboring grass species such as sorghum and brachypodium. The nutritional profiling, therapeutic potential, and stress physiology of little-, kodo-, and barnyard millets have been studied, however, the genetic determinants controlling these traits are not being identified due to lack of their draft genome sequences. Once the genome sequence information of these crops will release, ample opportunities would be created for the discovery of novel genes/alleles, resolving the genetic complexity of abiotic stress tolerance and molecular regulation of C_4 photosynthesis. Rice and maize are the only cereals where pangenome analysis has been done. Pangenome analysis of all the available small millet accessions may provide complete genomic information to further explore their genetic complexity and identification of overall gene pool.

Acknowledgments The authors' work in this study is supported by J.C. Bose National Fellowship Grant of Department of Science and Technology (File No.: JCB/2018/000001) and DBT Multi-institutional project entitled "Germplasm Characterization and Trait Discovery in Wheat using Genomics Approaches and its Integration for Improving Climate Resilience, Productivity and Nutritional Quality" under mission program of "Characterization of Genetic Resources" [BT/Ag/Network/Wheat/2019-20]. The authors are also thankful to DBT-eLibrary Consortium (DeLCON) for providing access to the e-resources.

References

Abrouk M, Ahmed HI, Cubry P, Šimoníková D, Cauet S, Pailles Y, Bettgenhaeuser J, Gapa L, Scarcelli N, Couderc M, Zekraoui L, Kathiresan N, Čížková J, Hřibová E, Doležel J, Arribat S, Bergès H, Wieringa JJ, Gueye M, Kane NA, Leclerc C, Causse S, Vancoppenolle S, Billot C, Wicker T, Vigouroux Y, Barnaud A, Krattinger SG (2020) Fonio millet genome unlocks African orphan crop diversity for agriculture in a changing climate. Nat Commun 11(1):4488. https://doi.org/10.1038/s41467-020-18329-4

Bandyopadhyay T, Singh RK, Ramesh P, Maurya J, Prasad M (2023) The promise of millets in the twenty-first century: emphasis on breeding, nutrition, food security and sustainability. J Soil Sci Plant Nutr 23:628–637

Barton L, Newsome SD, Chen FH, Wang H, Guilderson TP, Bettinger RL (2009) Agricultural origins and the isotopic identity of domestication in northern China. Proc Natl Acad Sci USA 106:5523–5528

Bennetzen JL, Schmutz J, Wang H et al (2012) Reference genome sequence of the model plant *Setaria*. Nat Biotechnol 30:555–561

Cannarozzi G, Plaza-Wüthrich S, Esfeld K, Larti S, Wilson YS, Girma D, de Castro E, Chanyalew S, Blösch R, Farinelli L, Lyons E, Schneider M, Falquet L, Kuhlemeier C, Assefa K, Tadele Z (2014) Genome and transcriptome sequencing identifies breeding targets in the orphan crop tef (Eragrostis tef). BMC Genomics 15(1):581. https://doi.org/10.1186/1471-2164-15-581

Dhaka A, Singh RK, Muthamilarasan M, Prasad M (2021) Genetics and genomics interventions for promoting millets as functional foods. Curr Genomics 22:154–163

Gimode D, Odeny DA, de Villiers EP, Wanyonyi S, Dida MM, Mneney EE, Muchugi A, Machuka J, de Villiers SM (2016) Identification of SNP and SSR markers in finger millet using next generation sequencing technologies. PLoS One 11:e0159437

Goff SA, Ricke D, Lan TH et al (2002) A draft sequence of the rice genome (*Oryza sativa* L. ssp. japonica). Science 296:92–100

Guo L, Qiu J, Ye C, Jin G, Mao L, Zhang H, Yang X, Peng Q, Wang Y, Jia L, Lin Z, Li G, Fu F, Liu C, Chen L, Shen E, Wang W, Chu Q, Wu D, Wu S, Xia C, Zhang Y, Zhou X, Wang L, Wu L, Song W, Wang Y, Shu Q, Aoki D, Yumoto E, Yokota T, Miyamoto K, Okada K, Kim DS, Cai D, Zhang C, Lou Y, Qian Q, Yamaguchi H, Yamane H, Kong CH, Timko MP, Bai L, Fan L (2017) Echinochloa crus-galli genome analysis provides insight into its adaptation and invasiveness as a weed. Nat Commun 8(1):1031. https://doi.org/10.1038/s41467-017-01067-5

Hatakeyama M, Aluri S, Balachadran MT et al (2018a) Multiple hybrid de novo genome assembly of finger millet, an orphan allotetraploid crop. DNA Res 25:39–47

Hatakeyama M, Aluri S, Balachadran MT, Sivarajan SR, Patrignani A, Grüter S, Poveda L, Shimizu-Inatsugi R, Baeten J, Francoijs KJ, Nataraja KN, Reddy YAN, Phadnis S, Ravikumar RL, Schlapbach R, Sreeman SM, Shimizu KK (2018b) Multiple hybrid de novo genome assembly of finger millet, an orphan allotetraploid crop. DNA Res 25(1):39–47. https://doi.org/10.1093/dnares/dsx036

He Z, Chen M, Ling B et al (2023) Overexpression of the autophagy-related gene SiATG8a from foxtail millet (*Setaria italica* L.) in transgenic wheat confers tolerance to phosphorus starvation. Plant Physiol Biochem 196:580–586

Hittalmani S, Mahesh HB, Shirke MD, Biradar H, Uday G, Aruna YR, Lohithaswa HC, Mohanrao A (2017) Genome and transcriptome sequence of finger millet (*Eleusine coracana* (L.) Gaertn.) provides insights into drought tolerance and nutraceutical properties. BMC Genomics 18(1):465

Lata C, Gupta S, Prasad M (2013) Foxtail millet: a model crop for genetic and genomic studies in bioenergy grasses. Crit Rev Biotechnol 33:328–343

Li S, Cui Y, Liu D, Zhao Z, Zhang J, Liu Z (2022) Transcriptome analysis and characterization of genes associated to leaf tannin content in foxtail millet [*Setaria italica* (L.) P. Beauv.]. BMC Genom 23:512

Liu H, Shi J, Cai Z, Huang Y, Lv M, Du H, Gao Q, Zuo Y, Dong Z, Huang W, Qin R, Liang C, Lai J, Jin W (2019) Evolution and domestication footprints uncovered from the genomes of Coix. Mol Plant 13(2):295–308. https://doi.org/10.1016/j.molp.2019.11.009

Muthamilarasan M, Prasad M (2017) Advances in *Setaria* genomics for genetic improvement of cereals and bioenergy grasses. Theor Appl Genet 128(1):1–14

Muthamilarasan M, Venkata Suresh B, Pandey G, Kumari K, Parida SK, Prasad M (2014) Development of 5123 intron-length polymorphic markers for large-scale genotyping applications in foxtail millet. DNA Res 21:41–52

Muthamilarasan M, Dhaka A, Yadav R, Prasad M (2016) Exploration of millet models for developing nutrient rich graminaceous crops. Plant Sci 242:89–97

Muthamilarasan M, Shweta S, Prasad M (2017) Foxtail millet genome sequencing, assembly, annotation, and application. In: Prasad M (ed) The foxtail millet genome, pp 11–22

Pandey G, Misra G, Kumari K, Gupta S, Parida SK, Chattopadhyay D, Prasad M (2013) Genome-wide development and use of microsatellite markers for large-scale genotyping applications in foxtail millet [Setaria italica (L.)]. DNA Res 20:197–207

Puranik S, Sahu PP, Mandal SN, Venkata Suresh B, Parida SK, Prasad M (2013) Comprehensive genome-wide survey, genomic constitution and expression profiling of the NAC transcription factor family in foxtail millet (*Setaria italica* L.). PLoS One 8:e64594

Rajasekaran NS, Nithya M, Rose C, Chandra TS (2004) The effect of finger millet feeding on the early responses during the process of wound healing in diabetic rats. Biochim Biophys Acta 1689:190–201

Shan Z, Jiang Y, Li H et al (2020) Genome-wide analysis of the NAC transcription factor family in broomcorn millet (*Panicum miliaceum* L.) and expression analysis under drought stress. BMC Genomics 21:96

Singh RK, Jaishankar J, Muthamilarasan M, Shweta S, Dangi A, Prasad M (2016) Genome-wide analysis of heat shock proteins in C4 model, foxtail millet identifies potential candidates for crop improvement under abiotic stress. Sci Rep 6:32641

Singh RK, Muthamilarasan M, Prasad M (2021) Biotechnological approaches to dissect climate-resilient traits in millets and their application in crop improvement. J Biotechnol 327:64–73

Singh RK, Muthamilarasan M, Prasad M (2022) SiHSFA2e regulated expression of SisHSP21.9 maintains chloroplast proteome integrity under high temperature stress. Cell Mol Life Sci 79:580

VanBuren R, Man Wai C, Wang X, Pardo J, Yocca AE, Wang H, Chaluvadi SR, Han G, Bryant D, Edger PP, Messing J, Sorrells ME, Mockler TC, Bennetzen JL, Michael TP (2020) Exceptional subgenome stability and functional divergence in the allotetraploid Ethiopian cereal teff. Nat Commun 11(1):884. https://doi.org/10.1038/s41467-020-14724-z

Wang T, Song H, Li P et al (2020) Transcriptome analysis provides insights into grain filling in foxtail millet (*Setaria italica* L.). Int J Mol Sci 14:5031

Wang H, Han Y, Wu C et al (2022) Comparative transcriptome profiling of resistant and susceptible foxtail millet responses to *Sclerospora graminicola* infection. BMC Plant Biol 22:567

Wang H, Zhang H, Liu J et al (2024) Transcriptome analysis reveals the mechanism of nitrogen fertilizers in starch synthesis and quality in waxy and non-waxy proso millet. Carbohydr Polym 323:121372

Washburn JD, Schnable JC, Davidse G, Pires JC (2015) Phylogeny and photosynthesis of the grass tribe Paniceae. Am J Bot 102:1493–1505

Xu C, Luo M, Sun X, Yan J, Shi H, Yan H, Yan R, Wang S, Tang W, Zhou Y, Wang C, Xu Z, Chen J, Ma Y, Jiang Q, Chen M, Sun D (2022) SiMYB19 from foxtail millet (*Setaria italica*) confers transgenic rice tolerance to high salt stress in the field. Int J Mol Sci 23(2):756

Yadav CB, Muthamilarasan M, Pandey G, Sadullabax YK, Prasad M (2014) Development of novel microRNA-based genetic markers in foxtail millet for genotyping applications in related grass species. Mol Breed 34:2219–2224

Yadav CB, Bonthala VS, Muthamilarasan M, Pandey G, Khan Y, Prasad M (2015) Genome-wide development of transposable elements-based markers in foxtail millet and construction of an integrated database. DNA Res 22:79–90

Ye CY, Wu D, Mao L, Jia L, Qiu J, Lao S, Chen M, Jiang B, Tang W, Peng Q, Pan L, Wang L, Feng X, Guo L, Zhang C, Kellogg EA, Olsen KM, Bai L, Fan L (2020) The genomes of the allohexaploid Echinochloa crus-galli and its progenitors provide insights into Polyploidization-driven adaptation. Mol Plant 13(9):1298–1310. https://doi.org/10.1016/j.molp.2020.07.001

Yu J, Hu S, Wang J et al (2002) A draft sequence of the rice genome (*Oryza sativa* L. ssp. indica). Science 296:79–92

Zhang G, Liu X, Quan Z et al (2012) Genome sequence of foxtail millet (*Setaria italica*) provides insights into grass evolution and biofuel potential. Nat Biotechnol 30:549–554

Zhang S, Yu Y, Song T et al (2022) Genome-wide identification of foxtail millet's TRX family and a functional analysis of SiNRX1 in response to drought and salt stresses in transgenic Arabidopsis. Front Plant Sci 13:946037

Zhang Y, He Z, Qi X et al (2023) Overexpression of MYB-like transcription factor SiMYB30 from foxtail millet (*Setaria italica* L.) confers tolerance to low nitrogen stress in transgenic rice. Plant Physiol Biochem 196:731–738

Zou C, Li L, Miki D et al (2019) The genome of broomcorn millet. Nat Commun 10:436

Technology for the Development of Value-Added Products in Minor Millets

Archana Kushwaha, Krati Goel, Ayushi Joshi, and Anil Kumar

Abstract

Small millets, such as foxtail, little, barnyard, proso, kodo, finger, and browntop, are notable for their resilience and nutritional richness, offering high levels of protein, fiber, iron, and bioactive compounds. Despite these benefits, their underutilization is attributed to processing challenges and the dominance of staple grains such as wheat and rice. Advanced processing technologies can enhance the palatability, digestibility, and shelf life of millets, facilitating their conversion into diverse products such as flour, ready-to-eat snacks, and breakfast cereals. Primary processing methods such as dehulling and milling improve grain usability, while secondary processes such as parboiling, germination, puffing, extrusion, fermentation, and baking add further value. These techniques not only preserve the nutritional integrity of millets but also diversify dietary options, meeting the global demand for nutritious super foods. Integrating millets into value-added products aligns with sustainable food systems, promoting the "Farm to Fork" concept. By addressing processing complexities and highlighting their nutritional benefits, millets can play a significant role in enhancing public health and supporting sustainable agriculture worldwide. The development of value-added products from minor millets, including composite flour, chapatti, upma, papad, vermicelli, idli, dosa, baked goods, ready-to-eat snacks, beverages, and functional foods, demonstrates their versatility and potential to improve global dietary patterns.

A. Kushwaha (✉) · K. Goel · A. Joshi
Department of Food and Nutrition, College of Community Science, G.B. Pant University of Agriculture & Technology, Pantnagar, Uttarakhand, India

A. Kumar
Department of Plant Breeding & Genetics, Bihar Agriculture College, Bihar Agricultural University, Sabour, Bihar, India

D. C. Joshi et al. (eds.), *Minor Millets*,
https://doi.org/10.1007/978-981-96-4265-6_12

Keywords

Small millets · Barnyard miller · Proso millet · Kodo millet · Nutrient demand · Value-added products · Processing techniques

1 Introduction

Small millets, known for their resilience and nutritional richness, are crucial in combating malnutrition and ensuring food security globally. These millet grains, including foxtail millet, little millet, barnyard millet, proso millet, kodo millet, finger millet, and browntop, thrive in diverse ecological conditions with minimal inputs (Devi et al. 2014). Despite their nutritional superiority—high in protein, fiber, iron, and bioactive compounds like polyphenols—millets face underutilization due to processing challenges and dominance of wheat and rice (Saleh et al. 2013).

To enhance millets' palatability and utility, advanced processing technologies play a pivotal role. Processing improves digestibility, extends shelf life, and enhances consumer acceptance by converting raw grains into diverse products such as flour, ready-to-eat snacks, and breakfast cereals (Suma and Asna 2017). Primary processing involves cleaning and sorting grains, while secondary processing adds value through cooking, baking, or fortifying products with additional nutrients (Yadav et al. 2020). Techniques such as milling, grinding, and extrusion not only make millets easier to cook but also preserve their nutritional integrity (Arabska 2021). These strategies not only diversify dietary options but also meet global demand for nutritious superfoods, ensuring economic viability for farmers and processors alike (Gowda et al. 2022).

Moreover, integrating millets into value-added products aligns with the "Farm to Fork" concept, promoting sustainable food systems (Arabska 2021). Therefore, leveraging advanced processing technologies is essential to mainstreaming millets in global diets. By overcoming processing complexities and promoting their nutritional benefits through value addition, millets can contribute significantly to improving public health and sustainable agriculture worldwide.

2 Processing Technologies of Minor Millets

Processing encompasses a range of techniques that turn raw millet grains to palatable form while improving their quality. Processing methods may exhibit positive and negative impacts on the nutrient as well as antinutrient components. Using different processing methods, small millet grains can be transformed into a variety of consumable forms, including rice/grain, flour, sprouts, ready-to-eat, flaked, popped, porridge, and fermented items (Fig. 1).

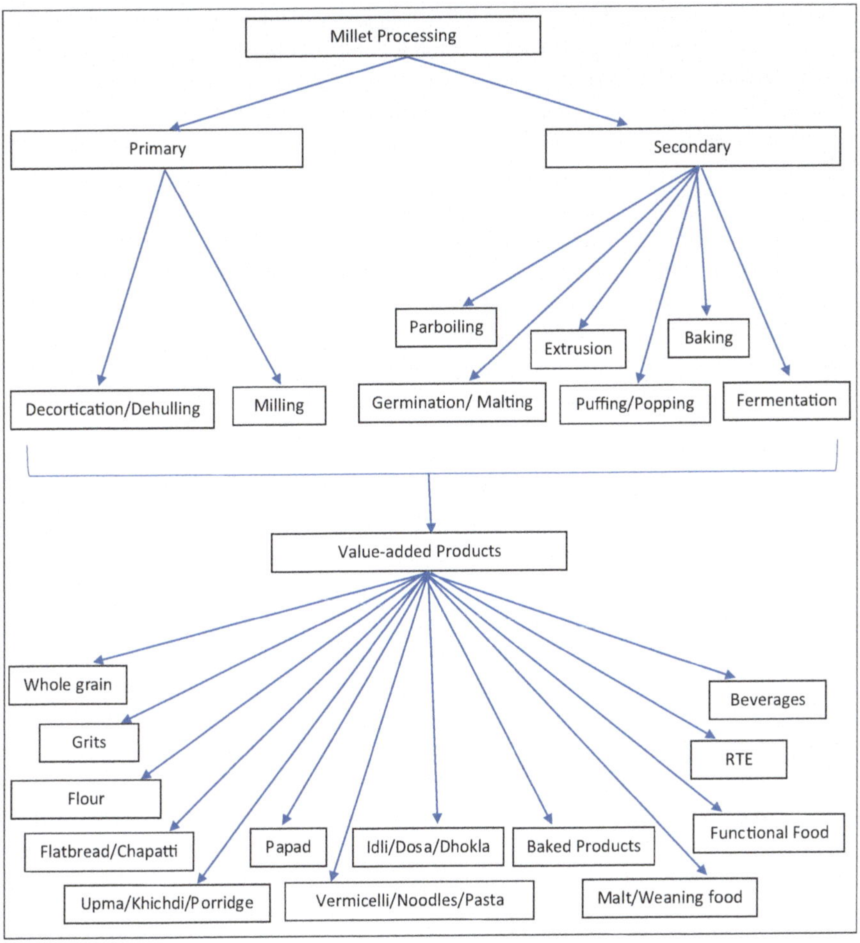

Fig. 1 Schematic representation of technologies for value-addition of minor millet

2.1 Primary Processing Methods

Grain unit operations that enhance grain quality convert it into a more usable form and are performed on the grain at the producer level or in the farm's area (Malleshi 2014).

2.1.1 Dehulling

Dehulling or dehusking, also known as decortication, is the most important stage in millet processing because of its extremely hard seed coat. Millets are traditionally decorated by hand pounding. A tiny stone that is carried in the hand and a larger flat stone that is laid on the ground make up the traditional mill stones used for dehusking and grinding millets. The millet grains are ground and dehulled by the friction between these grinding stones. Typically, this process takes four to five rounds to yield fully dehusked seeds. The conventional dehulling process is time-consuming, labor-intensive, and only produces a 50% dehulling yield. To address the issues with conventional systems, new automated dehulling technologies, such as centrifugal dehulling systems, have been introduced. The proportion of dehulled grains obtained using the conventional compounding approach is significantly lower than the dehulling yield (72–76%). The feed rate, the impeller's peripheral speed, and the moisture content of the grain are the main factors that determine the dehulling yield produced by a centrifugal system (Gupta and Das 1999). Another low-cost technique for dehulling millets is soaking. Seed coat of millets is removed when millets are decorticated. This results in a notable drop in fiber content but an increase in mineral bioaccessibility because fewer inhibitory factors, such as phytic acid and polyphenols, are present (Chethan and Malleshi 2007; Krishnan et al. 2012).

2.1.2 Milling

Millet's tough, fibrous seed coat, colored pigments, astringent flavor, and low processing product shelf life are the main obstacles to its widespread use (Desikachar 1975). Because there is a lack of appropriate milling technology, the majority of millets produced in India are utilized as staple foods and less of them are made into easy, ready-to-eat foods. For milling of minor millets, a variety of mills were created utilizing aspiration, friction, and other methods.

The key factors in milling that affect head yield and subsequent processing are milling efficiency and shelling index. Little millet, kodo, and kutki had milling efficiency values of 83.38, 76.43, and 74.10%, respectively. However, upon polishing, their corresponding head yields were 71.5, 56.25, and 54.57% (Fellows 2017). Bolder grains and varieties with lower protein concentrations break more easily during milling, as seen by the strong and positive association found between milling breakage and thousand kernel weight. B-vitamin content was decreased by milling and sieving, especially thiamin and riboflavin (Oghbaei and Prakash 2012; Oghbaei and Prakash 2016).

2.2 Secondary Processing Technologies

Grain-related unit processes that convert grain into products are typically intended for direct consumption, either before or after primary processing. Typically, they are carried out in organized or unorganized sectors distant from farms (Malleshi 2014).

2.2.1 Parboiling

Parboiling is a conventional processing method of millets. It improves the quality of milling (Shrestha 1972), hardens the endosperm, lessens sliminess during cooking, and ease production of grits (Desikachar 1975). It is commonly recognized that parboiling lessens loss of thiamine and is another way to produce a precooked, ready-to-eat millet product. By preventing phenolics from being lost during processing, parboiling millets raises their overall phenolic content and antioxidant activity (Bora et al. 2019).

2.2.2 Germination and Malting

Germination is a conventional method used to boost the nutritional content of millet grains and soften their kernel structure (Pushparaj and Urooj 2011). In the past, the germination process was conducted within environmental chambers; however, specialized grain sprouters are now available that allow for the adjustment of environmental conditions, such as light source, illumination wavelength, and so on, to produce a high-quality product with adequate safety and shelf life (Kate and Singh 2021). Since it increases digestibility through the production of hydrolytic enzymes that designate sprouts as "*functional foods*," it is an economical and efficient way to improve the overall nutritional quality of millets (Benincasa et al. 2019). Millets germinate more readily and have higher protein content, digestibility, and efficiency ratios (Sharma et al. 2015; Mbithi-Mwikya et al. 2000). Moreover, antinutrients that bind minerals and prevent their absorption are diminished by germination, including tannins, oxalates, and phytic acid (Sharma et al. 2015; Morah and Etukudo 2017). Finger millet is widely used for malting in parts of Tamil Nadu and Karnataka due to its good malting qualities. For low dietary bulk and calorie-dense weaning foods, supplementary foods, health foods, and amylase-rich foods, millet malt is utilized as a cereal foundation.

2.2.3 Puffing or Popping

Popping or puffing is a basic processing technique for ready-to-eat goods. Packed with flavor, popping grain is a crunchy, precooked, and porous product (Malleshi 2014). In both rich and developing nations, eating popped and puffed products is a prevalent practice. Starch is the main constituent that undergoes physicochemical and structural alterations most notably in the thermal treatment of grains that cause the grains to expand and result in a puffed product. Expanded flakes and other popped products can be easily made using the popular and practical high treatment short time (HTST) method (Jaybhaye et al. 2014). For the purpose of producing large quantities of popped and puffed millet grains, air blowing puffing technology have been developed (Verma and Patel 2013). Popping millet increases bioaccessible minerals such as iron and zinc due to high-temperature treatment (Krishnan et al. 2012; Nazni and Shobana 2016). According to Krishnan et al. (2012), popping millets lower the overall phenol content but may also lower phytic acid, increasing the availability of minerals.

2.2.4 Extrusion

Extrusion cooking is a high-temperature, short-time (HTST) cooking method that can be applied to both proteinaceous and starchy materials (Dahlin and Lorenz 1993). Extrusion cooking is achieved by applying heat either directly through the injection of steam, indirectly through the jacket, or through the dissipation of mechanical energy through the shearing process that takes place within the mixture. Kodo millet flour blend is used to make millet-based extruded snack items using a twin-screw extruder (Geetha et al. 2014). Using microwave cold extrusion, puffed barnyard millet-based RTE fast foods with good sensory quality can be made (Jaybhaye et al. 2014). In contrast to RTE foods, pasta and noodle products are the most extensively consumed extruded food items in developed and emerging nations across all age categories. Pasta and noodles are more economically significant and have a longer keeping quality or shelf life than other items.

2.2.5 Fermentation

Millets are an overall decent protein source, but their essential amino acid composition makes them low-quality protein (Jaybhaye et al. 2014). Fermented food products such as idli and dosa are two popular and ubiquitous breakfast items consumed in many parts of India (Antony et al. 1996; Mbithi-Mwikya et al. 2000). This is one of the least expensive and most efficient ways to dramatically lower phytic acid and polyphenols while also increasing the bioavailability of minerals; nevertheless, sequential fermentation yields the greatest reduction in these parameters. In order to improve vitamin bioavailability and mineral absorption, fermentation lowers antinutrients and enhances vitamin content (Carter and Carpenter 1981; Onyango et al. 2013).

2.2.6 Baking

Millet can be transformed into baked goods with remarkable sensory qualities through the process of baking. Thus, the primary attributes that customers assess about bakery items are their flavor, appearance, and surface color (Purlis 2010). Worldwide, baked goods are well-liked by people of all ages due to their extended shelf life, ease of marketing, and attractive packaging but the many wheat products with added value are what really drive the baking sector. People with celiac disease, especially those in industrialized nations, are allergic to gluten and thus cannot consume gluten-based products. Bakery products made from millet are gluten free and also provide better nutrition (Verma and Patel 2013). For preparing biscuits and muffins, foxtail millet and finger millet flour and for preparing cakes, barnyard millet is preferable.

3 Value-Added Products from Minor Millets

Millets have garnered global attention for their nutritional quality and resilience to drought, prompting research efforts to enhance their varieties and utilization in food (Kumar et al. 2018). In India, millets have long been staples, integrated into regional cuisines such as *ragi roti*, *mudde* in various states and *jhangora kheer* in Uttarakhand and Uttar Pradesh (ICAR-IIMR 2018). Their versatility extends to sweets, crispy products, *khichdi*, and noodles, catering to diverse nutritional needs across life stages (Thathola and Srivastava 2002). Millets are particularly suitable as weaning foods due to their nutrient-rich profiles and enhanced bioavailability (Sihag et al. 2016). Scientific evidence underscores their benefits in childhood nutrition, improving growth rates and hemoglobin levels in adolescents (Anitha et al. 2022; Karkada et al. 2019). Moreover, systematic reviews highlight millets' role in combating iron-deficiency anemia and reducing the risk of noncommunicable diseases such as cardiovascular ailments and cancer (Anitha et al. 2021; WHO 2022). As demand grows for nutritious, sensory-rich foods, millets stand poised to contribute significantly to global dietary diversification and health improvement initiatives. The above discussion gives the clarity that millets can be incorporated during all stages of life providing health benefits and can easily fit to one's food plate/*thali*. Various research works has been carried out on the utilization of millet crops using different processing methods (Table 1) as discussed in this section.

3.1 Composite Flour

Despite being an excellent source of both macro- and micronutrients, millets are not used very often. Blending them with staple flours such as wheat is one strategy to increase their use as food (Singh et al. 2005). It is feasible to substitute 10–20% of wheat flour with millet (finger, foxtail, proso, barnyard, and tiny millet), which lowers the proportion of damaged protein and starch and raises the ash content (Kamaraddi and Shanthakumar 2003; Singh et al. 2005). To create high-quality extrudates with tolerable sensory qualities, composite flour utilizing whole flour of foxtail millet and additional flours, such as rice flour, chick pea flour, amaranth seed flour, and cow pea flour (60:05:05:20:10), might be utilized (Deshpande and Poshadri 2011). Such studies showed that blending of millets to make composite flour can be considered one of the best ways to include millets as part of daily dietary regimes.

3.2 Flatbread or Chapatti or Roti

Chapatis, traditional unleavened flatbreads made from whole wheat flour (*atta*), are staples in the Indian subcontinent and the Middle East. In the Northwest Himalayan region of India, coarse cereals such as finger millet are also consumed as roti, despite lacking gluten for puffing upon baking. In Uttarakhand, a method using dried bark of *Boehmeria rugulosa*, locally known as "gethi," softens and enables finger millet rotis to puff-like rotis made from wheat and retain softness even during cold (Khulbe et al. 2014). Supplementation of wheat flour with salt-treated soy dhal, sorghum, rice, maize, and pearl millet enhances protein content and produces composite flour *chapatis* with favorable sensory attributes (Khetarpaul and Goyal 2009; Panghal et al. 2019; Walde et al. 2021).

3.3 Upma or Khichdi or Porridge

Khichdi, a balanced meal combining pulses and cereals, provides nutrition while *upma* serves as a breakfast in southern India. Millet khichdi and upma's nutritional

Table 1 Utilization of millets as food using different processing methods

Processing	Value-added product	Description	Reference
Decortication + baking	Cake	80% decorticated finger millet +20% refined wheat flour	Patel and Thorat (2019)
Decortication + soaking + gelatinization	Papad	60% sorghum +40% finger millet flour	Prabhakar et al. (2016)
Decortication + gelatinization	Papad	15–20% finger millet flour + black or green gram + rice + spices	Gull et al. (2014)
Milling	Chapatti	7:3 (wheat: Finger millet)	Gull et al. (2014)
Milling	Chapatti	10–30% finger millet + wheat flour	Panghal et al. (2019)
Milling	Chapatti	10–15% finger millet +10–15% pearl millet + wheat	Walde et al. (2021)
Roasting	Khichdi	70% little millet +25% mix of legumes (green gram, yellow and red lentil)	Chhabra and Kaur (2022)
Roasting	Upma	85% foxtail millet + Bengal gram	Chhabra and Kaur (2022)
Roasting	Laddu	60% finger millet +40% gram flour	Sharma and Yamer (2022)
Roasted + pressure cooked	Chapatti	10–100% barnyard millet flour + wheat flour	Devi and Nazni (2016)
Germination + baking	Biscuits	30–50% germinated finger millet +50–70% wheat flour	Shimray et al. (2012)
Germination + baking	Cookies	0–30% germinated finger millet +50–100% wheat + African yam bean	Abioye et al. (2018)
Malting	Weaning food	24% millet mix (kodo millet, finger millet, foxtail millet, little millet) + 76% other ingredients	Prasanna et al. (2020)
Malting and roasting	Weaning food	30% foxtail millet +30% barnyard millet +25% soybean flour +15% skim milk powder	Thathola and Srivastava (2002)
Germination + fermentation	Dosa mix	33% germinated millet +33% rice +33% dehusked black gram	Krishnamoorthy et al. (2013)
Germination + fermentation	Idli mix	33% germinated millet +33% rice +33% dehusked black gram	Krishnamoorthy et al. (2013)
Fermentation	Pizza base	25% foxtail millet +25% barnyard millet +50% refined wheat flour	Veena et al. (2020)
Fermentation	Dhokla mix powder	55% foxtail and barnyard millet +35% legumes +10% fenugreek seeds	Pathak et al. (2000)

(continued)

Table 1 (continued)

Processing	Value-added product	Description	Reference
Dehydration	Instant khichdi mix	22–28% foxtail millet +28–35% green gram dal +4–6% oats	Negi et al. (2021)
Extrusion (cold)	Noodles	50% finger millet +50% wheat	Gull et al. (2014)
Extrusion (cold)	Vermicelli	64.25% barnyard millet +22.32% rice +2.36% xanthan gum	Goel et al. (2021)
Extrusion (cold)	Vermicelli	40% barnyard millet +58% wheat +2% Ekanayakam root bark powder	Chandraprabha et al. (2017)
Puffing	Puffed millet	100% finger millet	Gull et al. (2014)
Baking	Waffles	30% finger millet +30% pearl millet +40% wheat	Chaitra et al. (2020)
Baking	Cake	40% finger millet +60% wheat	Adanse et al. (2021)
Baking	Buns	10% finger millet +82.5% refined wheat flour +7.5% moringa leaf	Boria et al. (2021)
Baking	Muffins	50% finger millet +50% refined wheat flour	Sharma and Yamer (2022)
Frying	Doughnuts	50% finger millet, quinoa flour +50% wheat	Joshi et al. (2023)

profile was improved by substituting rice with little, foxtail, and other minor millets. Millet-based products resulted in increased protein and dietary fiber content (Chhabra and Kaur 2022). Additionally, studies noted that porridge made from soaked and germinated millets exhibited improved nutritional characteristics including protein content, ash, and water absorption capacity (Ocheme and Chinma 2008).

3.4 Papad

In Indian cuisine, *papads* are a popular thin, crispy, disc-shaped dish that are enjoyed as a significant component of the meal (Sharma et al. 2020). Finger millet papad can be made using gelatinized dough with other necessary components such as rice, spices, and either black or green gram (Gull et al. 2014). Because the pericarp and starch are combined, millet lends the papad a slightly darker hue. However, after cooking or roasting, the starch turns lighter and is more palatable to consumers.

3.5 Vermicelli or Noodles or Pasta

Millets are a good choice for producing extruded pasta products such as noodles, vermicelli, and so on because they have a larger proportion of inaccessible carbohydrates, which reduces the release of glucose and lowers the glycemic index (Rathore et al. 2016). Due to its inherent amber color and ability to retain a firm, desired texture after cooking, durum wheat has historically been used to make these items. Addition of millets to vermicelli, noodles, and pasta improves their nutritional value and functionality (Devaraju et al. 2008; Shukla and Srivastava 2014; Goel et al. 2021).

3.6 Idli or Dosa

Fermented foods are essential for human nutrition and are therefore quite popular and consumed worldwide (Mugocha et al. 2000; Gotcheva et al. 2001). In India, the most popular fermented meals are *idli*, *dosa*, and *dhokla*. The process of fermentation alters the chemical makeup of millet grains and, consequently, the food products made from them. Compared to traditional idli and dosa produced from rice, millets provide higher quantities of protein, fat, and carbohydrates and are easier to make (Krishnamoorthy et al. 2013).

3.7 Baked Products

Bakery products including biscuits, *nan-khatai*, chocolate, cakes, muffins, and so on can be made with finger millet and foxtail millet flour (Desai et al. 2010). There have been attempts to add malted finger millet flour to cakes in order to increase their nutritional value in terms of fiber and mineral content (Desai et al. 2010). According to Kamaraddi and Shanthakumar (2003), cake and icing agents can be prepared using a combination of minor millet flour in the desired quantity. According to Misra and Tiwari (2014), biscuits are the most widely consumed bakery item. Without significantly compromising most baking and sensory qualities, millet could be utilized to make biscuits with higher nutritional value. Biscuits made with 40% replacement finger millet flour exhibited satisfactory sensory qualities (Shimray et al. 2012). Incorporation of millets in biscuits increased calcium and crude fiber contents as well as improve the sensory attributes of the product (Kaur et al. 2020).

3.8 Ready-to-Eat Popped or Puffed Millet

Cooking whole grains for use as cereal or a snack involves an ancient custom known as puffing or popping. Popped millet is consumed regularly in numerous African countries, including India. If native grains (12% moisture content) are treated using HTST, the cereal processing technologies can be successfully used for foxtail millet to produce ready-to-eat products such as flaked, extruded, roller dried, decorated, and popped grains (Ushakumari et al. 2004). Novel microwave puffing process for producing ready-to-eat (RTE) foods that are based on minor millet such as finger millet and barnyard millet have been used.

3.9 Beverage

Millets are also utilized in the preparation of certain fermented drinks. A common, wholesome, traditional cuisine in many parts of Zimbabwe is fermented composite beverages made from milk and finger millet (Mugocha et al. 2000). A thick, syrupy drink known for its versatility as both a food and a drink, *maschusvu* is produced by heating malted finger millet grain and water. It has a light brown hue. Millets have cultural and dietary significance, as demonstrated by the unique flavors and nutritional benefits of traditional millet-based ferments such as Bushera, Malwa (Uganda), and Mahewu (South Africa), as well as other millet-based beverages such as Kunn-zaki (Northern Nigeria). Millets can also be used to make alcoholic drinks. In the Eastern Himalayas, a traditional mild-alcoholic beverage known as "*Kodo ko jaanr*" is made from finger millet seeds through fermentation (Thapa and Tamang 2004). Fermented beverages made from millet have a high nutritional content and a high concentration of lactic acid bacteria, which gives them their probiotic qualities (Satish Kumar et al. 2010; Ilango and Antony 2014).

3.10 Functional Foods

Functional foods have become widely used to describe diets that improve health by preventing certain degenerative diseases such as diabetes, cancer, Parkinson's disease, cataracts, and so on. Because of its health-promoting phytochemicals, millets have drawn interest for being useful as functional foods. Those with gluten sensitivity and celiac disease can safely consume millet. According to Saleh et al. (2013), millets are easy to digest because they do not generate acids or allergies. The low glycemic index (40–50) of minor millets includes foxtail and barnyard millet. These tiny millets can be used to make a variety of low-GI foods, such as finger millet vermicelli, foxtail millet burfi, and barnyard millet crackers.

3.11 Malt and Weaning Foods

Malt made from millet is traditionally used to feed infants. Finger millet is widely used for malting in parts of Tamil Nadu and Karnataka due to its good malting qualities. Millet malt is used as a cereal base for low dietary bulk, calorie-dense weaning foods, health foods, supplementary foods, and amylase-rich diets (Jaybhaye et al. 2014).

4 Conclusion

The utilization of advanced processing technologies is crucial for enhancing the value of minor millets throughout their transformation into nutritionally intensive food products. From primary methods such as decortication and milling to secondary techniques such as parboiling, extrusion, and fermentation, these processes not only improve palatability and shelf life but also enhance nutritional content and consumer acceptance. These advancements enable the creation of diverse value-added products including flour, ready-to-eat snacks, breakfast cereals, and fermented beverages, meeting varied dietary preferences globally. Moreover, by promoting local agricultural diversity and reducing reliance on resource-intensive crops, such innovations support sustainable food systems. Enhanced with higher protein, fiber, and essential minerals, processed millet products contribute significantly to combating malnutrition and improving public health outcomes worldwide, fostering healthier and more sustainable diets.

References

Abioye VF, Ogunlakin GO, Taiwo G (2018) Effect of germination on anti-oxidant activity, total phenols, flavonoids and anti-nutritional content of finger millet flour. J Food Process Technol 9(2):1–5. https://doi.org/10.4172/2157-7110.1000719

Adanse J, Bıgson K, Asante Maureen N, Dorothy A (2021) Quality characteristics and sensory evaluation of cakes produced from composite blends of wheat (*Triticum aestivum L.)* and finger millet (*Pennisetum glaucum*). Flour Eurasian J Food Sci Technol 5(2):190–204

Anitha S, Botha R, Kane-Potaka J, Givens DI, Rajendran A, Tsusaka TW, Bhandari RK (2021) Can millet consumption help manage hyperlipidemia and obesity?: a systematic review and meta-analysis. Front Nutr 8:700778. https://doi.org/10.3389/fnut.2021.700778

Anitha S, Givens DI, Subramaniam K, Upadhyay S, Kane-Potaka J, Vogtschmidt YD, Bhandari RK (2022) Can feeding a millet-based diet improve the growth of children?—a systematic review and meta-analysis. Nutrients 14(1):225. https://doi.org/10.3390/nu14010225

Antony U, Sripriya G, Chandra TS (1996) Effect of fermentation on the primary nutrients in finger millet (*Eleusine coracana*). J Agric Food Chem 44(9):2616–2618. https://doi.org/10.1021/jf950787q

Arabska E (2021) From farm to fork: human health and well-being through sustainable agri-food systems. J Life Econ 8(1):11–27. https://doi.org/10.15637/jlecon.8.1.02

Benincasa P, Falcinelli B, Lutts S, Stagnari F, Galieni A (2019) Sprouted grains: a comprehensive review. J Nutr 11(2):421. https://doi.org/10.3390/nu11020421

Bora P, Ragaee S, Marcone M (2019) Characterisation of several types of millets as functional food ingredients. Int J Food Sci Nutr 70(6):714–724. https://doi.org/10.1080/09637486.2019.1570086

Boria A, Gour S, Dubey AD, Shukla SS (2021) Chemical composition and shelf-life of burger buns from ragi and moringa leaf powder. J Pharma Innov 10(10):2179–2185

Carter EGA, Carpenter KJ (1981) Bound niacin in sorghum and its availability. Nutr Res 1(6):571–579. https://doi.org/10.1093/jn/112.11.2091

Chaitra U, Abhishek P, Sudha ML, Vanitha T (2020) Impact of millets on wheat based. Belgian waffles: quality characteristics and nutritional composition. LWT Food Sci Technol 124:1–7. https://doi.org/10.1016/j.lwt.2020.109136

Chandraprabha LS, Sharon CL, Panjikkaran ST, Aneena ER, Beena C (2017) Development and nutritional qualities of vermicelli prepared from barnyard millet and Ekanayakam root bark. J Pharmacogn Phytochem 6(6):2359–2362

Chethan S, Malleshi NG (2007) Finger millet polyphenols: optimization of extraction and the effect of pH on their stability. Food Chem 105(2):862–870. https://doi.org/10.1016/j.foodchem.2007.02.012

Chhabra I, Kaur A (2022) Development of a convenient, nutritious ready to cook packaged product using millets with a batch scale process development for a small-scale enterprise. J Food Sci Technol 59(2):488–497. https://doi.org/10.1007/s13197-021-05031-6

Dahlin K, Lorenz K (1993) Protein digestibility of extruded cereal grains. Food Chem 48(1):13–18. https://doi.org/10.1016/0308-8146(93)90214-Z

Desai AD, Kulkarni SS, Sahoo A, Ranveer R, Dandge P (2010) Effect of supplementation of malted ragi flour on the nutritional and sensorial quality characteristics of cake. Adv J Food Sci Technol 2(1):67–71

Deshpande HW, Poshadri A (2011) Physical and sensory characteristics of extruded snacks prepared from foxtail millet based composite flours. Int Food Res J 18(2):751–756

Desikachar SR (1975) Processing of maize, sorghum and millets for food uses. J Sci Ind Res 34:231–237

Devaraju B, Begum JM, Begum S, Vidya K (2008) Finger millet pasta fortified with plant and animal protein and their sensory qualities. Asian J Dairy Food Res 27(3 and 4):193–195

Devi RS, Nazni P (2016) Sensory characteristics, total polyphenol content and in vitro antioxidant activity of value added processed barnyard millet flour chapattis. J Food Process Technol 7(6):1000595. https://doi.org/10.4172/2157-7110.1000595

Devi PB, Vijayabharathi R, Sathyabama S, Malleshi NG, Priyadarisini VB (2014) Health benefits of finger millet (*Eleusine coracana* L.) polyphenols and dietary fiber: a review. J Food Sci Technol 51:1021–1040. https://doi.org/10.1007/s13197-011-0584-9

Fellows PJ (2017) Properties of food and principles of processing. J Food Process Technol 4:1

Geetha R, Mishra HN, Srivastav PP (2014) Twin screw extrusion of kodo millet-chickpea blend: process parameter optimization, physico-chemical and functional properties. J Food Sci Technol 51:3144–3153. https://doi.org/10.1007/s13197-012-0850-5

Goel K, Goomer S, Aggarwal D (2021) Formulation and optimization of value-added barnyard millet vermicelli using response surface methodology. Asian J Dairy Food Res 40(1):55–61. https://doi.org/10.18805/ajdfr

Gotcheva V, Pandiella SS, Angelov A, Roshkova Z, Webb C (2001) Monitoring the fermentation of the traditional Bulgarian beverage boza. Int J Food Sci Technol 36(2):129–134. https://doi.org/10.1046/j.1365-2621.2001.00429.x

Gowda NN, Siliveru K, Prasad PV, Bhatt Y, Netravati BP, Gurikar C (2022) Modern processing of Indian millets: a perspective on changes in nutritional properties. Food Secur 11(4):499. https://doi.org/10.3390/foods11040499

Gull A, Jan R, Nayik GA, Prasad K, Kumar P (2014) Significance of finger millet in nutrition, health and value added products: a review. Magnesium (mg) 130(32):120

Gupta RK, Das SK (1999) Performance of centrifugal dehulling system for sunflower seeds. J Food Eng 42(4):191–198. https://doi.org/10.1016/S0260-8774(99)00119-3

ICAR-Indian Institute of Millets Research (2018) Millets annual report 2016–17; 2017. http://millets.res.in/annual_report/ar16-17.pdf. Accessed 27 Mar 2018

Ilango S, Antony U (2014). Assessment of the microbiological quality of koozh, a fermented millet beverage. Afr J Microbiol Res. 8(3):308–312

Jaybhaye RV, Pardeshi IL, Vengaiah PC, Srivastav PP (2014) Processing and technology for millet-based food products: a review. J Ready-To-Eat Food 1(2):32–48

Joshi A, Srivastava S, Shahi NC, Kushwaha A, Acharya R (2023) Formulation and numerical optimization of finger millet and quinoa incorporated eggless doughnuts. Asian J Dairy Food Res DR-2072:1–9. https://doi.org/10.18805/ajdfr.DR-2072

Kamaraddi V, Shanthakumar G (2003) Effect of incorporation of small millet flour to wheat flour on chemical, rheological and bread characteristics. Recent trends in millet processing and utilization. CCS Hisar Agricultural University, pp 74–81

Karkada S, Upadhya S, Upadhya S, Bhat G (2019) Beneficial effects of ragi (finger millet) on hematological parameters, body mass index, and scholastic performance among anemic adolescent high-school girls (AHSG). Compr Child Adolesc Nurs 42(2):141–150

Kate A, Singh A (2021) Processing technology for value addition in millets. In: Kumar A, Tripathi MK, Joshi D, Kumar V (eds) Millets and millet technology. Springer, Singapore. https://doi.org/10.1007/978-981-16-0676-2_11

Kaur A, Kumar K, Dhaliwal HS (2020) Physico-chemical characterization and utilization of finger millet (*Eleusine coracana* L.) cultivars for the preparation of biscuits. J Food Process Preserv 44(9):e14672

Khetarpaul N, Goyal R (2009) Effect of composite flour fortification to wheat flour on the quality characteristics of unleavened bread. Br Food J 111(6):554–564

Khulbe RK, Sood S, Sharma A, Agrawal K, Bhatt JC (2014) Value addition and nutritional fortification of finger millet [*Eleusine coracana* (L.) *Gaertn.*] using bark of Gethi (*Boehmeria regulosa* Wedd.) tree. Indian J Tradit Knowl 13(3):519–524

Krishnamoorthy S, Kunjithapatha S, Manickam L (2013) Traditional Indian breakfast (Idli and Dosa) with enhanced nutritional content using millets. Nutr Diet 70(3):241–246

Krishnan R, Dharmaraj U, Malleshi NG (2012) Influence of decortication, popping and malting on bioaccessibility of calcium, iron and zinc in finger millet. LWT-Food Sci Technol 48(2):169–174. https://doi.org/10.1016/j.lwt.2012.03.003

Kumar A, Tomer V, Kaur A, Kumar V, Gupta K (2018) Millets: a solution to agrarian and nutritional challenges. Agric Food Secur 7(1):1–15. https://doi.org/10.1186/s40066-018-0183-3

Malleshi NG (2014) Post-harvest processing of millets for value addition. https://isites.harvard.edu/fs/docs/icb.topic868074

Mbithi-Mwikya S, Van Camp J, Yiru Y, Huyghebaert A (2000) Nutrient and antinutrient changes in finger millet (*Eleusine coracana*) during sprouting. LWT-Food Sci Technol 33(1):9–14. https://doi.org/10.1006/fstl.1999.0605

Misra NN, Tiwari BK (2014) Biscuits. In: Bakery products science and technology, pp 585–601. https://doi.org/10.1002/9781118792001.ch33

Morah F, Etukudo UP (2017) Effect of sprouting on nutritional value of Panicium miliaceum (proso millet) Edorium. J Nutr Diet 4:1–4. https://doi.org/10.5348/N09-2017-4-SR-1

Mugocha PT, Taylor JRN, Bester BH (2000) Fermentation of a composite finger millet-dairy beverage. World J Microbiol Biotechnol 16:341–344. https://doi.org/10.1023/A:1008929403215

Nazni P, Shobana DR (2016) Effect of processing on the characteristics changes in barnyard and foxtail millet. J Food Process Technol 7(3):1–9

Negi N, Srihari SP, Wadikar DD, Sharma GK, Semwal AD (2021) Optimization of instant foxtail millet based khichdi by using response surface methodology and evaluation of its shelf stability. J Food Sci Technol 58:4478–4485. https://doi.org/10.1007/s13197-020-04929-x

Ocheme OB, Chinma CE (2008) Effects of soaking and germination on some physicochemical properties of millet flour for porridge production. J Food Sci Technol 6(5):185–188

Oghbaei M, Prakash J (2012) Bioaccessible nutrients and bioactive components from fortified products prepared using finger millet (*Eleusine coracana*). J Sci Food Agric 92(11):2281–2290. https://doi.org/10.1002/jsfa.5622

Oghbaei M, Prakash J (2016) Effect of primary processing of cereals and legumes on its nutritional quality: a comprehensive review. Cogent Food Agric 2(1):1136015. https://doi.org/10.1080/23311932.2015.1136015

Onyango CA, Ochanda SO, Mwasaru MA, Ochieng JK, Mathooko FM, Kinyuru JN (2013) Effects of malting and fermentation on anti-nutrient reduction and protein digestibility of red sorghum, white sorghum and pearl millet. J Food Res 2(1):41. https://doi.org/10.5539/jfr.v2n1p41

Panghal A, Chhikara N, Khatkar BS (2019) Characterisation of Indian wheat varieties for chapatti (flat bread) quality. J Saudi Soc Agric Sci 18(1):107–111. https://doi.org/10.1016/j.jssas.2017.02.005

Patel P, Thorat SS (2019) Studies on chemical, textural and color characteristics of decorticated finger millet (*Eleusine coracana*) fortified sponge cake. J Pharm Innov 8(3):64–67

Pathak P, Srivastava S, Grover S (2000) Development of food products based on millets, legumes and fenugreek seeds and their suitability in the diabetic diet. Int J Food Sci Nutr 51(5):409–414. https://doi.org/10.1080/096374800427019

Prabhakar B, More DR, Shivashankar S, Mallesh S, Babu GN (2016) Physico-chemical and sensory evaluation of sorghum-finger millet papad. Int J Food Ferment Technol 6(2):387–395. https://doi.org/10.5958/2277-9396.2016.00064.7

Prasanna MS, Sowjanya VS, Jaya E, Rajender G (2020) Development of millet based instant weaning mix. J Pharmacogn Phytochem 9(4):1908–1913

Purlis E (2010) Browning development in bakery products–a review. J Food Eng 99(3):239–249. https://doi.org/10.1016/j.jfoodeng.2010.03.008

Pushparaj FS, Urooj A (2011) Influence of processing on dietary fiber, tannin and in vitro protein digestibility of pearl millet. Food Nutr Sci 2(8):895–900. https://doi.org/10.4236/FNS.2011.28122

Rathore S, Singh K, Kumar V (2016) Millet grain processing, utilization and its role in health promotion: a review. Int J Nutr Food Sci 5(5):318–329. https://doi.org/10.11648/j.ijnfs.20160505.12

Saleh AS, Zhang Q, Chen J, Shen Q (2013) Millet grains: nutritional quality, processing, and potential health benefits. Compr Rev Food Sci Food Saf 12(3):281–295. https://doi.org/10.1111/1541-4337.12012

Satish Kumar R, Ragu Varman D, Kanmani P, Yuvaraj N, Paari KA, Pattukumar V, Arul V (2010) Isolation, characterization and identification of a potential probiont from south Indian fermented foods (Kallappam, Koozh and Mor Kuzhambu) and its use as biopreservative. Probiotics Antimicrob Proteins 2:145–151

Sharma D, Yamer P (2022) Value addition of products from finger millet (*Eleusine coracana*). J Postharvest Technol 10(2):120–138

Sharma S, Saxena DC, Riar CS (2015) Antioxidant activity, total phenolics, flavonoids and antinutritional characteristics of germinated foxtail millet (*Setaria italica*). Cog Food Agric 1(1):1081728. https://doi.org/10.1080/23311932.2015.1081728

Sharma L, Garg S, Yadav A (2020) Functional and sensory properties of papads developed with semolina and chia seeds. Int J Scient Technol Res 9(1):3607–3612

Shimray CA, Gupta S, Venkateswara Rao G (2012) Effect of native and germinated finger millet flour on rheological and sensory characteristics of biscuits. Int J Food Sci Technol 47(11):2413–2420. https://doi.org/10.1111/j.1365-2621.2012.03117.x

Shrestha KB (1972) Dehusking of Varagu (*Paspalum scrobiculatum*) and its utilization for edible purposes (Doctoral dissertation, Central Food Technological Research Institute)

Shukla K, Srivastava S (2014) Evaluation of finger millet incorporated noodles for nutritive value and glycemic index. J Food Sci Technol 51:527–534. https://doi.org/10.1007/s13197-011-0530-x

Sihag MK, Sharma V, Goyal A, Arora S, Kapila R (2016) In vivo assessment of iron bioavailability from fortified pearl millet based weaning food. J Sci Food Agric 96(13):4410–4415. https://doi.org/10.1002/jsfa.7651

Singh P, Singh G, Srivastava S, Agarwal P (2005) Physico-chemical characteristics of wheat flour and millet flour blends. J Food Sci Technol 42(4):340–343

Suma F, Asna U (2017) Impact of household processing methods on the nutritional characteristics of pearl millet (*Pennisetum typhoideum*): a review. Food Process Technol 4(1):1–5. https://doi.org/10.15406/MOJFPT.2017.04.00082

Thapa S, Tamang JP (2004) Product characterization of *kodo ko jaanr*: fermented finger millet beverage of the Himalayas. Food Microbiol 21(5):617–622. https://doi.org/10.1016/j.fm.2004.01.004

Thathola A, Srivastava S (2002) Physicochemical properties and nutritional traits of millet-based weaning food suitable for infants of the Kumaon hills, Northern India. Asia Pac J Clin Nutr 11(1):28–32. https://doi.org/10.1046/j.1440-6047.2002.00251.x

Ushakumari SR, Latha S, Malleshi NG (2004) The functional properties of popped, flaked, extruded and roller-dried foxtail millet (*Setaria italica*). Int J Food Sci Technol 39(9):907–915. https://doi.org/10.1111/j.1365-2621.2004.00850.x

Veena T, Nidoni U, Devi GS (2020) Millet incorporated pizza base. Int J Curr Microbiol App Sci 9(10):2830–2837. https://doi.org/10.20546/ijcmas.2020.910.341

Verma V, Patel S (2013) Value added products from nutri-cereals: finger millet (*Eleusine coracana*). Emir J Food Agric 25(3):169. https://doi.org/10.9755/ejfa.v25i3.10764

Walde SG, Agrawal S, Mittal S (2021) Development of multigrain chapatti with spices mix to enhance the nutritional values and their storage study. J Food Sci Technol 58:1132–1142. https://doi.org/10.1007/s13197-020-04627-8

Yadav SK, Kauldhar BS, Sandhu PP, Thakur K, Sharma TR (2020) Retrospect and prospects of secondary agriculture and bioprocessing. J Plant Biochem Biotechnol 29(1):1–14. https://doi.org/10.1007/s13562-020-00550

Value Chain Development of Minor Millets: Challenges and Opportunities

Sangappa, Rafi Dudekula, Laxmi B, and Meghana G

Abstract

The cultivation and integration of minor millets into agricultural systems present a compelling yet intricate landscape characterized by distinct challenges and promising opportunities. To boost the cultivation of nutritious millets and raise awareness among the people, the year 2023 has been designated as the "Global Year for Millets." The nutritional richness of minor millets presents an opportunity to position them as health-oriented choices, and diversifying product offerings can cater to varied consumer preferences. Additionally, creating inclusive value chains, empowering smallholder farmers, especially women, and securing governmental support through subsidies and policies can contribute to the sustainable development of minor millet cultivation. In summary, navigating the challenges and leveraging the opportunities in minor millet cultivation necessitates a comprehensive and collaborative approach, ensuring resilience in agricultural systems and advancing broader food security goals. Addressing the key challenges in the millet value chain requires a comprehensive approach involving government policies, market incentives, supply chain improvements, and continued awareness efforts. Coordination between various stakeholders is essential to create a sustainable and inclusive millet industry in India.

Keywords

Minor millets · Value chain · Millet cultivation · Climate · Food security

Sangappa (✉) · R. Dudekula · G. Meghana
ICAR-IIMR, Hyderabad, India
e-mail: sangappa@millets.res.in

B. Laxmi
College of Forestry, KSNUAHS, Ponnampet, India

D. C. Joshi et al. (eds.), *Minor Millets*,
https://doi.org/10.1007/978-981-96-4265-6_13

1 Introduction

Millets indeed have a long history of cultivation in India and have played a crucial role in the country's food and agricultural systems. Millets are highly nutritious and gluten-free, making them a valuable addition to diets, especially for individuals with gluten sensitivities (Assefa et al. 2011). They are well-suited to grow in challenging environmental conditions and have the ability to thrive in harsh weather conditions and poor soil quality makes them climate resilient crops. They can be vital in regions where other crops might struggle due to changing weather patterns (Burkill 1985). India stands out as one of the largest producers and consumers of millets globally. The widespread cultivation and consumption of millets in India contribute to the country's food diversity and security. Millets are not only important in India but also in other parts of Asia and Africa where they contribute to food security in regions facing climate challenges. Millets are of two types that includes major ones such as pearl millet (bajra), sorghum (jowar), finger millet (ragi), as well as minor millets such as foxtail millet (kangni), little millet (kutki), proso millet (cheena), kodo millet (koden), barnyard millet (sanwa), and browntop millet (andu korralu) (Gebretsadik et al. 2014). Each of these millets has its own nutritional profile and culinary uses. To boost the cultivation of nutritious millets and raise awareness among the people, the year 2023 has been designated as the "Global Year for Millets." Owing to millet attributes, beneficial uses, and the growing public interest in health, food scientists and researchers have developed an array of enhanced products derived from millets (Sheahan 2014). These value-added products hold significant market appeal, as consumers perceive millet-based products as highly advantageous for their well-being (Table 1).

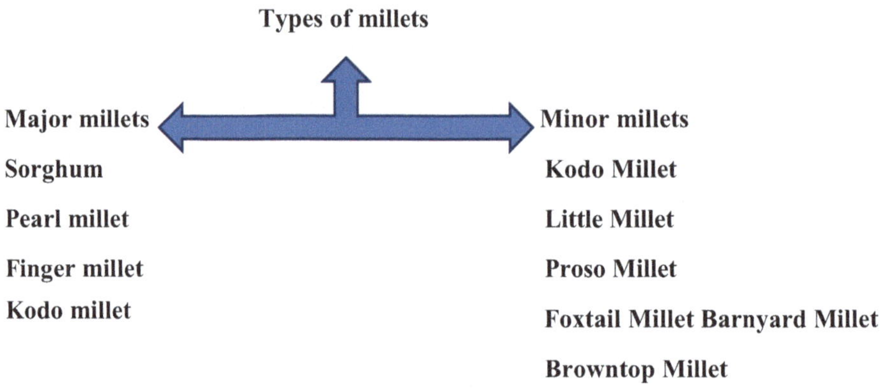

Table 1 General cultivation practices of minor millets

S. no.	Parameters	Crop-wise recommendations of minor millets					
		Little millet (*Panicum sumatrense* L.)	Proso millet (*Panicum miliaceum*)	Foxtail millet (*Setariaitalia*)	Barnyard millet (*Echinochloa utilis*)	Kodo millet (*Paspalum scrobiculatum*)	Browntop millet (*Brachiariaramosa* (L.))
1	Varieties	CLMV 1 (Jaicar Sama 1), Sree Neelima, OLM 203, OLM 217, CO 4, DHLM 36–3, JK 8, LMV 518	PR 18, TNAU 202, TNAU 145, GPUP 21, Bhawna, TNAU 164, TNAU 151, Sagar, Nagarjuna, CO 4, CO 3, ATL 1 (TNPm 230), GPUP 25	SiA 3156, SiA 3085, SiA 3088, SiA 2644, SiA 2593	DHBM 93–3, DHBM 23–3, CO 1, CO 2, VL 29, VL 172, VL 207	JK 155, GPUK 3, TNAU 86, JK 41, RK 390–25, RBK 155, ATL-2, BK-36	IIMR AK 2, HBr 2
2	Season	*Kharif*: June–July *Rabi*: September–October	*Kharif*: June–July *Rabi*: September–October	*Kharif*: June–July *Rabi*: September–October	*Kharif*: June–July	*Kharif*: June–July *Rabi*: September–October	Mid-April until mid-august
3	Seed rate (kg/ha)	For line sowing: 6–8 kg/ha For broadcasting: 10–12 kg/ha	For line sowing: 10–12 kg/ha For broadcasting: 15 kg/ha	For line sowing: 8–10 kg/ha For broadcasting: 15 kg/ha	For line sowing: 10 kg/ha For broadcasting: 15 kg/ha	For line sowing: 8–10 kg/ha For broadcasting: 12–15 kg/ha	For line sowing: 3–5 kg/ha
4	Spacing (cm)	22.5 cm row to row × 10 cm plant to plant is recommended	25 cm row to row × 10 cm plant to plant is recommended	25 cm row to row × 10 cm plant to plant is recommended	25 cm row to row × 10 cm plant to plant is recommended	22.5 cm row to row × 10 cm plant to plant is recommended	25 cm row to row × 10 cm plant to plant is recommended
5	Maturity (days)	75–90 days	70–90 days	80–90 days	85–100 days	80–90 days	90–100 days
6	Yield	Grain: 12–15 q/ha Fodder: 20–25 q/ha	Grain: 20–23 q/ha Fodder: 50–60 q/ha	Grain: 15–18 q/ha Fodder: 20–40 q/ha	Grain: 12–15 q/ha Fodder: 20–25 q/ha	Grain: 15–20 q/ha Fodder: 30–40 q/ha	Grain: 17–18 q/ha Fodder: 30–40 q/ha

2 Steps Involved in Cultivation and Harvesting of Minor Millets

The cultivation and harvesting of minor millets involved several steps, from seed quality and availability to reaching the consumer. The overview of each step is shown in Table 2:

Seed quality and availability:

- Ensure the availability of high-quality millet seeds adapted to the local environment.
- Encourage the use of certified millet seeds and provide support for seed quality testing.

Training and extension services:

- Provide training to millet farmers on modern and sustainable cultivation practices of minor millets.
- Extend agricultural extension services to disseminate knowledge about improved techniques, pest management, and resource optimization.

Input supply:

- Ensure timely and adequate supply of essential inputs such as fertilizers, pesticides, and water for millet cultivation.

Table 2 Steps in cultivation of minor millets

Seed quality and availability
↓
Input supply
↓
Cultivation and harvesting
↓
Training and extension services
↓
Postharvest handling
↓
Milling and processing
↓
Product diversification
↓
Quality control
↓
Market access and market research
↓
Market linkages
↓
Certification and delivery
↓
Consumer

Cultivation and harvesting:

- Prepare the land and sow millet seeds by following recommended practices as suggested by agriculture experts.
- Implement proper irrigation and weed control measures during the growing season.
- Harvest the millets at the right stage of maturity to ensure optimal yield and quality.

Postharvest handling:

- Proper handling and storage of harvested minor millets is essential to prevent contamination and to maintain quality.
- Implement postharvest practices to reduce losses and preserve nutritional content.

Milling and processing:

- Mill the harvested minor millets to obtain grains or flour.
- Invest in suitable processing machinery to enhance efficiency.
- Encourage the adoption of standardized processing methods for consistency in quality.

Product diversification:

- Explore opportunities for diversifying millet products and cater the consumer preferences.
- Encourage the development of millet value-added products such as snacks, cereals, and beverages.

Quality control:

- Implement quality control measures at various stages, from cultivation to processing.
- Establish quality standards and certifications to build consumer confidence.

Market access and market research:

- Conduct market research to understand consumer preferences and trends.
- Facilitate access to markets for farmers and processors by identifying potential buyers.

Market linkages:

- Establish linkages between millet producers, processors, and distributors to create ideal supply chain.

- Encourage collaborations with retailers, restaurants, and other outlets to increase market reach.

Certification and delivery:

- Obtain relevant certifications for organic or sustainable farming practices and ensure timely and reliable delivery of millet products to meet market demands.

Consumer:

- Create awareness among consumers about the nutritional benefits of minor millets and promote consumption through educational campaigns and marketing strategies.
- By addressing each of these steps, stakeholders can contribute to the development of a robust and sustainable minor millets industry, benefiting both farmers and consumers.

3 Understanding the Millet Value Chain

The traditional supply chain for millets faces several challenges that impact the overall efficiency and sustainability of the system (Duke 1983). The presence of numerous intermediaries, including brokers and middlemen, in the marketing process resulted in a fragmented supply chain, leading to increased costs and reduced profits for millet growers. Additionally, the lack of direct linkages between farmers and end consumers hampers transparency and traceability in the supply chain. This lack of transparency can contribute to quality issues and inconsistencies in product standards. Furthermore, the traditional supply chain may not adequately address the unique characteristics and nutritional benefits of millets, potentially limiting consumer awareness and demand (Ejeta 2007). Inadequate infrastructure, including storage and transportation facilities, poses another challenge, particularly in remote or rural areas where millets are often cultivated. These challenges collectively contribute to the vulnerability of millet farmers and hinder the growth potential of the millet industry. Addressing these issues requires innovative approaches, such as the intervention of Farmer Producer Organizations (FPOs) or the adoption of technology to streamline the supply chain and create a more sustainable and equitable market for millets (Bikkina et al. 2018). Here are some key differences and advantages associated with the involvement of FPOs in the marketing process:

Traditional marketing/supply chain:

1. **Dependence on middlemen:**
   (a) Millet farmers often rely on intermediaries or middlemen for selling their produce.
   (b) This dependence has led to lower prices for millet farmers and reduced bargaining power.

2. **Limited economies of scale:**
   (a) Individual millet grower has limited economies of scale in terms of production, processing, and marketing due to increased costs in value chain of minor millets.
3. **Lack of market access:**
   (a) In traditional supply chain, individual millet growers were not linked to markets directly, thus leading to limited exposure to market channels.
4. **Limited value addition:**
   (a) Value addition activities, such as processing and packaging, may be limited, reducing the overall value of the produce.

FPO intervention in millet value chain:

1. **Elimination of middlemen:**
   (a) Millet FPOs act as a collective entity by eliminating the need for middlemen in the marketing process.
   (b) Direct linkages between millet FPOs and retailers or consumers reduce the number of intermediaries involved in the value chain.
2. **Economies of scale:**
   (a) Millet FPOs aggregate the millet produce from multiple farmers, enabling economies of scale in procurement, processing, and marketing.
   (b) Bulk purchasing of inputs and collective bargaining enhance the financial benefits for farmers.
3. **Enhanced market access:**
   (a) Millet FPOs as organized entities can establish direct market linkages and negotiate better prices for their members.
   (b) Collaboration with retailers and participation in value chains provide broader market access.
4. **Value addition and processing:**
   (a) Millet FPOs may invest in processing units to add value to the produce, such as milling millets into flour or producing value-added products.
   (b) Processing and value addition activities contribute to higher returns for farmers.
5. **Input support and training:**
   (a) Millet FPOs provide support to farmers by facilitating access to quality inputs, agricultural training, and capacity building programs.
   (b) Knowledge transfer and skill development enhance the overall productivity and quality of the produce.
6. **Quality control:**
   (a) FPOs can implement quality control measures, ensuring that the millet produce meets industry standards.
   (b) Quality assurance enhances the marketability of the produce.
7. **Direct sale of value-added products:**
   (a) Millet FPOs can directly sell value-added products to retailers or consumers, capturing a larger share of the value chain.
   (b) Direct sales contribute to increased revenue for both farmers and the FPO.

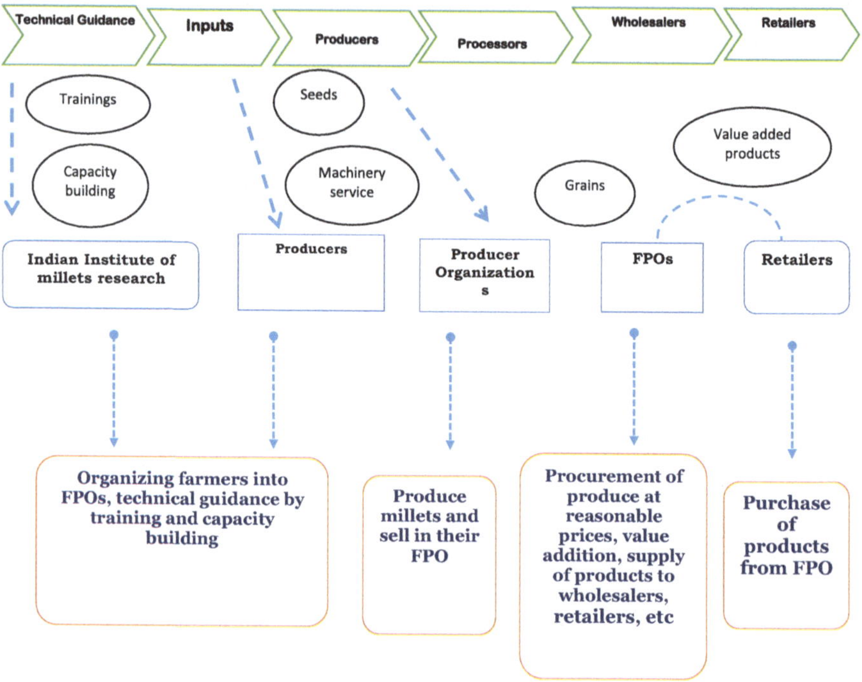

Fig. 1 Value chain of small millets and FPOs

The interlinkage between millet growers and FPOs, as described in your statement, represents a more organized and efficient approach to agricultural marketing as shown in Fig. 1. It helps empower farmers by providing them with collective strength, improved market access, and opportunities for value addition. Overall, FPO interventions contribute to the development of a more sustainable and inclusive agricultural value chain.

4 Millet Processing

The processing of minor millets involves a series of steps to make them fit for consumption and to create a variety of value-added products (Fig. 2). Primary processing methods are employed to clean and prepare the millets for further use. These methods include dehulling, soaking, germination, roasting, drying, polishing, and milling (size reduction). Dehulling is the removal of the outer husk or hull from the millet grains, while soaking and germination are often used to enhance nutritional content (Sangappa et al. 2023). Roasting and drying help improve flavor, reduce moisture, and increase shelf life. Polishing involves removing the outer layer to create a smoother texture, and milling reduces the size of the grains. Simultaneously, millet-based value-added processed food products are manufactured using secondary processing techniques (Hariprasanna and Rakshit

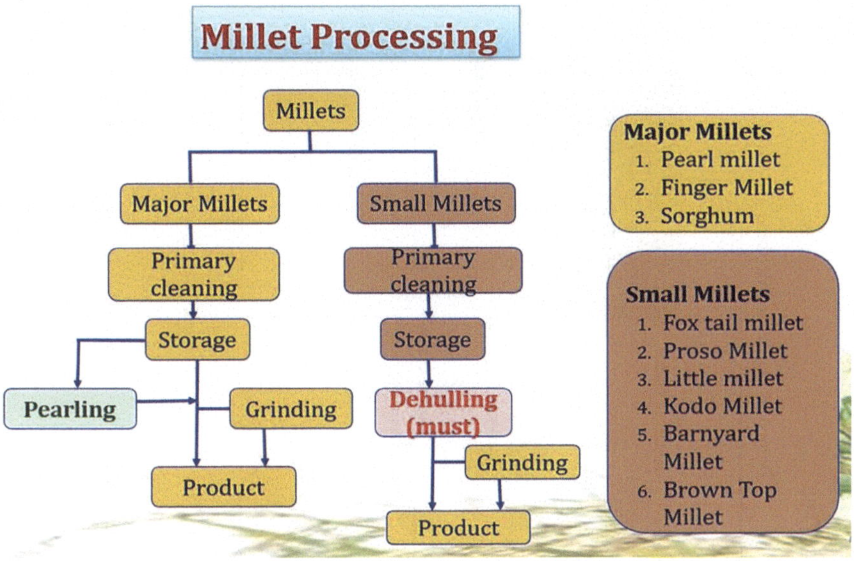

Fig. 2 Flowchart of millet processing unit

2016). These techniques include fermentation, parboiling, cooking, puffing, popping, malting, baking, flaking, extrusion, among others. Fermentation can enhance the nutritional profile and digestibility of millet-based products, while parboiling involves partially boiling the grains before further processing. Cooking, baking, and extrusion are methods used to prepare a wide range of millet-based food items, from bread and snacks to cereals and pasta. Puffing and popping involve the expansion of millet grains, creating light and crunchy textures in snacks. The overall processing of millets can be categorized into three stages: primary processing, where the raw millet is cleaned and prepared; secondary processing, where value-added products are created using various modern techniques; and tertiary processing, which involves additional steps such as packaging and labeling to prepare the final products for distribution and consumption (Pandey and Bolia 2023).

5 Primary Processing of Minor Millets

Primary processing of millets is a vital step to convert the grain into edible form and thereby enhancing its quality (Doggett 1989). Although processing millets without husk (naked grains), i.e., sorghum, pearl, and finger millets are easy, whereas processing of minor millets with husk, i.e., little, proso, kodo, barnyard, and foxtail millets is difficult (Gupta et al. 2009). These have an inedible husk, which needs to be removed through processing (Figs. 3 and 4). The major challenges in the processing of small millets are as follows.

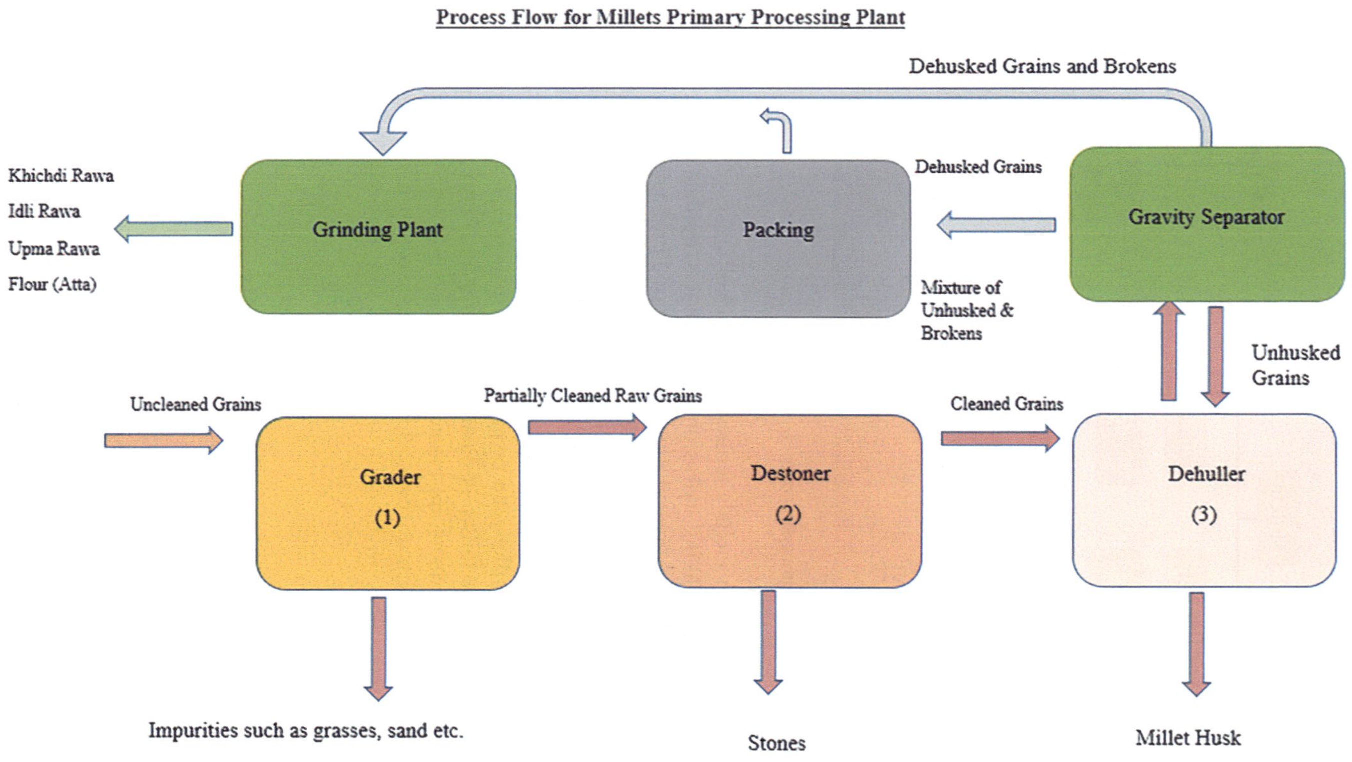

Fig. 3 Process flow for millets primary processing plant

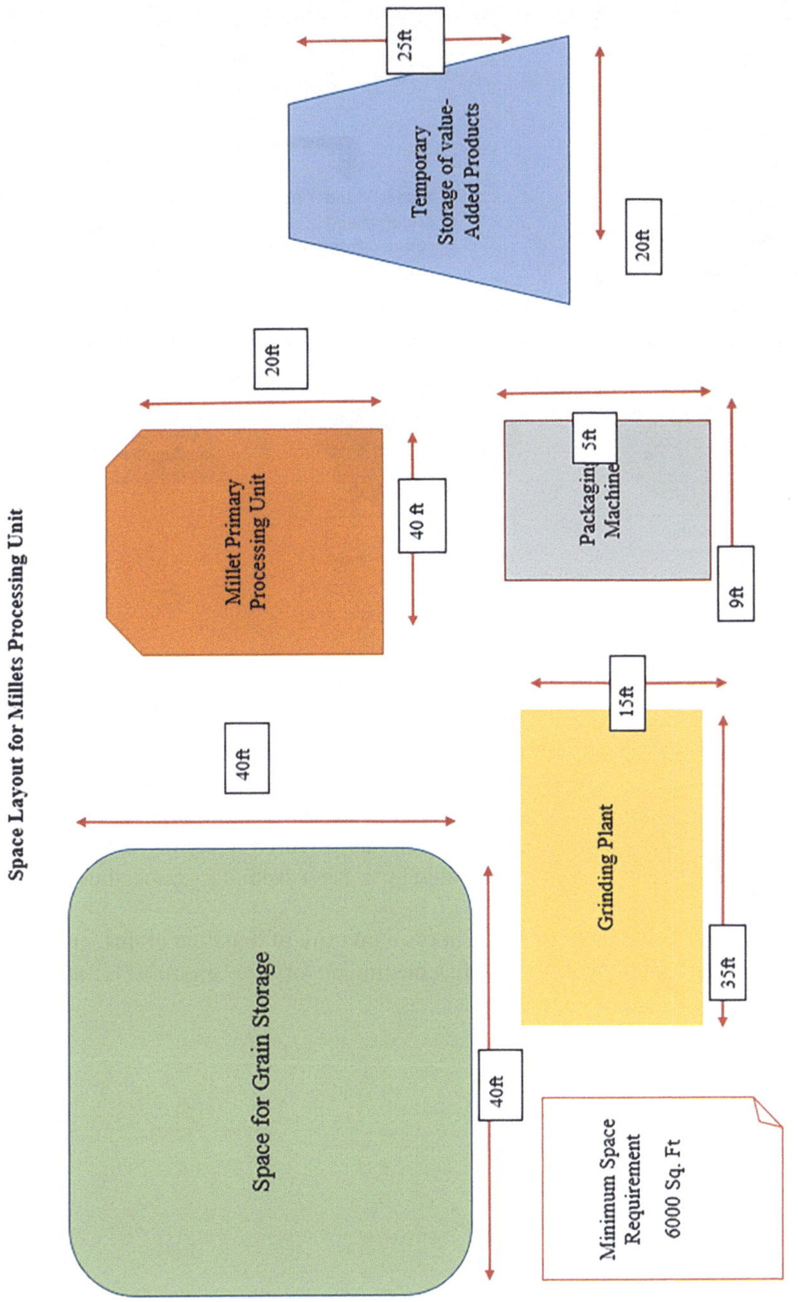

Fig. 4 Space layout for millets processing units

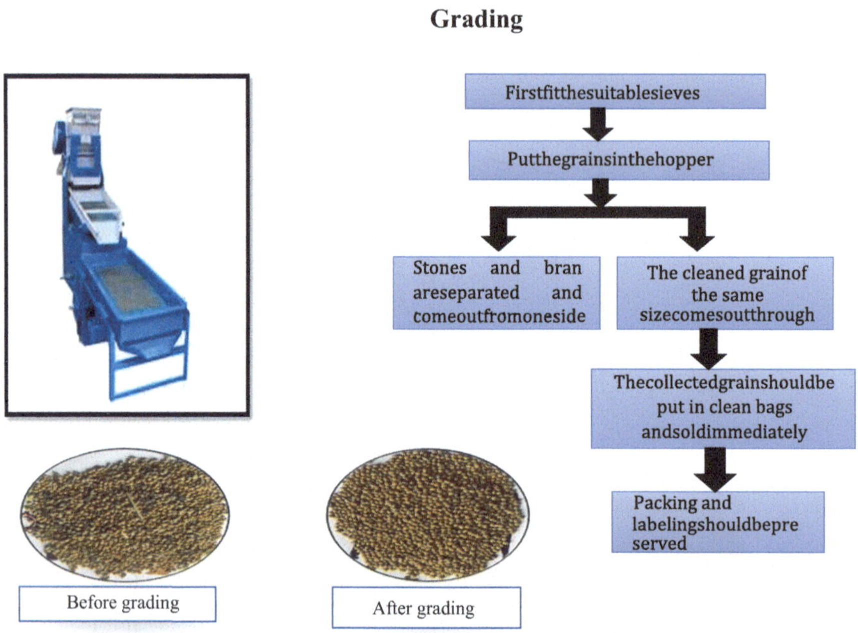

Fig. 5 Grading mechanism of minor millets

5.1 Steps Involved in Primary Processing of Millets

- *Cleaning*: Procured grains from farmers come as a mixture of grains, stones, dust, grasses, and so on. So, the first step will be to get rid of all these impurities. Destoner-cum-grader-cum-aspirator is the machine used for this work.
- *Dehulling*: Cleaned raw grains are then subjected to dehulling, "which is the procedure to remove outer indigestible husk layer from the grains, thus improving its overall digestibility."
- *Separation*: Output from the dehuller is a mixture of dehulled grains, un grains, and brokens. In this step each constituent of the mixture is separated (Figs. 5 and 6).

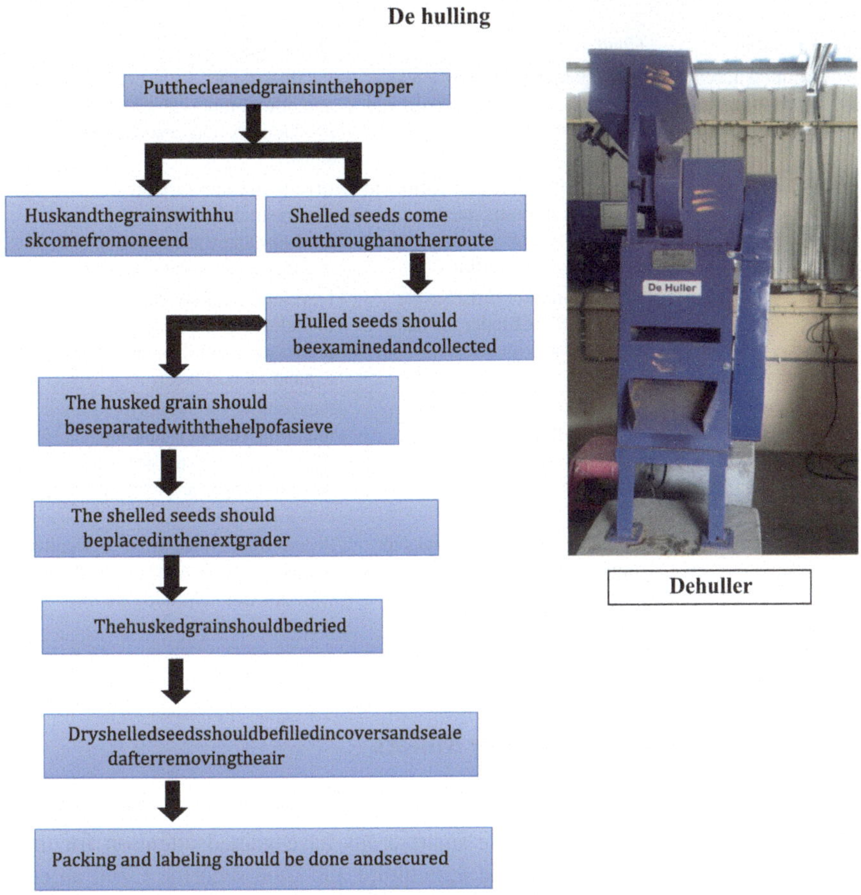

Fig. 6 Grading mechanism of minor millets

6 Challenges in Minor Millets Cultivation

S. no.	Particulars	Challenge	Opportunity
1	Limited awareness and demand	There is often limited awareness among consumers about the nutritional benefits of minor millets, leading to low demand	Educating consumers about the nutritional value and health benefits can create increased demand and market acceptance
2	Fragmented supply chain	The supply chain for minor millets is often fragmented, with smallholder farmers producing limited quantities	Developing efficient aggregation systems and promoting collective farming can enhance the scale of production
3	Lack of processing infrastructure	Inadequate processing facilities hinder the development of value-added products	Investments in processing infrastructure and technology can improve efficiency, reduce losses, and diversify product offerings

S. no.	Particulars	Challenge	Opportunity
4	Limited access to credit and inputs	Smallholder farmers often face difficulties accessing credit and quality inputs	Facilitating access to credit, seeds, and other inputs can enhance productivity and ensure a stable supply
5	Postharvest losses	Minor millets are susceptible to postharvest losses due to inadequate storage and transportation facilities	Investment in postharvest management, including storage facilities and transportation networks, can reduce losses
6	Market linkages	Limited market linkages for minor millets hinder farmers' ability to access broader markets	Developing strong market linkages, including partnerships with retailers, processors, and exporters, can expand market reach
	Policy support	Limited policy support for minor millets in terms of subsidies, research, and extension services	Advocating for supportive policies, including research funding, subsidies, and extension services, can encourage farmers to cultivate minor millets
	Climate change risks	Climate change poses risks to minor millet cultivation	Promoting climate-smart agriculture practices and developing resilient varieties can help mitigate the impact of climate change

7 Opportunities in Minor Millets

Minor millets present a host of opportunities that extend beyond their nutritional benefits. These resilient and climate-smart crops have gained attention for their potential to contribute to sustainable agriculture and food security. From an economic perspective, there are opportunities for diversification in agriculture, providing farmers with alternative crops that are well-suited to varying agro-climatic conditions. Minor millets are known for their ability to grow in marginal lands, requiring fewer inputs such as water and fertilizers, making them suitable for resource-constrained environments (Seetharam 2015). As consumer awareness about health and nutrition continues to rise, there is an opportunity for the marketing and promotion of minor millets as nutritious superfoods. The growing interest in gluten-free and natural foods also positions minor millets as attractive options for those with dietary preferences or restrictions (Sood et al. 2015). Moreover, the value chain associated with minor millets offers opportunities for entrepreneurship, including processing and value addition, leading to the development of a range of innovative and marketable products. Embracing these opportunities can not only contribute to the well-being of farmers but also foster sustainable agricultural practices and promote dietary diversity, aligning with global goals for a resilient and nutritious food future.

8 Conclusion

The value chain development of minor millets presents a dynamic landscape marked by both challenges and opportunities. Addressing the low productivity associated with traditional farming practices is a pivotal challenge. However, overcoming this hurdle through the introduction of high-yielding varieties and improved agronomic practices can significantly enhance the productivity of minor millet cultivation. Creating robust market linkages, coupled with investments in storage and transportation infrastructure, is essential to bridge the gap between producers and consumers. Additionally, there is a need for concerted efforts to raise awareness and alter consumer perceptions regarding the nutritional benefits of minor millets. Challenges such as the impact of climate change on cultivation can be mitigated through the development of climate-resilient varieties and the promotion of sustainable agricultural practices. Leveraging the nutritional richness of these grains and positioning them as health-conscious choices can tap into a growing market segment. Diversifying product offerings, including flour, flakes, and snacks, can cater to diverse consumer preferences. Moreover, fostering inclusive value chains by empowering smallholder farmers, particularly women, and garnering government support through subsidies and policy incentives, can contribute to the overall success and sustainability of minor millet value chains. In conclusion, a strategic and collaborative approach is paramount to realizing the full potential of minor millets, overcoming challenges, and capitalizing on the multitude of opportunities for sustainable agricultural development.

References

Assefa K, Yu JK, Zeid M, Belay G, Tefera H, Sorrells ME (2011) Breeding tef [Eragrostis tef(Zucc.) trotter]: conventional and molecular approaches. Plant Breed 130:1–9

Bikkina N, Turaga RM, Bhamoriya V (2018) Farmer producer organizations as farmer collectives: a case study from India. Dev Policy Rev 36(6):669–687. https://doi.org/10.1111/dpr.12274

Burkill HM (1985) The useful plants of west tropical Africa, vol 2. Royal Botanic Gardens, p 648

Doggett H (1989) Small millets—a selective overview. In: Seetharam A et al (eds) Small millets in global agriculture. Oxford & IBH Publishing Co., Delhi, pp 59–70

Duke JA (1983) Handbook of energy crop. http://www.hort.purdue.edu/newcrop/dukeenergy/arundodonax

Ejeta G (2007) The Striga scourge in Africa–a growing pandemic and integrating new technologies for striga control towards ending the witch-hunt. World Scientific Publishing Co, Singapore, pp 3–16

Gebretsadik R, Shimelis H, Laing MD, Tongoona P, Mandefro N (2014) A diagnostic appraisal of the sorghum farming system and breeding priorities in striga infested agro-ecologies of Ethiopia. Agric Syst 123:54–61

Gupta A, Mahajan V, Kumar M, Gupta HS (2009) Biodiversity in the barnyard millet (Echinochloafrumentacea link: Poaceae) germplasm in India. Genet Resour Crop Evol 56:883–889

Hariprasanna K, Rakshit S (2016) Economic importance of millets. Springer International Publishing, Singapore, pp 1–25

Pandey A, Bolia NB (2023) Millet value chain revolution for sustainability: a proposal for India. Socio Econ Plan Sci 101592:101592

Sangappa, Dudekula R, Laxmi B, Charishma E, Babu KS (2023) Strengthening millet value chain through farmer producer organizations. Indian J Extension Educ 59(3):26–31. https://epubs.icar.org.in/index.php/IJEE/article/view/136293

Seetharam A (2015) Genetic improvement in small millets: ensuring climate resilience and nutritional security. Daya Publishing House, New Delhi, p 649

Sheahan CM (2014) Plant guide for minor millets USDA-Natural Resources Conservation Service. Cape May Plant Materials Center, Cape May

Sood S, Khulbe RK, Gupta AK, Agrawal PK, Upadhyaya HD, Bhatt JC (2015) Barnyard millet—a potential food and feed crop of future. Plant Breed 134:135–147

Frontline Demonstrations in Small Millets: Dissemination and Impact Assessment

T. S. Sukanya, Dinesh Chandra Joshi, K. Sathya, A. L. Narayanan, Ajay Kumar, C. Chaithra, Jyotsna Kiramnai, H. E. Patil, Nagendra Kumar Nag, Manisha Shyam, and Yogesh Gangadhar Ben

Abstract

Small millets productivity in India is much lesser than potential yield for the reason of inadequate crop management. Frontline demonstrations (FLDs) program is one of the approaches and an effective tool for disseminating farm tech-

T. S. Sukanya (✉) · C. Chaithra
Project Coordinating Unit, ICAR-AICRP on Small Millets, GKVK, Bengaluru, Karnataka, India

D. C. Joshi
ICAR-Vivekananda Institute of Hill Agriculture, Almora, Uttarakhand, India

K. Sathya
ICAR-AICRP on Small Millets, Centre of Excellence in Millets, Athiyandal, Tamil Nadu, India

A. L. Narayanan
Pandit Jawaharlal Nehru College of Agriculture and Research Institute, Govt. of Puducherry, Karaikal, Puducherry, India

A. Kumar
ICAR- AICRP on Small Millets, Uttarakhand University of Horticulture and Forestry, Ranichauri, Uttarakhand, India

J. Kiramnai
ICAR-AICRP on Small Millets, Nandyal, Andhra Pradesh, India

H. E. Patil
ICAR-AICRP on Small Millets, Waghai, Gujarat, India

N. K. Nag
ICAR-AICRP on Small Millets, Indira Gandhi Agricultural University, Raipur, Chhattisgarh, India

M. Shyam
ICAR-AICRP on Small Millets, Dindori, Madhya Pradesh, India

Y. G. Ben
ICAR-AICRP on Small Millets, Kolhapur, Maharashtra, India

D. C. Joshi et al. (eds.), *Minor Millets*,
https://doi.org/10.1007/978-981-96-4265-6_14

nologies generated by research system to bridge the yield gap. The "Cluster Frontline Demonstration on Small Millets" was conducted mostly during *kharif* season of 2010–11 to 2022–23 in different states, namely Andhra Pradesh, Gujarat, Karnataka, Chhattisgarh, Maharashtra, Madhya Pradesh, Tamil Nadu, and Uttarakhand under ICAR-AICRP on Small Millets, India, stated a wide yield gap between yields of potential and demonstration owing to extension and technology yield gaps. Over the base of 13 years data, demonstrations gave a mean of 41.16% (barnyard millet), 43.22% (finger millet), 43.74% (foxtail millet), 62.16% (little millet), 62.47% (kodo millet), and 68.42% (proso millet) higher yield than farmers practices. Extension yield gaps varied among different small millets (463, 475, 566, 569, 607, 632, and kg/ha in barnyard millet, little millet, foxtail millet, kodo millet, proso millet, and finger millet, respectively) during the study period. Further, technology index in finger millet (5.68–35.40%) and other small millets (11.99% in foxtail millet; 39.83% in kodo millet; 16.89% in barnyard millet; 12.72% in proso millet; 14.29% in little millet) has denoted that demonstrated technology in small millets is rather viable under prevailing agriculture situations in different states, however, requires to be disseminated in concentrated means to communicate farmers to enhance small millets productivity and farm profitability. Upgraded technology package resulted in enhanced profitability in small millets in respect of gross and net returns besides enhancing benefit–cost ratio. In general, it is implied that adoption of improved production technology has an exceptional potential in enhancing the small millets productivity via frontline demonstration strategy.

Keywords

Small millets · Frontline demonstration · Productivity · Technology gap · Extension gap · Technology index · Profitability

1 Introduction

India is the principal millets producer and consumer, crops that are habitually designated as coarse cereals. Nowadays, these are further popular compared to any other cereals due to their distinctive qualities of adaptability to array of marginal growing settings, nutritious, and survivability under severe abiotic and biotic stresses (Sukanya et al. 2023). Yet, apprehending the nutrient formulations of these grains and their special adaptations to rainfed areas, they are nowadays considered as nutricereals and recently as "Sree Anna," which means the best among all the food grains. The comprehensive millets cataloguing divides crops into two groups as major and small or minor millets. The small millets cover finger millet, foxtail millet, little millet, proso millet, barnyard millet, kodo millet, and browntop millet, and these are the key food sources for large number of people, exclusively as principal food for the people who are tribes living in warm and drier zones (Amadou et al. 2013).

2 Scenario of Small Millets

These small millets farming spreads from the height of sea level on the coastal Andhra Pradesh to more than 2000 m above mean level in Uttarakhand and North-Eastern states. Altogether of 9.95 m ha of millets, finger millet and other small millets occupy 1.65 m ha in which, finger millet takes a major share of 1.22 m ha (Ministry of Agriculture and Farmers Welfare 2023). The finger millet area was 2.19 m ha during 1969–70 upsurged towards 2.78 m ha but abruptly declined to 1.22 m ha in 2021–22 (Fig. 1). While, the area of other small millets during 1951–52 was 4.76 m ha but dropped down to 0.43 m ha; however, the productivity is interestingly higher from 402 kg/ha to 857 kg/ha (Fig. 2).

Small millets are grown over varied soils, rolling landscape or topography, erratic rainfall regimes, and extensively contrary in photoperiods and air temperature. The crop resilience displayed by millets is supportive in fine-tuning to diverse biological niche. Meanwhile, their eco-friendly nature and comparatively free from key diseases and pests are included under organic cultivation. Hence, these are crucial element of dryland agriculture, tribal farming, and hill farming. Despite these advantages, these crops are ignored for cultivation under larger area. However, with improved agricultural techniques, the circumstances are rapidly changing in many developing countries.

The finger millet cultivating states are Karnataka, Maharashtra, Uttarakhand, Tamil Nadu, Odisha, Madhya Pradesh, Jharkhand, Gujarat, Andhra Pradesh, West

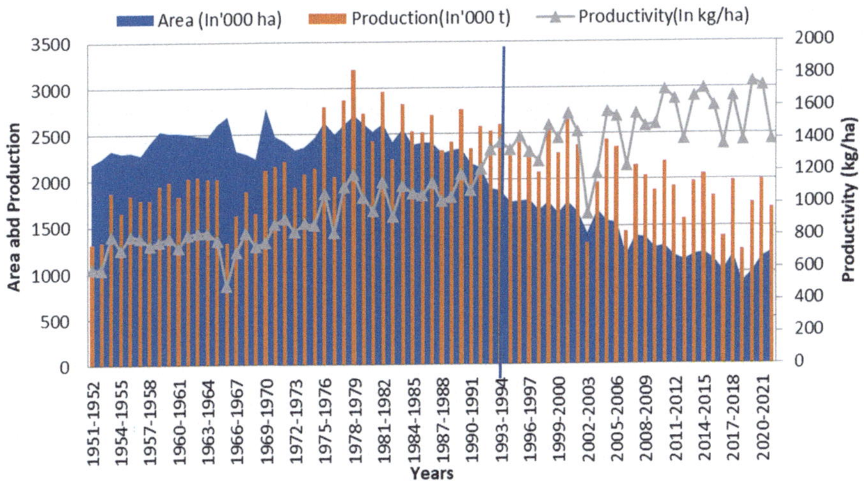

Fig. 1 The finger millet status (area, production, and productivity) in India

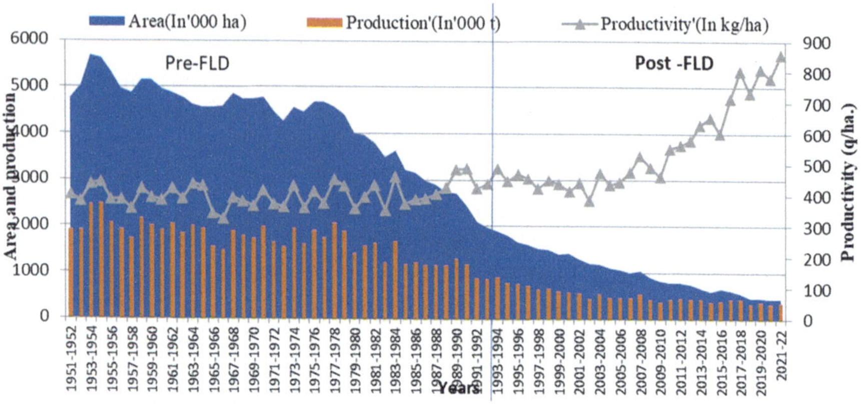

Fig. 2 The area, production, and productivity of small millets (excluding finger millet) in India

Bengal, and Chhattisgarh, whereas Chhattisgarh, Uttarakhand, Maharashtra, Odisha, Karnataka, Arunachal Pradesh, Andhra Pradesh, Tamil Nadu, and Gujarat are major other small millets growing states. Being eco-friendly, these crops are appropriate for fragile and susceptible ecologies (Michaelraj and Shanmugam 2013). The use of these crops is mostly for food meant for human consumption. Nevertheless, straw is often a valuable feed for bovines and for bird's feed. The promotion of these crops could lead to effective natural resource supervision and holistic approach in nourishing precious agro-biodiversity (Sukanya et al. 2024a, b).

The reviews have shown that the low yield of dryland crops such as small millets is due to lack of cultivation information among farmers on high yielding varieties and other improved cultivation practices. The main factors directly influence yield are (1) cultivating of improved varieties, (2) plant nutrients application in optimum extent, and (3) effective crop management. A critical task for agrarian extension deceits with conveying the main advance in agriculture technologies to farmers proficiently and in a timely manner. Small millets production is less owing to incessant usage of poor unrestored landraces that are generally liable to diseases, especially low yielding and blast tolerant, inadequate information on improved varieties, scanty distribution of seeds, and postharvest management related to crop (Degu et al. 2009; Molla et al. 2020; Sukanya et al. 2022) (Plate 1).

3 Frontline Demonstrations

Frontline demonstration (FLD) is the approach that shows the value of improved crop production know-hows and its management applications and further convinces the farmers about their capabilities for extra spread. However, the acceptance of the

Plate 1 Imparting training to FLD farmers in the demonstration village

technologies in most of crops by farmers is not high (Kalita et al. 2019; Kumar and Yadav 2007). The field demos are navigated with the close direction of researchers or scientists and verified for the foremost time by scientists beforehand being pushed into the core extension system of the State Agriculture Department. The need and necessity of demonstration consistently be grounded on the emerging issues. FLDs are implemented in an orderly means and this is an efficient knowledge transfer means for improved technology adoption that links the yield gaps (Plate 2).

Frontline demonstrations are shown in a cluster so as to have well influence of the demonstrated technologies on growers and extension officials. Under the scheme, only critical inputs and capacity-building programs are imported and the most prerequisite is the training of the farmers connected through the FLDs budget, the remaining inputs are provided by the farmers themselves. The target audiences of the FLD are farmers and the extension officers, and the main purpose is to persuade extension officials and farmers together round the technology potentials for further extensive scale dissemination. The purpose of FLDs is not only to examine the effectiveness of the technology under real farm circumstances but also to persuade farmers, subject matter specialists, and field extension officials, as these technologies are organized under the straight direction of who generated the technology. Adoption.

4 Operational Details

AICRP—All India Coordinated Research Project on Small Millets—is a multi-crop project. The "Cluster Frontline Demonstration on Small Millets" was conducted by "AICRP on Small Millets" to recognize the yield and to find out the technological gap, extension gap, and economics during the year 2010 and 2022. The FLD

Plate 2 Scientists and extension officials along with farmers at the FLD plot

concept was introduced in small millets in the year 1993–94. Finger millet or ragi and five other small millets, namely foxtail millet, kodo millet, little millet, barnyard millet, and proso millet, have been encompassed under this movement in different states based on their local status. The size of individual demonstration was not less than 0.4 hectare or 1 acre and clustering of two or more demos by choosing adjoining fields of farmers was attempted including mostly small and marginal farmers. Other equal size plots of the representing farmers or the equal size of plot of neighboring FLD farmers under the similar farming situation represent the control plots for objective of results comparison. Rigorous trainings were given to the FLD farmers and field day and group meetings were held at site of demonstration to attain extensive dissemination of skills and technologies. A more or less participatory approach was invariably implemented by collaborating with Krishi Vigyan Kendra, State Department of Agriculture, State Agriculture Universities, and Civil Society Organizations. The brief about the technologies implemented in the farmer's field is given in Table 1 and Fig. 3 and the varieties cultivated under FLDs are represented in Tables 2 and 3, Plates 3 and 4.

After harvesting the crop, the information obtained from both demonstration and farmer's plot were analyzed and the economical parameters such as cost of production, net return, and benefit–cost ratio were calculated. In all, 5395 ha demonstration were conducted involving 17,099 farmers in the year 2010–2022. Data on expenditure from 2018 to 2022 incurred by demonstration plots and farmer (farmer's practice) were analyzed. Based on local market prices of small millets, the gross income was calculated and net income was obtained by deducting the total cost of production from gross return. By dividing gross returns with cost of production in all small millets, the benefit–cost proportion was computed. The extension,

Table 1 The technologies adopted in the frontline demonstrations

Operations	Particulars
Sowing	2nd fortnight of June to first fortnight of August
Seed amount or seed rate	10 kg/ha (4 kg/acre)
Plant geometry or spacing	22.5–30 cm row spacing and 7.5–10 cm plant to plant spacing
Plant population	3.33–4.44 lakh per ha
Seed treatment	*Azospirillum brasilense* and *Aspergillus awamori* @ 25 g/kg of seeds
Inorganic nutrient or fertilizer dose ($N:P_2O_5:K_2O$ kg/ha):	50:40:25 $N:P_2O_5:K_2O$ kg/ha for rainfed situation and 100:50:50 kg/ha $N:P_2O_5:K_2O$ kg/ha for irrigated crop. 40:20:20 $N:P_2O_5:K_2O$ kg/ha for kodo millet, foxtail millet, little millet, barnyard millet, and proso millet (50% nitrogen and complete P_2O_5 and K_2O as initial or basal dose and remaining nitrogen at 30–35 days after sowing as topdressing). The regional fertilizer recommendation based on soil test is to be adopted
Irrigation	If available when crops experience longer dry spell at critical periods of moisture first at germination, second at initiation of panicle, and third at grain filling period
Weed control or management	Isoproturon Spray @ 0.5 kg a.i./ha instantly after sowing. The intercultivation operations at two times around 25 and 35–40 DAS and one hand weeding
Plant protection	Necessity-based plant protection measures

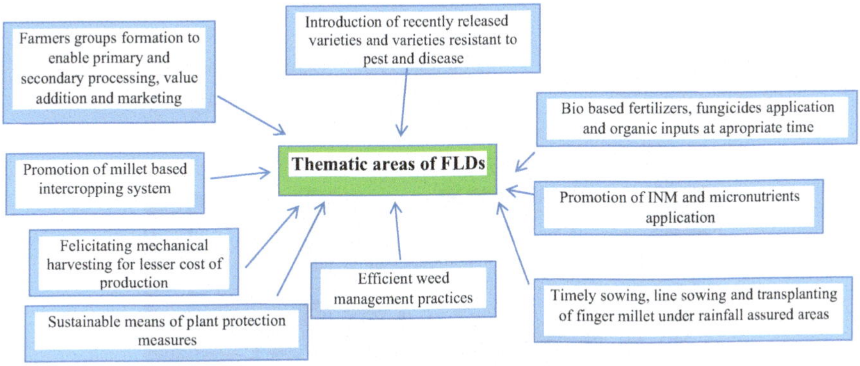

Fig. 3 Frontline demonstrations technologies demonstrated

technology index, and technology gap were computed by means of the formulas as mentioned below as proposed by Kadian et al. (1997) and Samui et al. (2000).

Table 2 List of varieties used in frontline demonstrations

State	Varieties
Finger millet	Arjuna (OEB-526), DHFM-78-3, GPU 45, GPU 48, GPU 28, GPU 66, GPU 67, Indaf 7, KMR 204, KMR 301, KMR-316, KMR 340, KMR 630, MR 1, MR 6, ML 365, PR 202, VL Mandua 352 (VL 352), VL Mandua 376 (VL 376), Vakula (PPR 2700)
Foxtail millet	DHFt 109–3, GPUF 3, HMT 100–1, HN-46, Narasimharaya, PS 4, SiA 326, SiA 3088, SiA 3085, SiA 3156
Little millet	BL-41-3, Chhattisgarh Kutki 1 (BL-6), DHLM 36–3, DHLM 14–1, GPUL 6, GPUL 11, JK 8, LMV 518, OLM 203
Proso millet	ATL 1 (TNPm 230), DHPM-2769, GPUP 8, GPUP 21, GPUP 25, GPUP 28, GPUP 32, HB 1, TNAU 145, TNAU 151, TNAU 202, TNAU 164
Kodo millet	ATL-2, BK-36, RK 390–25, GPUK 3, RBK 155, TNAU 86
Barnyard millet	DHBM 23–3, DHBM 93–3, DHBM 93–2, RAU 11, VL 172, VL 181
Browntop millet	HBR 2, GPUBt 2

Table 3 The formulas for calculating extension gap, technology gap, and technology index

Sl. no.	Particulars	Formula
1	% yield increase over farmer's practice	(Demonstration Yield – Farmers yield/ Farmers yield) × 100
2	The extension gap	Demonstration yield – Farmers yield
3	The technology gap	Potential yield – Demonstration yield
4	The technology index	{(Potential yield – Demonstration yield) × 100}/ Potential yield

Plate 3 Women participation in FLD program in Uttarakhand state

Plate 4 Scientists visit to the FLD in Chhattisgarh state

5 Selection of Site for FLD

The demonstrations site must be at a place effortlessly nearby and at fundamental point to appeal large audience/farmers number for maximum impact, effective monitoring, and feedback. The technology designated for demonstration must be of paramount status, and in order to form good and evident impact of a technology, demonstrations of at least 10.0 ha in cluster form must be implemented. One demonstration plot should never be less than 0.4 ha but not exceeding 1 ha keeping in view the technical prerequisite for consistency. Participatory approach is required in following demonstrations collaborating farm scientists, extension workers, and representing farmers to have operative operation, leading to better acceptance and dissemination of technology.

6 Critical Inputs Supply to Demonstrations and Other Requirements at Conduction of FLD

Always preceding to the beginning of crop season, meetings will be apprehended in the villages to educate implementers or beneficiaries about the objectives and drive of the demonstrations. This would help in launching moral understanding between farmers and scientists and aid in measuring the socioeconomic situations, in addition to current level of technologies adoption and farmers receptivity to current technologies. Further, it also serves as yardstick for judging the effect after demonstration. The critical inputs cover mostly seeds that are usually seeds of high-yielding improved varieties, nutrient mixtures preferably organic based, necessary plant protection chemicals, and machineries require for taking field operations and other inputs that are found critical. Essential monitoring was completed at recurrent intervals by speaking to farmers and extension workers, visiting demonstration plots with the eventual aim of attaining as much response as probable. The data on yield and other necessary evidence were gained from distinct demonstrations via a proforma specially intended for the purpose. This was analyzed, consolidated, and the outcomes were deliberated at the annual workshop or group meeting and know-hows demanding refinement were acknowledged. The data generated was spread in various ways through communication media and methods.

7 Results of Frontline Demonstrations

Across countries, namely Andhra Pradesh, Karnataka, Maharashtra, Gujarat, Chhattisgarh, Tamil Nadu, Madhya Pradesh, and Uttarakhand on farmer's fields, frontline demonstrations on small millets remained organized to show the productivity potential and profitability. The demos on whole package conducted from 2010–11 to 2022–23 were compiled. The frontline demonstration total area was 5396 ha in which 2430 ha in finger millet, 923 ha in little millet, 853 ha in foxtail millet, 696 ha in kodo millet, 410 ha in barnyard millet, and 84 ha in proso millet (Fig. 4), besides farmers' number involved under FLDs on Small Millets from the year 2010 to 2022 are depicted in Fig. 5 (Plate 5).

Among different small millets demonstrations, the highest FLD area for finger millet was found in Karnataka, while Andhra Pradesh for foxtail millet, Madhya Pradesh for little millet and kodo millet, Uttarakhand for barnyard millet, and Tamil Nadu for proso millet were observed.

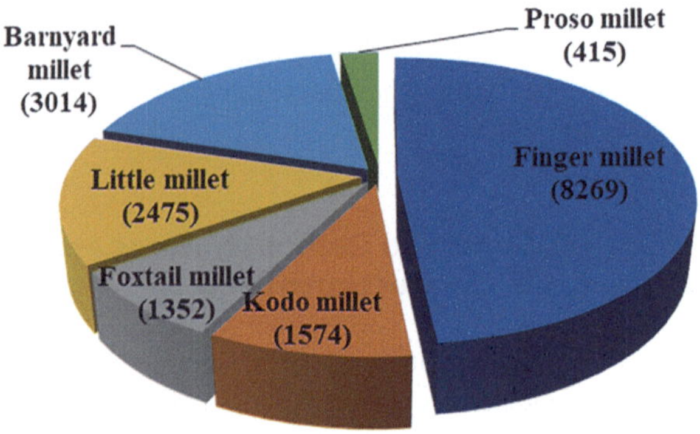

Fig. 4 Crop-wise division of area covered (ha) under FLDs on Small Millets from 2010 to 2022

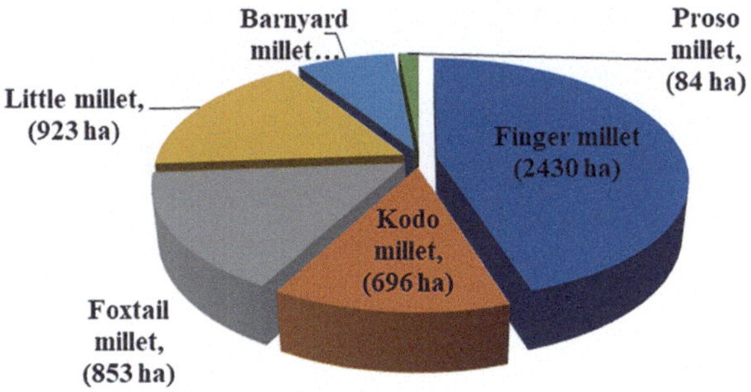

Fig. 5 Crop-wise farmers' number involved under FLDs on Small Millets from 2010 to 2022

Plate 5 FLD in Waghai, Gujarat state

8 Grain Yield and Percent Increase Yield

Facts on grain yield (Table 4) exposed that the technologies transfer under frontline demonstrations in small millets caused in outstanding higher grain yield under demonstrations (1689–2681 kg/ha in finger millet; 1664–2342 kg/ha in foxtail millet; 897–1875 kg/ha in little millet; 1243–2333 kg/ha in kodo millet; 1495–1675 kg/ha in barnyard millet; 1382 kg in proso millet) than farmers' plot yield (677–2087 kg/ha in finger millet; 1097–1831 kg/ha in foxtail millet; 840–1466 kg/ha in kodo millet; 658–1217 kg/ha in little millet; 1058–1200 kg/ha in barnyard millet; 1537–1840 kg in proso millet), which credited to the implementation of suggested technologies in FLDs all through study period. It is likewise perceived that an average percent yield (2010–2022) increase over farmer's practice was 43.22, 43.74, 62.16, 62.47, 41.16, and 68.42 for finger millet, foxtail millet, little millet, kodo millet, barnyard millet, and proso millet, respectively (Table 5). The outcomes are in homogeneity with the results of

Table 4 Grain yield (pooled over the years from 2010 to 2022) recorded at FLD on Small Millets

States/crops	Finger millet		Foxtail millet		Little millet		Kodo millet		Barnyard millet		Proso millet	
	FLD	Farmers practice	FLD	Farmers practice	FLD	Farmers practice	FLD	Farmers practice	FLD	Farmers practice	FLD	Farmers practice
Andhra Pradesh	2340	1636	2342	1831	1578	783	–	–	–	–	–	–
Chhattisgarh	1689	677	–	–	897	842	1243	840	–	–	–	–
Jharkhand	2237	1798	–	–	–	–	–	–	–	–	–	–
Karnataka	2649	1923	1664	1097	1794	568	1764	1198	–	–	–	–
Madhya Pradesh	–	–	–	–	1110	953	1454	916	–	–	–	–
Maharashtra	1892	1409	–	–	1410	860	–	–	–	–	–	–
Odisha	2226	1115	–	–	1206	995	–	–	–	–	–	–
Pondicherry	2470	2087	–	–	–	–	–	–	–	–	–	–
Tamil Nadu	2681	1245	1750	1172	1423	615	2333	1466	1495	1058	1537	949
Uttarakhand	1813	1076	–	–	–	–	–	–	1675	1200	1840	1153
Gujarat	1941	1636	–	–	1875	371	–	–	–	–	–	–

Table 5 The % yield increase, extension yield gap, technology gap, and technology index observed in FLD on small millets from the year 2010 to 2022

	Finger millet				Foxtail millet				Kodo millet			
Year	% yield increase	Extn. gap (kg/ha)	Tech. gap (kg/ha)	Tech. gap Index (%)	% yield increase	Extn. gap (kg/ha)	Tech. gap (kg/ha)	Tech. gap index (%)	% yield increase	Extn. gap (kg/ha)	Tech. gap (kg/ha)	Tech. gap index (%)
2010	59.6	836	135	5.68	47.98	843	−477	−22.47	57.1	751	684	24.87
2011	43.8	610	521	20.63	77.26	862	273	12.13	72.15	550	1389	51.44
2012	62.6	713	1015	35.4	47.88	491	732	32.53	69.64	588	1507	64.13
2013	30.7	528	521	18.84	61.37	759	−509	−22.62	22.7	316	613	26.39
2014	39.9	639	734	24.67	21.17	440	−216	−9.39	42.84	478	706	30.7
2015	37.5	565	478	18.75	31.66	614	−237	−10.24	40.68	485	581	25.73
2016	55.1	754	279	11.63	5.05	85	532	23.13	125.99	709	1278	50.12
2017	44.7	698	730	24.41	39.16	512	556	23.41	66.72	575	1292	47.33
2018	10.0	100	835	28.6	35.91	362	1380	50.18	104.61	757	1219	45.15
2019	59.4	839	816	26.6	54.98	702	772	28.07	83.06	685	1165	43.55
2020	39.8	617	676	23.78	57.15	533	634	30.19	51.17	566	1037	38.27
2021	37.5	631	719	23.73	52.01	638	62	3.22	39.01	485	989	36.4
2022	41.2	682	438	15.78	36.99	512	410	17.77	36.48	454	865	33.74

Contd.................................

	Barnyard millet				Proso millet				Little millet			
Year	% yield increase	Extn. gap (kg/ha)	Tech. gap (kg/ha)	Tech. gap index (%)	% yield increase	Extn. gap (kg/ha)	Tech. gap (kg/ha)	Tech. gap index (%)	% yield increase	Extn. gap (kg/ha)	Tech. gap (kg/ha)	Tech. gap index (%)
2010	63.79	629	175	9.78	–	–	–	–	72.19	549	140	9.66
2011	33.87	411	175	9.72	–	–	–	–	69.58	499	214	14.97
2012	30.52	346	156	10.28	43.17	497.5	250	13.16	68.96	467	281	19.72

(continued)

Table 5 (continued)

Contd………………………………

Year	Barnyard millet				Proso millet				Little millet			
	% yield increase	Extn. gap (kg/ha)	Tech. gap (kg/ha)	Tech. gap index (%)	% yield increase	Extn. gap (kg/ha)	Tech. gap (kg/ha)	Tech. gap index (%)	% yield increase	Extn. gap (kg/ha)	Tech. gap (kg/ha)	Tech. gap index (%)
2013	22.21	235	714	35.52	13.11	153	580	30.53	31.74	276	251	17.97
2014	32.16	325	554	29.31	–	–	–	–	53.95	401	292	20.32
2015	20.03	210	500	28.43	18.37	203	550	29.6	42.33	300	137	11.97
2016	26.87	289	613	30.96	41.71	297	841	45.46	80.31	445	100	9.09
2017	36.2	425	605	27.43	154.32	1162	−115	−6.39	74.94	542	34	2.62
2018	52.71	575	345	17.16	72.35	764	580	24.17	80.51	535	563	46.92
2019	56.67	686	138	6.78	78.36	829	−27	−1.45	75.15	623	134	9.22
2020	60.5	690	59	3.12	113.26	897	−389	−29.92	45.79	459	114	7.8
2021	65.62	742	138	6.87	98.62	784	271	14.65	57.31	557	270	17.65
2022	33.94	462	93	4.85	53.05	479	476	25.62	55.26	528	−32	−2.16

Plate 6 Bountiful crop - FLD on Finger millet at Kolar, Karnataka state

Madhushekar et al. (2021) and Tankodara et al. (2018), who noticed average growth in yield that varied from 9.10% to 13.50% among farmers of FLD in different crops, while Sagar and Chandra (2004), Sharma et al. (2012), and Choudhary et al. (2009) too conveyed yield improvement by recommended agro-technologies under FLDs (Plate 6).

9 Technology Gap

The variance between potential yield and demonstration plots yield represents the technology gap, in finger millet it ranged from 135 kg/ha (in 2010) to 1015 kg/ha (in 2012) (Table 5). Technology gap on an average for a 13-year FLD program was 607 kg/ha in finger millet. However, 301 kg/ha in foxtail millet, 192 kg/ha in little millet, 1025 kg in kodo millet, 302 kg/ha in proso millet, 328 kg/ha in barnyard millet were observed over the years from 2010 to 2022. The gap is very widespread and needs to be lessened over several interventions and the gap might be owed to soil fertility variation, cultivation on marginal lands, improper nutrient management, nonuniform distribution of rainfall, managerial skills of individual farmers, non-congenial climate, local-specific crop management problems faced by FLD farmers (Sagar and Chandra 2004; Vaghasia et al. 2005; Choudhary et al. 2009; Tunvar et al. 2017; Shankar et al. 2022) who all have witnessed the wide technology gap under FLD in different crops. These all notes indicate that the location-specific crop management is needed to fill the gap (Vedna et al. 2007).

10 Extension Gap

The extension yield gaps of small millets were calculated for 13 years from *kharif* 2010 to 2022 under FLD program in different Indian states, which revealed that a wide extension yield gap was existed. Extension yield gaps are the gauges towards the awareness dearth for enhanced farm technologies adoption by farmers (Choudhary et al. 2009).

Extension yields gaps differed to an extent of 100–836 kg/ha in finger millet, 85–862 kg/ha in foxtail millet, 276–528 kg/ha in little millet, 316–757 kg/ha in kodo millet, 210–742 kg/ha in kodo millet, and 153–1162 kg/ha in proso millet during the study period. The FLDs conducted in small millets gave an average (from the year 2010 to 2022) extension gap of 632 kg/ha in finger millet, 566 kg/ha in foxtail millet 475 kg/ha in little millet, 569 kg/ha in kodo millet, 607 kg/ha in proso millet 463 kg/ha in barnyard millet; however, 552 kg/ha extension gap was found as overall extension gap of all small millets put together. This highlighted the essential to train the farmers through the use of novel information and communication tools (ICT) for more implementation of good agronomic production techniques to converse wide extension gap trend. More usage of up-to-date production technologies along with improved varieties would subsequently alter the trend (Morwal et al. 2018; Shah et al. 2019).

The wider extension yield gap in the current study (Table 5) specifies that there is a strong necessity to persuade the demonstration farmers for upgraded farm technologies adoption over prevailing local practices of small millets. The refinement in farmers' practices for more implementation of location exact farm technology for sustaining crop efficiency is an added choice open for scientists.

11 Technology Index

The viability or feasibility of generated technologies in farmers' fields under prevailing agro climatic conditions designates the technology index (Vedna et al. 2007; Choudhary et al. 2009). Lower technology index indicates the higher viability of generated farm technology under farmers' fields and vice versa. The technology index outcomes from Table 5 shows the technology index in finger millet varied from 5.68% to 35.40%. The technology index of 21.42% (average) was accounted during the 13 years of FLD program, which illustrates the efficacy, efficiency, and comfort of acceptance of integrated crop management segment for increase yield in finger millet. However, technology index in other small millets also followed the similar trend (11.99% in foxtail millet; 39.83% in kodo millet; 16.89% in barnyard millet; 12.72% in proso millet; 14.29% in little millet). The outcomes found are in unity with the results of Kumar et al. (2020), Singh et al. (2020), and Shankar et al. (2022), who observed yield increase amid FLD farmers showing comfort of different technologies adoption. The result showed that demonstrated crop practices in small millets is fairly feasible under prevailing situations of farming in the country. Again, there is a prerequisite to be promoted in an exhaustive means to teach millet farmers to implement technologies to enhance small millets productivity and uplift their farm profitability.

12 Economic Analysis

The highest gross (Rs. 74,383/ha) and the net-returns (Rs. 47,641/ha) in finger millet were obtained during 2021–22 and 2022–23, respectively (Tables 6 and 7), while for other small millets, years 2021 and 2022 have noticed higher return, but for proso millet during 2020 accounted with more gross returns. The net returns also almost followed the similar inclination. The monetary returns variations might be ascribed to the small millets variable performance in terms of grain yield with upgraded technology under FLDs. Benefit–cost ratio found between 2.02 and 2.91 in finger millet; 2.34 and 3.25 in foxtail millet; 2.15 and 2.60 in kodo millet; 1.86 and 2.68 in little millet; 1.40 and 3.16 in barnyard millet; 1.83 and 2.50 in proso millet) (Table 8). The package of upgraded technology also enhanced the small millets profitability with respect to gross and net returns besides enhancing benefit–cost ratio.

13 Conclusion

Overall analysis of yield gap identified that the technology gaps observed were lesser than extension gap in small millets except kodo millet. Among small millets, proso millet (68.63%), kodo millet (62.47%), and little millet (62.17%) observed with more yield gap than barnyard millet (41.16%), finger millet (43.22), and foxtail millet (43.74%). The main production associated constraints perceived in all the implemented states mainly was low market value/low profit and next in the order were high inputs cost, labor nonavailability, lack of inputs subsidy, noncredit facilities, and lack of information sources. The extension interventions are required to decrease the gap in both technological and extension in small millets at farmers' field. The distinct location-specific package of practices need to be established for seed rate, seed treatment, soil testing, plant population, nutrient management, moisture conservation methods, and concentrated demonstrations efforts at farmers field as a technological intervention. Technology transfer need to be more supported in a farmer participatory mode with energetic participation of multidisciplinary researchers' team and all stakeholders. Furthermore, government needs to ensure processing, storage facilities, and to establish custom hiring centers to small millets growers and must encourage value addition, packaging, grading, and selling of small millets products and mechanization in group line. The minimum support price (MSP) needs to be enhanced for all small millets to assure better price to the farmers. The "Seed Hub" for recommended variety seed production of location-specific area has to be strengthened. The middlemen/commission agents need to be curtailed in the marketing channel to declare better price to the consumers and farmer. A need to provide special packages for small millets growers to ensure timely availability of all critical inputs obligatory in the small millets cultivating areas to enhance the sustainability of production and productivity.

Table 6 Gross returns (Rs/ha) of small millets under the demonstrations of FLD and farmers practice (FP)

	Finger millet		Foxtail millet		Kodo millet		Little millet		Barnyard millet		Proso millet	
	FLD	FP	FLD	FP	FLD	FP	FLD	FP	FLD	FP	FLD	FP
2017	56,754	36,766	37,957	26,020	34,291	20,713	36,921	21,309	49,083	25,040	58,697	23,193
2018	54,350	33,870	37,740	25,233	41,776	21,677	38,970	22,443	48,135	34,881	55,873	32,348
2019	59,573	39,096	53,333	31,248	44,202	24,228	45,297	25,100	50,859	32,503	57,884	32,396
2020	61,047	43,629	44,376	25,116	51,663	32,622	48,441	31,524	61,751	39,373	69,195	32,472
2021	74,383	51,207	58,152	38,289	55,099	36,442	53,019	32,353	64,916	40,900	56,872	28,633
2022	72,626	53,010	58,267	42,870	55,280	41,088	51,432	33,175	65,815	47,513	49,427	32,255

Note: *FLD* frontline demonstration, *FP* farmers practice

Table 7 The net returns (Rs/ha) of small millets in the demonstrations of FLD and farmers practice (FP)

	Finger millet		Foxtail millet		Kodo millet		Little millet		Barnyard millet		Proso millet	
	FLD	FP	FLD	FP	FLD	FP	FLD	FP	FLD	FP	FLD	FP
2017	32,763	18,576	21,795	12,485	18,977	85,65	19,846	7924	20,334	7189	35,197	3542
2018	32,955	14,131	24,693	13,579	24,867	9041	21,402	8177	24,229	11,642	32,373	12,698
2019	38,119	20,485	35,619	16,261	25,152	10,649	24,720	10,256	28,916	12,216	34,384	12,596
2020	41,910	24,875	25,231	9369	28,638	13,515	27,886	14,328	38,360	18,297	41,353	9857
2021	46,421	28,831	39,138	21,974	33,994	19,736	31,517	15,815	43,431	19,400	32,372	7633
2022	47,641	30,174	37,221	24,434	33,880	23,030	32,232	17,525	44,160	27,847	22,486	8705

Note: *FLD* frontline demonstration, *FP* farmers practice

Table 8 The benefit–cost ratios of small millets in the demonstrations of FLD and farmers practice (FP)

Year	Finger millet		Foxtail millet		Kodo millet		Little millet		Barnyard millet		Proso millet	
	FLD	FP	FLD	FP	FLD	FP	FLD	FP	FLD	FP	FLD	FP
2017	2.02	1.47	2.37	1.88	2.15	1.61	1.86	1.28	1.40	0.6	2.5	1.18
2018	2.63	1.84	3.14	2.17	2.57	1.67	2.28	1.56	2.05	1.5	2.38	1.65
2019	2.88	2.07	3.04	2.13	2.24	1.81	2.2	1.61	2.32	1.6	2.46	1.64
2020	2.61	1.98	2.34	1.71	2.48	1.78	2.18	1.59	2.66	1.92	2.48	1.44
2021	2.66	2.29	3.25	2.54	2.60	2.13	2.51	2.09	3.16	2.18	2.32	1.36
2022	2.91	2.32	2.77	2.33	2.58	2.28	2.68	2.12	3.04	2.42	1.83	1.37

Note: *FLD* frontline demonstration, *FP* farmers practice

References

Amadou I, Ganga ME, Le GW (2013) Millets: Nutritional composition, some health benefits and processing-a review. Emeritus J Food Agric 25(7):501–508. https://doi.org/10.9755/ejfa.v25i7.12045

Choudhary AK, Yadav DS, Singh A (2009) Technological extension yield gaps in oilseeds in Mi district of Himachal Pradesh. Indian J Soil Conserv 37(3):224–229

Degu E, Debello A, Belete K (2009) Combining ability study for grain yield and yield-related traits of grain sorghum [Sorghum bicolor (L.) Moench] in Ethiopia. Acta Agron Hungarica 57(2):175–184

Kadian KS, Sharma R, Sharma AK (1997) Evaluation of frontline demonstration trials on oilseeds in Kangra valley of Himachal Pradesh. Ann Agric Res 18(1):40–43

Kalita SK, Chhonkar DS, Kanwat M (2019) Assessment of cluster front line demonstrations on rapeseed (Brassica campestris L.) in Tirap district of Arunachal Pradesh. Indian J Extension Educ 55(3):17–22

Kumar H, Yadav DS (2007) Effect of phosphorus and sulphur levels on growth, yield and quality of Indian mustard (Brassica juncea) cultivars. Indian J Agron 52(2):154–157

Madhushekar BR, Narendar G, Avil Kumar K (2021) Impact of front–line demonstrations on extent of adoption and horizontal spread of direct seeding in rice with drum seeder in Nalgonda district of Telangana. Pharm Innov J 10(9):784–788

Michaelraj SJ, Shanmugam A (2013) Study on millets cultivation in Karur district of Tamil Nadu. Int J Manag Res Rev 4:2167–2179

Molla A, Beuving J, Ruben R (2020) Risk aversion, cooperative membership, and path dependences of smallholder farmers in Ethiopia. Rev Dev Econ 24(1):167–187

Morwal BR, Pagaria P, Kanthwa SL, Das S (2018) Performance of frontline demonstration on yield enhancement of cumin in Barmer district of Rajasthan. J Krishi Vigyan 6(2):176–178

Sagar RL, Chandra G (2004) Frontline demonstrations on sesame in West Bengal. Agric Extension Rev 16(2):7–10

Samui SK, Maitra S, Roy DK, Mondal AK, Saha D (2000) Evaluation of front line demonstration on groundnut (Arachis hypogea L.) in Sundarbans. J Indian Soc Coastal Agric Res 18(2):180–183

Shah KA, Nayak P, Lad AN (2019) Impact of cluster frontline demonstration of scientific cultivation practices of major pulses in Navsari district of Gujarat. Gujarat J Extension Educ 30(1):38–41

Shankar A, Prabhakar Reddy T, Jaganmohan Reddy M, Afifa J, Rajashekhar M, Rajashekhar B, Ramakrishna K, Bhatt PS (2022) Impact of front-line demonstration on integrated management of brinjal shoot and fruit borer in Nagarkurnool district, Telangana state. Int J Bio-Resou Stress Manag 13(3):292–298

Sharma PK, Kumar S, Ishar AK, Parkash S, Jamwal SS (2012) Economic impact of front line demonstrations (FLD's) in Poonch district of Jammu & Kashmir. Econ Aff 57(1):99–106

Sukanya TS, Prabhakar KGKT, Swarna R, Hariprasanna K, Tonapi VA (2022) Good agronomic practices for higher yield in small millets. ICAR-Indian Institute of Millets Research, Hyderabad. https://www.millets.res.in/pub/2022/Good_Agronomic_Practices_2022.pdf

Sukanya TS, Ajay Kumar, Sathya K, Chaithra C, Narayanan AL, Anand MR, Kaushal Kishore, Manisha Shyam, Narendra Kumar Nag (2023) Nutricereals role in indian agriculture, food and nutritional security: A review. Mysore J Agric Sci 57(2):33–42. https://www.uasbangalore.edu.in/images/2023-2nd-Issue/1.pdf

Sukanya TS, Sneha MA, Chaithra C, Ragimasalawada M (2024a) Optimizing productivity and resource use efficiency under a finger millet-based cropping system. Sustain 16:11046. https://doi.org/10.3390/su162411046

Sukanya TS, Kumar A, Sathya K, Narayanan AL, Kishore K, Shyam M, Nag NK, Chaithra C, (2024b) Millet-based cropping systems for enhanced productivity. In: Mishra S, Kumar S, Srivastava RC (Eds.) Genetic Improvement of Small Millets. Springer Nature Singapore, pp. 63–86. https://doi.org/10.1007/978-981-99-7232-6_4

Tankodara KD, Gohil GR, Thakar DS (2018) Impact of training programme on knowledge level of farmers regarding scientific cultivation technologies of horticultural crops. Gujarat J Extension Educ 29(1):69–71

Tunvar MA, Patel AJ, Prajapati VV (2017) Impact of Front–line demonstration on groundnut conducted by Krishi Vigyan Kendra, Deesa. Gujrat J Extension Educ (Special Issue on National Seminar):56–58

Vaghasia PM, Savalia RL, Kelaiya GR (2005) Evaluation of frontline demonstrations on groundnut in Saurashtra region of Gujarat. J Oilseeds Res 22(1):238–239

Vedna, Kumar A, Bhateria S (2007) Demonstration an effective tool for increasing productivity of rapeseed-Niger in Kangra district of Himachal Pradesh. J Oilseeds Res 33(2):257–261

Advances in Genetic Transformation Studies of Small Millets

T. P. Ajeesh Krishna, T. Maharajan,
and Stanislaus Antony Ceasar

Abstract

Millets have been important cereal crops for the poor people of Asia and Africa for the past several hundred years. There are two main reasons why these crops are the center of attraction for the cross-sectoral community around the world: (1) Millets possess excellent climate resilience features for adapting to harsh climates. (2) They are a rich source of mineral nutrients and vitamins, conferring several health benefits. Millets are considered future food security crops due to these two features. It is the priority of the researchers to dissect and understand the mechanisms behind the climate resilience and nutrient enrichment features of the millets. Genetic engineering and genome editing are important tools in molecular genetics used by plant breeders to determine the role of a specific gene. Although main crops such as rice have received much attention during the past decades on the use of these tools, millets have been neglected for these studies until recent years. This chapter reviews the present status of genetic engineering and genome editing reports available for millets. Among the minor millets, foxtail millet and finger millet have considerable reports on genetic engineering studies, and other millets have a few or no reports to date. The *Agrobacterium*-mediated system was most frequently applied for the genetic engineering of millets. Genome editing is another important tool used by plant biotechnologists to study genes and improve upon specific traits. However, except for foxtail millet, other millets have had no reports on genome editing studies to date. The genetic engineering and genome editing studies in millets might help study specific gene functions and transfer key traits within and outside the millets.

T. P. A. Krishna · T. Maharajan · S. A. Ceasar (✉)
Division of Plant Molecular Biology and Biotechnology, Department of Biosciences, Rajagiri College of Social Sciences, Kochi, Kerala, India

D. C. Joshi et al. (eds.), *Minor Millets*,
https://doi.org/10.1007/978-981-96-4265-6_15

Keywords

Genetic engineering · Minor millets · Genome editing · *Agrobacterium*-mediated system · CRISPR/Cas9

Abbreviations

Act1	rice actin gene promoter
bar	phosphinothricin resistance
BRI1	brassinosteroid insensitive 1
BZR1	brassinazole resistant 1
chi11	chitinase gene from rice
Ft5	promoter of C4 isoform of phosphoenolpyruate carboxylase gene from *Flaveria trinervia*
gus	β-glucuronidase
HptII	hygromycin phosphotransferase II
mtlD	mannitol-1-phosphate dehydrogenase
PcSrp	*Porteresia coarctata* serine-rich-protein
RbcS	rice ribulose 1,5-biphosphate carboxylase small subunit gene promoter
SiPf40	*Setaria italica* ZIP-like gene
SiPHT1	*Setaria italica* phosphate transporter 1
Srp	serine-rich-protein
TUAm1	α1-tubulin
TUB1	β1-tubulin
UidA	beta-glucuronidase
Uq1	ubiquitin gene promoter from maize
Zm401	maize pollen-specific gene
VPPase	vacuolar H^+-pyrophosphatase.

1 Introduction

Millets are a group of small-seeded coarse cereal crops majorly adopted by the resource-poor and rural farmers in Asia and Africa. Sorghum (*Sorghum bicolor* L.) and pearl millet (*Cenchrus americanus* (L.) Morrone) are usually called major millets. The finger millet (*Eleusine coracana* (L.) Gaertn), proso millet (*Panicum miliaceum* L.), kodo millet (*Paspalum scrobiculatum* L.), foxtail millet (*Setaria italica* (L.) Beauv), little millet (*Panicum sumatrense* Roth. ex Roem. and Schuktz), and barnyard millet (*Echinochloa esculenta* (A. Braun) H. Scholz) are considered minor millets, although mixed in the classifications generally noted. Millets are some of the oldest domesticated crops in the world. They are usually grown as rainfed crops

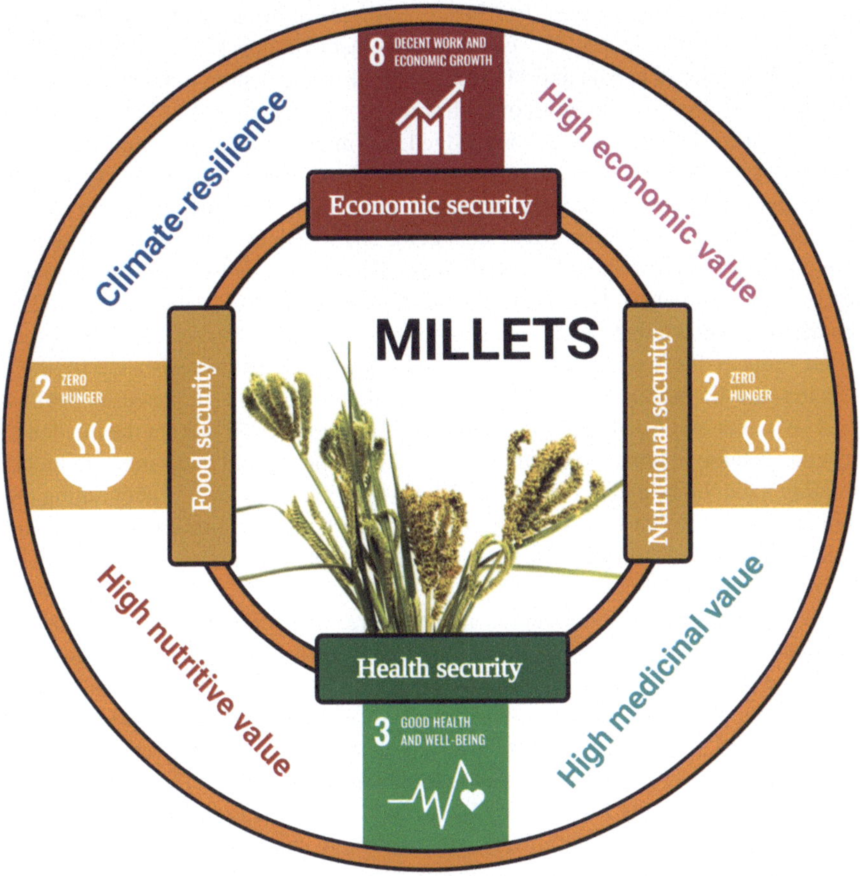

Fig. 1 Millet's contribution to the United Nations' Sustainable Development Goals (SDGs). Millets are a multipurpose crop providing solutions to future global food, nutritional, health, and economic challenges

on unimproved soils with low input systems (Vetriventhan et al. 2020). Millets are considered a sustainable solution to many global challenges, such as food insecurity, malnutrition, and climate change. Due to these properties, millets also contribute to the United Nations' Sustainable Development Goals (SDGs) (Fig. 1). Because of their many health benefits and high nutritional content, minor millets are becoming more and more popular these days. In the arid and semiarid regions of the emerging nation, they are regarded as a staple food (Upadhyaya and Vetriventhan 2018). Due to their high fiber, vitamin, and mineral content, minor millets are a highly recommended choice for a diet that is well-rounded (Amadou et al. 2013; Dayakar Rao et al. 2017; Nithiyanantham et al. 2019). Millet-based value-added food items are also becoming popular among urban people in recent years. Each millet has one or more unique nutrients, which make it more attractive to people. High fiber content and a low glycemic index also make the millets more suitable for people with lifestyle diseases. For instance, finger millet contains the following nutrients: 5–8% protein, 65–75% carbs, 1–2% ether extractives, 2.5–3.5% minerals, and 15–20% fiber (Chethan and Malleshi 2007). It is interesting to know that finger millet has the highest calcium concentration of any cereal grain (344 mg/100 g) (Maharajan et al. 2021). It is acknowledged that minor millets have more nutrients than major cereal crops such as maize, wheat, and rice (Bhat et al. 2018). Numerous studies have emphasized the advantages of small millets for human health (Saleh et al. 2013; Sarita and Singh 2016; Bhat et al. 2018). Small millets have a low glycemic index; therefore, people with diabetes or those trying to control their blood sugar levels can benefit from eating them (Ugare et al. 2014; Anitha et al. 2021). Studies have shown that regular consumption of food products derived from finger millet reduces fasting glucose by 32% and eliminates insulin resistance by 43% (Geetha and Easwaran 1990). Due to being gluten-free and marketed as healthy food (Woomer and Adedeji 2021), millet foods are believed to prevent the incidence of celiac disease in humans (Caeiro et al. 2022). People should focus more on incorporating a diet high in millet-based foods into their everyday lives. Encouraging the consumption of millet can aid in the fight against noncommunicable diseases and malnutrition. It will contribute to lowering health risks and improving the health of upcoming generations.

In order to maintain food security in the future, agriculture in dynamic climates must cultivate crops that are climate-resilient. In addition to their nutritional and medicinal properties, millet crops are resistant to extreme climate conditions (Singh et al. 2022). They can adapt to a wide range of ecological conditions (Goron and Raizada 2015). Minor millets thrive in the semiarid tropics and produce satisfactory yields within a short growing season (Swindale et al. 1981; Thilakarathna and Raizada 2015). Millets also exhibit high rates of seed reproduction and may be stored for many years without suffering damage from storage pests (Ajeesh Krishna et al. 2021). Minor millets are considered "smart crops" due to their low water requirements for their grain production. Despite having exceptionally high nutrient content and climate resilience, millet research is not keeping up with that of other important crops. Only a few labs have focused on studying and improving millets using modern genetic and genomic tools. Genetics and genomic tools play a crucial role in crop improvement. They help to create a new crop variety with desirable qualities. Genetic transformation is a promising approach to studying gene function and improving crop traits, especially yield and quality, and resistance to adverse environmental stresses (Chen et al. 2008; Ignacimuthu and Ceasar 2012; Hussain et al. 2014). We have reported a few genetic transformation and gene characterization studies for millets. Similarly, genome editing tools such as the CRISPR/Cas system could help advance functional genomics studies and impart key traits in millets. This chapter discusses the current status of minor millet research on genetic transformation and genome editing.

2 Genetic Transformation Studies in Minor Millets

The availability of robust regeneration and transformation protocols supports the genetic improvement of minor millets. Plant regeneration protocols are available for most of the minor millets. Many genetic transformation studies have been available on millets, including foxtail millet and finger millet, but few on other millets (Table 1). The *Agrobacterium*-mediated method was mainly used in millet transformation. However, the genetic transformation studies in millets are lagging compared to other major cereals such as rice (Ceasar and Ignacimuthu 2009). In the past few years, collective efforts have been made to improve millet transformation methods and make transgenic plants that express functionally active transgenes in millets. The details of genetic transformation studies available in millets are discussed below. Among minor millets, finger millet and foxtail millet have considerable reports on genetic transformation studies. However, few reports are available on the genetic transformation in barnyard millet, teff, and fonio. So far, no genetic transformation report is available for little millet and proso millet. In *planta*, genetic transformation is typically carried out via agrobacterium-mediated, biolistic, and electroporation approaches. Among these methods, *Agrobacterium*-mediated transformation was frequently employed in minor millets (Fig. 2). The *Agrobacterium* strain, plasmid, promoter, and selectable marker genes, as well as the amount of time for cocultivation and *Agrobacterium* infection, are all important parts of a successful genetic transformation. Additionally, regeneration of transformed cells/calluses/tissues is essential for successful genetic transformation.

2.1 Foxtail Millet

Genetic transformation research has demonstrated that foxtail millet is a top priority among minor millets. Its genome sequence is thoroughly annotated, and it is regarded as a model plant for C4 crops (Ceasar 2023). The development and optimization of an in vitro plant regeneration protocol is a crucial step for genetic transformation in foxtail millet. So far, many reports are available for the in vitro plant regeneration of foxtail millet, including direct and indirect methods. In genetic transformation studies, indirect plant regeneration protocols are widely used for genetic transformation in foxtail millet using various explants such as seed callus, immature inflorescence, and shoot tips. Further research is required to optimize each component in the transformation to develop a more efficient transformation protocol for foxtail millet. Numerous studies on the transformation of foxtail millet by *Agrobacterium* have been reported. Both direct and indirect plant regeneration techniques have been successfully applied for genetic transformation in foxtail millet. However, transgenic plants in foxtail millet were primarily created via the indirect plant regeneration method. The *Agrobacterium*-mediated transformations methodology for foxtail millet using seed callus via indirect plant generation was initially optimized by Yinghui et al. (2005). Following that, three groups reported transformation methods for foxtail millet employing immature inflorescences and

Table 1 Details of genetic transformation studies in minor millets. The name of the minor millet, explant, method of plant tissue culture, method of gene transfer, target gene, transformation efficiency, and observation are given

Name of the millet	Method of PTC	Explant	Method of delivery	Strain	Plasmid	Selection gene	Reporter gene	Candidate gene	Transformation efficiency	Achievements	Reference
Finger millet	Direct	Shoot apex	*Agrobacterium*-mediated	LBA4404	*pCAMBIA1301*	*HptII*	*uidA*	–	3.8% (stable)	Optimization of *Agrobacterium*-mediated transformation protocol	Antony Ceasar and Ignacimuthu (2011)
								chi11	3.9% (stable)	Successfully transformed rice *chi11* gene for improving leaf blast resistance	Ignacimuthu and Ceasar (2012)
				EHA105	*pCAMBIA1301*	*HptII*	*uidA*	–	11.8% (stable)	Optimization of *Agrobacterium*-mediated transformation protocol	Satish et al. (2017)
	Indirect	Seed callus	*Agrobacterium*-mediated	LBA4404	*pCAMBIA1301*	*HptII*	*uidA*	*SbVPPase*	–	Successfully transformed sorghum *VPPase* gene for salt tolerance	Anjaneyulu et al. (2014)
				LBA4404	*pBI-121*	*nptII*	*uidA*	–	44.4% (frequency)	Optimization of *Agrobacterium*-mediated transformation protocol	Sharma et al. (2011)
				EHA105	*pCNL-56*	*nptII*	*uidA*	–	44.4% (frequency)	Optimization of *agrobacterium*-mediated transformation protocol	Sharma et al. (2011)
				EHA105	*pCAMBIA1380*	*HptII*	*uidA*	*mtlD*	–	Transferred *mtlD* gene for multiple stress tolerance	Hema et al. (2014)
		Embryogenic calli	Biolistic method	–	*pPur*	*Bar*	*uidA*	*PIN*	–	Developed leaf blast disease-resistant plants	Latha et al. (2005)
		Shoot tip callus		–	*TG0063* of *pCAMBIA* series	–	–	*PcSRP*	–	Developed salt-tolerant plants	Mahalakshmi et al. (2006)
		Seed callus		–	–	–	*ActI/UqI/ RbcS/Ft /uidA*	–	–	Testing the efficiency of various promoters	Gupta et al. (2001)

Foxtail millet	Direct	Shoot apex	*Agrobacterium*-mediated	LBA4404	*pFGC*	*HptII*	–	*SiPHT1;1,2* and *3*	9% (frequency)	Characterization of phosphate transporters	Ceasar et al. (2017)
	Indirect	Seed callus		AGL1	*pB4NU*	*HptII*	*uidA*	–	~ 27%	Optimization of *Agrobacterium*-mediated transformation protocol	Sood et al. (2020)
	Indirect	Immature inflorescences		LBA4404	*pCAMBIA2300*	*nptII*	*uidA*	–	5.5%	Optimization of *Agrobacterium*-mediated transformation protocol	Wang et al. (2011)
	Indirect	Immature inflorescences		EHA105	*pCAMBIA2300*	*HptII*	*uidA*	–	5.5%	Optimization of *Agrobacterium*-mediated transformation protocol	Wang et al. (2011)
	Indirect	Seed callus		LBA4404	pCAMBIA1304	*HptII*	*uidA*	–	~ 27%	Optimization of *Agrobacterium*-mediated transformation protocol	Sood et al. (2020)
				EHA105	*p6i2x*	*HptII*	–	–	19.2%	Optimization of *Agrobacterium*-mediated transformation protocol	Santos et al. (2020)
		Shoot tip		LBA4404	*pCAMBIA1305*	*HptII*	–	*BRI1*	–	Showed stress tolerance	Zhao et al. (2021)
		Seed callus		LBA4404	*pBI121*	*nptII*	*uidA*	–	6.6%	Optimization of *Agrobacterium*-mediated transformation protocol	Yinghui et al. (2005)
				LBA4404	*pBIZm13Si401*	*nptII*	–	*Si401*	–	Confirmed the role of *Si401* for anther and pollen development	Qin et al. (2008)
		Florect calli	Biolistic method	–	*pNOV3604*	*manA*	–	*SiPf4*	–	Functional characterization of *SiPf4* gene	Liu et al. (2009)
Kodo millet	Indirect	Seed callus	*Agrobacterium*-mediated	EHA 105	*pCNL56*	*nptII*	*uidA*	–	15.5%	Optimization of *Agrobacterium*-mediated transformation protocol	Bhatt et al. (2021)
				LBA4404	*pCAMBIA2300*	*nptII*	*uidA*	–	14.4%	Optimization of *Agrobacterium*-mediated transformation protocol	Bhatt et al. (2021)

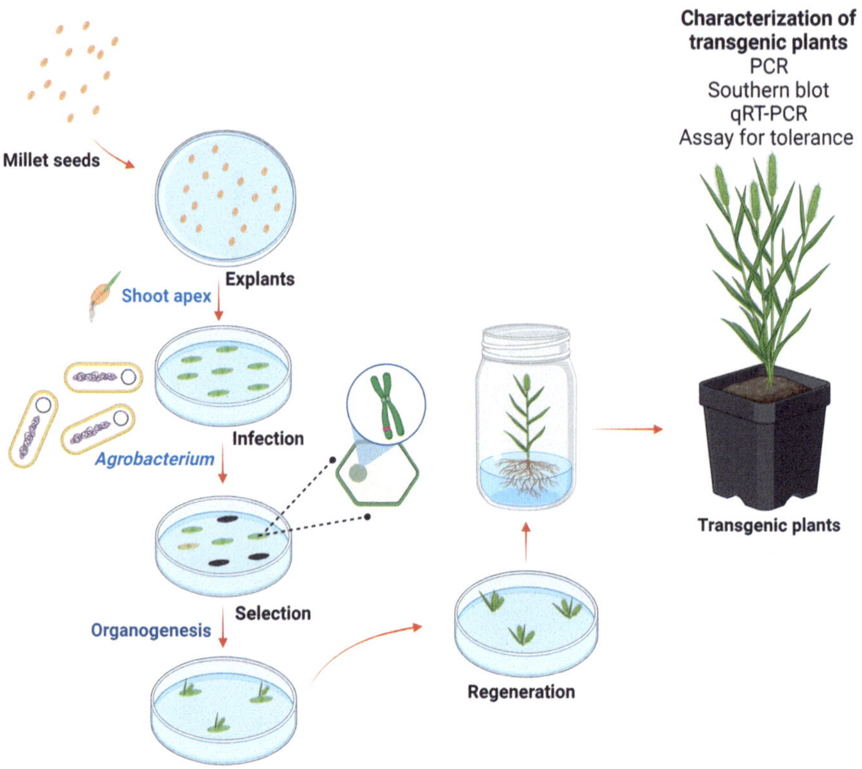

Fig. 2 Stepwise protocol followed for *Agrobacterium*-mediated transformation in minor millets. The *Agrobacterium* strain, plasmid, promoter, and selectable marker genes, as well as the amount of time for cocultivation and *Agrobacterium* infection, are all important parts of a successful genetic transformation

seed callus via an indirect plant regeneration technique (Wang et al. 2011; Sood et al. 2020; Santos et al. 2020). A few genes were introduced into the foxtail millet using the *Agrobacterium*-mediated transformations method. The function of the pollen-specific (*Si401*) gene in anther and pollen development was discovered by successfully transferring it into the foxtail millet utilizing the *Agrobacterium*-mediated transformations method (Qin et al. 2008). Ceasar et al. (2017) characterized the role of three *phosphate transporter1* genes (*SiPHT1;1*, *1;2*, and *1;3*) in foxtail millet by developing the knockdown lines through an RNAi strategy. The RNAi plasmids were used to transform the foxtail millet by the *Agrobacterium*-mediated transformation procedure. *Agrobacterium*-mediated transformation with shoot apex explants and transgenic plants was recovered through direct regeneration without any callus phase. Among three *PHT1* genes, downregulation of *SiPHT1;2* significantly reduced the phosphate transport. These studies proved the key roles of foxtail millet genes to overcome many physiological stresses. These genes may be utilized for improving the saline and drought tolerances in other minor millets by gene transfer.

2.2 Finger Millet

Finger millet is considered a nutraceutical crop due to its exceptional nutritional values and health benefits. Compared to other minor millets, research on genetic transformation has mostly focused on this particular millet. Both biolistic and *Agrobacterium*-mediated transformation protocols are available in finger millet. (Gupta et al. 2001; Latha et al. 2005; Anjaneyulu et al. 2014; Satish et al. 2017). Gupta et al. (2001) initially reported on the transformation study of finger millet. They carried out an initial investigation utilizing a biolistic method to compare the effectiveness of five different gene promoters for the production of the *GUS* reporter gene. Latha et al. (2005) used the prawn gene's antifungal protein (PIN) to enhance the genetic transformation of finger millet. To create leaf blast-resistant finger millet, they have altered the *PIN* gene using the biolistic (particle-inflow gun) technique. Significantly more, they developed a high-frequency, genotype-independent method of plant regeneration for the particle-inflow strategy of integrating *PIN* genes into finger millet (Latha et al. 2005). The genotype chosen, explants, media composition, and so on are important factors for successful plant regeneration in finger millet (Latha et al. 2005). Numerous studies have reported on finger millet regeneration protocols, and they have highlighted the significance of these elements for effective plant regeneration (Antony Ceasar and Ignacimuthu 2008; Pande et al. 2015; Satish et al. 2015; Atul Babu et al. 2018; Ngetich et al. 2018). *Agrobacterium*-mediated transformation techniques have been mostly utilized in finger millet. The technique for the first-ever *Agrobacterium*-mediated transformation (3.8% stable transformation frequency) of finger millet was reported by Ceasar and Ignacimuthu (2011) using shoot apex explants. They have used *hygromycin phosphotransferase* (*hptII*) as a selectable marker gene and *β-glucuronidase* (*GUS*) as a reporter gene to optimize the transformation in finger millet (Ceasar and Ignacimuthu 2011). By integrating the rice *chitinase* (*chi11*) gene into their transformation procedure, the same group created transgenic finger millet lines that were resistant to leaf blast disease caused by *Pyricularia grisea* (Ignacimuthu and Ceasar 2012). This indicates that the transformation protocol created by Ceasar and Ignacimuthu (2011) contributes to the introduction of numerous genes that are crucial for agronomy into the genome of finger millet. Sharma et al. (2011) evaluated the various factors influencing *Agrobacterium*-mediated genetic transformation in finger millet. They highlighted that the addition of a higher level of $CuSO_4$ in the regeneration medium has influenced the high recovery rate of transformed finger millet plants (Sharma et al. 2011). Later, Sathish et al. (2017) developed high-frequency transformation (11.8%) of finger millet by the *Agrobacterium*-mediated method via direct plant regeneration with the modification of various factors. In addition to these results, several researchers used the *Agrobacterium*-mediated approach to develop transgenic lines that are resistant/tolerant to different stressful conditions. For instance, Anjaneyulu et al. (2014) developed leaf blast-resistant finger millet lines by transferring the vacuolar *H+-pyrophosphatase* (*SbVPPase*) gene to finger millet using an *Agrobacterium*-mediated transformation. There are not many other papers available on the biolistic-mediated generation of transgenic finger millet plants. For instance, the biolistic approach was used to modify the *Porteresia coarctata serine-rich protein* (*PcSrp*) gene in order to create salt stress-tolerant transgenic finger millet (Mahalakshmi et al. 2006).

2.3 Kodo Millet

The kodo millet is among the most significant minor millets (Vetriventhan and Upadhyaya 2019), and most of it is grown in South Asian nations (Hegde et al. 2005). For the genetic transformation of kodo millet, there is now only one report available. Bhatt et al. (2021) established the *Agrobacterium*-mediated transformation mechanism in kodo millet. In this study, *Agrobacterium* tumefaciens strains EHA105 and LBA4404 harboring plasmids pCNL56 and pCAMBIA2300, respectively, were used for the development of a successful transformation protocol in kodo millet (Table 1). The transformation efficiency was shown to be higher with *Agrobacterium* strain EHA105 than LBA4404 (Bhatt et al. 2021). This study suggested that the infection time followed by various cocultivation times of EHA105 (3 days) and LBA4404 (5 days) strains improved the efficiency of transformation in kodo millet. It also revealed that successful genetic transformation efficiency depends on various factors (Bhatt et al. 2021). In this study, they have used an indirect regeneration protocol for the regeneration of transgenics. In order to develop kodo millet cultivars, efforts must be made to expand this process and transfer favorable genes to them.

2.4 Barnyard Millet, Tef, and Fonio Millet

Other minor millets such as little millet, barnyard millet, and proso millet are also considered future food and nutrient security crops (Nainwal and Verma 2018). Unfortunately, no information is available on the genetic transformation in these minor millets. Barnyardgrass (*Echinochloa crus-galli*) is a closely related species to barnyard millet (*Echinochloa esculenta*). In Barnyardgrass, the biolistic approach to genetic transformation is reported (Gupta et al. 2001). In this study, they analyzed the efficiency of five gene promoters (*ActI*, *UqI*, *RbcS*, *Ft*, and *uidA*) for the expression of the *GUS* reporter gene. They stated that the *GUS* gene expression responded well to promoters *Actl* and *Uql* (Gupta et al. 2001). This data could be useful in the future to choose the ideal promoter in barnyard millet to get the highest beneficial gene expression. Since then, there has been no other attempt to transform this millet, particularly through the *Agrobacterium*-mediated transformation method. A report is also available on the transformation of tef by an *Agrobacterium*-mediated system. To investigate its role in tef growth, the runner bean (*Phaseolus coccineus*) *gibberellin inactivation* (*PcGA2ox1*) gene was introduced into tef by an *Agrobacterium*-mediated system (Gebre et al. 2013). Embryogenic calluses obtained from immature embryos were used as explants for genetic transformation. The transgenic plants expressing the *PcGA2ox1* gene showed changes in growth characteristics. Notably, up to a 3.7-fold higher grain yield was seen in transformed plants. Ntui et al. (2017) reported on the *Agrobacterium*-mediated transformation of fonio. Calluses derived from stems were used as explants. Transformation was confirmed by histochemical GUS assay and PCR. No report is available to date on the transformation of other minor millets such as proso millet and little millet. More effort is needed to develop robust transformation protocols for introducing key genes to confer resistance or tolerance against various biotic or abiotic stresses.

3 Genome Editing in Minor Millets

Genome editing is a molecular genetic tool used to alter specific sequences/regions in the target organism to study its function and improve its performance. It has been widely applied to several organisms in the past decade. The genome editing approach needs programmable sequence-specific endonucleases that facilitate site-specific DNA double-stranded breaks (DSBs) at the targeted location, allowing the endogenous repair mechanisms to correct this break. The DSBs are repaired by utilizing two repair mechanisms, particularly homology-directed repair (HDR) and nonhomologous end-joining repair (NHEJ). To generate specific DNA breaks, varied genome editing technologies have emerged during the past few years, including zinc finger nucleases (ZFNs) and transcription activator-like effector nucleases (TALENs). These two tools were approached based on catalyst domains and DNA-binding proteins for inducing DSBs at target genomic loci. However, these tools need in-depth protein construction and cloning to make DSBs, which precludes them from being utilized for high-throughput applications. Most recently, a new class of Cas proteins related to clustered regularly interspaced short palindromic repeats (CRISPR/Cas), an adaptive system against phages and plasmids in nature, has been tailored as a robust RNA-guided DNA (gDNA) targeting tool for genome altering in eukaryotes. This has been easily adapted to function as a powerful tool for genetic engineering due to its feasibility in design, cost-efficiency, and labor intensity. The CRISPR/CRISPR-associated protein 9 (CRISPR-Cas) system has been widely used to study and increase crop plants' resistance or tolerance to a variety of biotic or abiotic stresses with great success.

The majority of minor millet crops, with the exception of foxtail millet, have not yet had genome editing using the CRISPR/Cas9 system documented. Foxtail millet is currently the sole millet that has undergone limited experimentation using genome editing techniques. The researchers Lin et al. (2018) have effectively devised a technique for gene editing in foxtail millet utilizing the CRISPR/Cas9 method, specifically targeting the *phytoene desaturase* (*PDS*) gene. The albino phenotype in foxtail millet plants is a direct consequence of the mutation of the *PDS* gene utilizing the CRISPR/Cas9 system (Lin et al. 2018). CRISPR/Cas9-mediated genome editing has been successfully applied in foxtail millet in order to develop double haploid (DH) lines by targeting the *matrilineal* (*SiMTL*) gene (Cheng et al. 2021). Two distinct gRNAs were used to target the exonic region of *SiMTL*. The expression of *SiMTL* was regulated by the OsU3 promoter, whereas the Uq promoter of maize controlled the expression of Cas9. The study utilized the foxtail millet genotype Ci846, and the transformation process was facilitated by *Agrobacterium*. The researchers documented a haploid induction frequency of 2.8% in the second-generation offspring. It is noteworthy that foxtail millet, which is a representative C4 crop, is used for genetic research through genome editing.

Other types of millets have not been utilized for high-resolution genetic investigations by genome editing. The complete genomic sequence is accessible for all minor millets, with the exception of kodo millet and little millet. It can facilitate comprehension of genomic research and gene discovery. Small millets have the potential to enhance the efficiency of genome editing. Research on genome editing in minor millets is now at a nascent phase when compared to other cereal crops. The possible reason for this could be the absence of an efficient transformation and regeneration strategy in small millets. A multitude of stress-responsive genes that confer tolerance or susceptibility have been discovered in minor millets. Therefore, employing the CRISPR/Cas9 technology to eliminate or introduce stress-related genes could be advantageous in elucidating the roles of crucial genes in millets (Fig. 3).

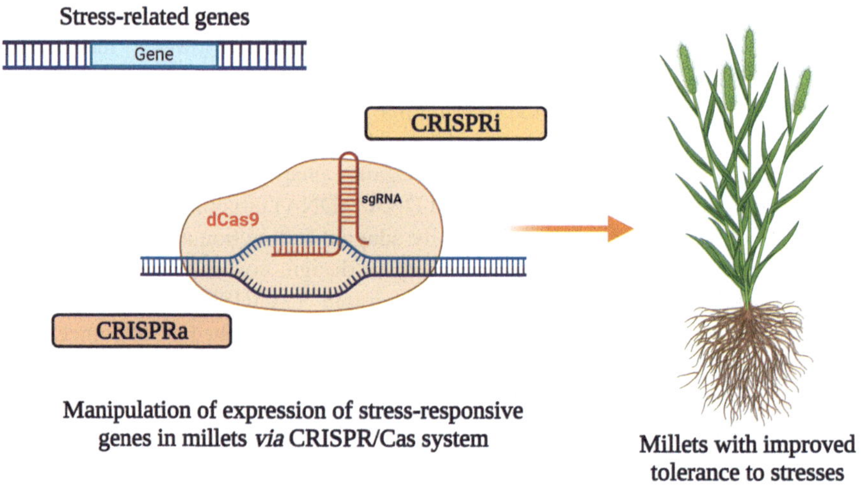

Fig. 3 Millet improvement through transcriptional regulation of genes by the CRISPR/Cas system. The variants of the CRISPR/Cas system, such as CRISPR interference (CRISPRi) and CRISPR activation (CRISPRa), were recently developed. In these systems, nuclease-dead mutants of Cas9 (dCas9), which lost their endonuclease function but retained DNA recognition activity with the guide RNA, offer gene interference and activation. Therefore, the CRISPRa or CRISPRi system has allowed the development of stress-resistant crop varieties either by activating of resistant genes or interference of susceptible genes

4 Conclusions and Future Prospects

An important study focus for small millets has been to comprehensively investigate climate resistance, nutrient enrichment characteristics, and crop enhancement potential. It is understood that hereditary variables play a role in an individual's ability to withstand various forms of stress. Currently, there is a substantial amount of genetic data accessible for small millets. The process of gene mining has been made easier by the complete sequencing of the genome, which has also aided in understanding the organization of the genome in many minor millets. Gaining a comprehensive understanding of the functional role of a gene is of the utmost importance for advancing future research. Developing a robust genetic transformation procedure and enhancing the current protocol are crucial for validating the gene function of minor millets. Currently, there is a scarcity of reports regarding genetic modification and the production of transgenic minor millets. *Agrobacterium*-mediated transformation was the primary method used to achieve genetic transformation in minor millets. Proso millet, little millet, and barnyard millet currently lack a published report on genetic modification. The advancement of effective gene transfer techniques relies on the in vitro tissue culture methodology of specific minor millet genotypes. From our perspective, it is imperative for researchers to enhance the efficacy of genetic transformation and in vitro regeneration techniques in cultivable minor millet genotypes. It can enhance the characteristics of minor millet by incorporating favorable qualities. In addition, genome editing has promise for enhancing minor millet crops in the future.

Acknowledgments We sincerely thank Rajagiri College of Social Science, Kochi, Kerala, for providing the research facilities and support.

Funding This work was financially supported by Rajagiri College of Social Sciences (Autonomous), Kerala, India, under Seed Money for Faculty Minor Research.

References

Ajeesh Krishna TP, Maharajan T, Ignacimuthu S, Antony Ceasar S (2021) Genomic-assisted breeding in finger millet (*Eleusine Coracana* (L.) Gaertn.) for abiotic stress tolerance. In: Genomic designing for abiotic stress resistant cereal crops. Springer, pp 291–317

Amadou I, Gounga ME, Le G-W (2013) Millets: nutritional composition, some health benefits and processing-a review. Emir J Food Agric 501–508:501

Anitha S, Kane-Potaka J, Tsusaka TW et al (2021) A systematic review and meta-analysis of the potential of millets for managing and reducing the risk of developing diabetes mellitus. Front Nutr 386

Anjaneyulu E, Reddy PS, Sunita MS et al (2014) Salt tolerance and activity of antioxidative enzymes of transgenic finger millet overexpressing a *vacuolar H+-pyrophosphatase* gene (*SbVPPase*) from Sorghum bicolor. J Plant Physiol 171:789–798

Antony Ceasar S, Ignacimuthu S (2008) Efficient somatic embryogenesis and plant regeneration from shoot apex explants of different Indian genotypes of finger millet (*Eleusine coracana* (L.) Gaertn.). In vitro Cell Dev Biol Plant 44:427–435

Antony Ceasar S, Ignacimuthu S (2011) *Agrobacterium*-mediated transformation of finger millet (*Eleusine coracana* (L.) Gaertn.) using shoot apex explants. Plant Cell Rep 30:1759–1770

Atul Babu G, Vinoth A, Ravindhran R (2018) Direct shoot regeneration and genetic fidelity analysis in finger millet using ISSR markers. Plant Cell Tiss Org Cult 132:157–164

Bhat S, Nandini C, Tippeswamy V (2018) Significance of small millets in nutrition and health-a review. Asian J Dairy Food Res 37:35–40

Bhatt R, Asopa PP, Jain R, Kothari-Chajer A, Kothari SL, Kachhwaha S (2021) Optimization of *Agrobacterium* mediated genetic transformation in *Paspalum scrobiculatum* L. (Kodo millet) Agronomy 11:1104

Caeiro C, Pragosa C, Cruz MC, Pereira CD, Pereira SG (2022) The role of pseudocereals in celiac disease: reducing nutritional deficiencies to improve well-being and health. J Nutr Metab 2022:1–8

Ceasar SA (2023) Foxtail millet (Setaria italica) as a model system to study and improve the nutrient transport in cereals. Plant Growth Regul 99:3–10

Ceasar SA, Ignacimuthu S (2009) Genetic engineering of millets: current status and future prospects. Biotechnol Lett 31:779–788

Ceasar SA, Ignacimuthu S (2011) Agrobacterium-mediated transformation of finger millet (*Eleusine coracana* (L.) Gaertn.) using shoot apex explants. Plant Cell Rep 30:1759

Ceasar SA, Baker A, Ignacimuthu S (2017) Functional characterization of the PHT1 family transporters of foxtail millet with development of a novel *Agrobacterium*-mediated transformation procedure. Sci Rep 7:1–16

Chen L-H, Zhang B, Xu Z-Q (2008) Salt tolerance conferred by overexpression of Arabidopsis vacuolar Na+/H+ antiporter gene *AtNHX1* in common buckwheat (*Fagopyrum esculentum*). Transgenic Res 17:121–132

Cheng Z, Sun Y, Yang S et al (2021) Establishing in planta haploid inducer line by edited *SiMTL* in foxtail millet (*Setaria italica*). Plant Biotechnol J 19(6):1089

Chethan S, Malleshi NG (2007) Finger millet polyphenols: optimization of extraction and the effect of pH on their stability. Food Chem 105:862–870

Dayakar Rao B, Bhaskarachary K, Arlene Christina GD et al (2017) Nutritional and health benefits of millets. ICAR_Indian Inst Millets Res Rajendranagar, Hyderabad

Gebre E, Gugsa L, Schlüter U, Kunert K (2013) Transformation of tef (Eragrostis tef) by Agrobacterium through immature embryo regeneration system for inducing semi-dwarfism. S Afr J Bot 1;87:9–17

Geetha C, Easwaran PP (1990) Hypoglycemic effect of millet incorporated breakfast items on selected non-insulin dependent diabetic patients. Indian J Nutr Diet 316–320

Goron TL, Raizada MN (2015) Genetic diversity and genomic resources available for the small millet crops to accelerate a new green revolution. Front Plant Sci 6:157

Gupta P, Raghuvanshi S, K Tyagi A (2001) Assessment of the efficiency of various gene promoters via biolistics in leaf and regenerating seed callus of millets, *Eleusine coracana* and *Echinochloa crusgalli*. Plant Biotechnol 18:275–282

Hegde PS, Rajasekaran NS, Chandra TS (2005) Effects of the antioxidant properties of millet species on oxidative stress and glycemic status in alloxan-induced rats. Nutr Res 25:1109–1120

Hema R, Vemanna RS, Sreeramulu S, Reddy CP, Senthil-Kumar M, Udayakumar M (2014) Stable expression of mtlD gene imparts multiple stress tolerance in finger millet. PloS One 9:e99110

Hussain Z, Ali S, Hayat Z, Zia MA, Iqbal A, Ali GM (2014) *Agrobacterium*-mediated transformation of '*DREB1A*' gene for improved drought tolerance in rice cultivars (*Oryza sativa* L.). Aust J Crop Sci 8:1114–1123

Ignacimuthu S, Ceasar SA (2012) Development of transgenic finger millet (*Eleusine coracana* (L.) Gaertn.) resistant to leaf blast disease. J Biosci 37:135–147

Latha AM, Rao KV, Reddy VD (2005) Production of transgenic plants resistant to leaf blast disease in finger millet (*Eleusine coracana* (L.) Gaertn.). Plant Sci 169:657–667

Lin C, Hsu C, Yang L et al (2018) Application of protoplast technology to CRISPR/Cas9 mutagenesis: from single-cell mutation detection to mutant plant regeneration. Plant Biotechnol J 16:1295–1310

Liu Y, Feng X, Xu Y, Yu J, Ao G, Peng Z, Zhao Q (2009) Overexpression of millet ZIP-like gene (SiPf40) affects lateral bud outgrowth in tobacco and millet. Plant Physiol Biochem 47:1051–1060

Mahalakshmi S, Christopher GSB, Reddy TP, Rao KV, Reddy VD (2006) Isolation of a cDNA clone (PcSrp) encoding serine-rich-protein from *Porteresia coarctata* T. and its expression in yeast and finger millet (*Eleusine coracana* L.) affording salt tolerance. Planta 224:347–359

Maharajan T, Antony Ceasar S, Ajeesh Krishna TP, Ignacimuthu S (2021) Finger millet [*Eleusine coracana* (L.) Gaertn]: an orphan crop with a potential to alleviate the calcium deficiency in the semi-arid tropics of Asia and Africa. Front Sustain Food Syst 5:258

Nainwal K, Verma O (2018) Conservation of minor millets for sustaining agricultural biodiversity and nutritional security. J Pharmacogn Phytochem 7:1576–1580

Ngetich A, Mweu C, Ngugi M et al (2018) Efficient plant regeneration protocol for finger millet [*Eleusine coracana* (L.) Gaertn.] via somatic embryogenesis. Afr J Biotechnol 17(21):660–667

Nithiyanantham S, Kalaiselvi P, Mahomoodally MF et al (2019) Nutritional and functional roles of millets—a review. J Food Biochem 43:e12859

Ntui VO, Uyoh EA, Nakamura I, Mii M (2017) Agrobacterium-mediated genetic transformation of Fonio (*Digitaria exilis* (L.) Stapf). Afr J Biotechnol, 16(23):1302–1307

Pande A, Dosad S, Chawla HS, Arora S (2015) In-vitro organogenesis and plant regeneration from seed-derived callus cultures of finger millet (Eleusine coracana). Rev Bras Bot 38:19–23

Qin FF, Zhao Q, Ao GM, Yu JJ (2008) Co-suppression of Si401, a maize pollen specific Zm401 homologous gene, results in aberrant anther development in foxtail millet. Euphytica 163:103–111

Saleh ASM, Zhang Q, Chen J, Shen Q (2013) Millet grains: nutritional quality, processing, and potential health benefits. Compr Rev Food Sci Food Saf 12:281–295

Santos CM, Romeiro D, Silva JP, Basso MF, Molinari HBC, Centeno DC (2020) An improved protocol for efficient transformation and regeneration of Setaria italica. Plant Cell Rep 39:501–510

Sarita ES, Singh E (2016) Potential of millets: nutrients composition and health benefits. J Scient Innov Res 5:46–50

Satish L, Ceasar SA, Shilpha J et al (2015) Direct plant regeneration from in vitro-derived shoot apical meristems of finger millet (*Eleusine coracana* (L.) Gaertn.). In vitro Cell Dev Biol Plant 51:192–200

Satish L, Ceasar SA, Ramesh M (2017) Improved agrobacterium-mediated transformation and direct plant regeneration in four cultivars of finger millet (*Eleusine coracana* (L.) Gaertn.). Plant Cell Tissue and Organ Culture (PCTOC) 131:547–565

Sharma M, Kothari-Chajer A, Jagga-Chugh S, Kothari SL (2011) Factors influencing *Agrobacterium tumefaciens*-mediated genetic transformation of *Eleusine coracana* (L.) Gaertn. Plant Cell Tissue and Organ Culture (PCTOC) 105:93–104

Singh RP, Qidwai S, Singh O et al (2022) Millets for food and nutritional security in the context of climate resilient agriculture: a review. Int J Plant Soil Sci:939–953

Sood P, Singh RK, Prasad M (2020) An efficient *Agrobacterium*-mediated genetic transformation method for foxtail millet (*Setaria italica* L.). Plant Cell Rep 39:511–525

Swindale LD, Virmani SM, Sivakumar MVK (1981) Climatic variability and crop yields in the semi-arid tropics. In: Food-climate interactions: proceedings of an international workshop held in Berlin (West), December 9–12, 1980. Springer, pp 139–166

Thilakarathna MS, Raizada MN (2015) A review of nutrient management studies involving finger millet in the semi-arid tropics of Asia and Africa. Agronomy 5:262–290

Ugare R, Chimmad B, Naik R et al (2014) Glycemic index and significance of barnyard millet (*Echinochloa frumentacae*) in type II diabetics. J Food Sci Technol 51:392–395

Upadhyaya HD, Vetriventhan M (2018) Underutilized climate-smart nutrient rich small millets for food and nutritional security

Vetriventhan M, Upadhyaya HD (2019) Variability for productivity and nutritional traits in germplasm of kodo millet, an underutilized nutrient-rich climate smart crop. Crop Sci 59:1095–1106

Vetriventhan M, Azevedo VCR, Upadhyaya HD et al (2020) Genetic and genomic resources, and breeding for accelerating improvement of small millets: current status and future interventions. Nucleus 63:217–239

Wang M-Z, Pan Y-L, Li C, Liu C, Zhao Q, Ao G-M, Yu J-J (2011) Culturing of immature inflorescences and agrobacterium-mediated transformation of foxtail millet (Setaria italica). Afr J Biotechnol 10:16466–16479

Woomer JS, Adedeji AA (2021) Current applications of gluten-free grains–a review. Crit Rev Food Sci Nutr 61:14–24

Yinghui L, Jingjuan Y, Qian Z, Dengyun Z, Guangming A (2005) Genetic transformation of millet (*Tetaria italica*) by agrobacterium-mediated. Nong ye Sheng wu ji shu xue bao= J Agric Biotechnol 13:32–37

Zhao Z, Tang S, Zhang Y, Yue J, Xu J, Tang W, Sun Y, Wang R, Diao X, Zhang B (2021) Evolutionary analysis and functional characterization of *SiBRI1* as a Brassinosteroid receptor gene in foxtail millet. BMC Plant Biol 21:291

The manufacturer's authorised representative in the EU is Springer Nature Customer Service Centre GmbH, Europaplatz 3, 69115 Heidelberg, Germany. If you have any concerns regarding our products, please contact ProductSafety@springernature.com

Printed and bound by CPI Group (UK) Ltd, Croydon, CR0 4YY
01/10/2026
02225929-0009